Triton and Pluto

The long lost twins of active worlds

Online at: https://doi.org/10.1088/2514-3433/ad5278

AAS Editor in Chief

Ethan Vishniac, Johns Hopkins University, Maryland, USA

About the program:

AAS-IOP Astronomy ebooks is the official book program of the American Astronomical Society (AAS) and aims to share in depth the most fascinating areas of astronomy, astrophysics, solar physics and planetary science. The program includes publications in the following topics:

GALAXIES AND COSMOLOGY

INTERSTELLAR MATTER AND THE LOCAL UNIVERSE

STARS AND STELLAR PHYSICS

EDUCATION, OUTREACH, AND HERITAGE

HIGH-ENERGY PHENOMENA AND FUNDAMENTAL PHYSICS

THE SUN AND THE HELIOSPHERE

THE SOLAR SYSTEM, EXOPLANETS, AND ASTROBIOLOGY

LABORATORY ASTROPHYSICS, INSTRUMENTATION, SOFTWARE, AND DATA

Books in the program range in level from short introductory texts on fast-moving areas, graduate and upper-level undergraduate textbooks, research monographs, and practical handbooks.

For a complete list of published and forthcoming titles, please visit iopscience.org/books/aas.

About the American Astronomical Society

The American Astronomical Society (aas.org), established 1899, is the major organization of professional astronomers in North America. The membership (~7,000) also includes physicists, mathematicians, geologists, engineers, and others whose research interests lie within the broad spectrum of subjects now comprising the contemporary astronomical sciences. The mission of the Society is to enhance and share humanity's scientific understanding of the universe.

Editorial Advisory Board

Triton and Pluto

The long lost twins of active worlds

Edited by

Adrienn Luspay-Kuti
The Johns Hopkins University Applied Physics Laboratory, Laurel, USA

Kathleen Mandt
NASA Goddard Space Flight Center, Greenbelt, USA

IOP Publishing, Bristol, UK

Adrienn Luspay-Kuti and Kathleen Mandt have asserted their right to be identified as the editors of this work in accordance with sections 77 and 78 of the Copyright, Designs and Patents Act 1988.

ISBN   978-0-7503-5618-3 (ebook)
ISBN   978-0-7503-5616-9 (print)
ISBN   978-0-7503-5619-0 (myPrint)
ISBN   978-0-7503-5617-6 (mobi)

DOI   10.1088/2514-3433/ad5278

Version: 20250101

AAS–IOP Astronomy
ISSN 2514-3433 (online)
ISSN 2515-141X (print)

British Library Cataloguing-in-Publication Data: A catalogue record for this book is available from the British Library.

Published by IOP Publishing, wholly owned by The Institute of Physics, London

IOP Publishing, No.2 The Distillery, Glassfields, Avon Street, Bristol, BS2 0GR, UK

US Office: IOP Publishing, Inc., 190 North Independence Mall West, Suite 601, Philadelphia, PA 19106, USA

This book is dedicated to Louise Prockter, who tirelessly worked to make a mission to Triton possible, and who positively impacted the careers of many scientists, including the Editors and several authors of this book. We will forever be grateful for your leadership and support.

Contents

Preface   xvi

Acknowledgements   xvii

Editor Biographies   xix

List of Contributors   xx

Introduction   xxiv

1 Triton and Pluto: Same Origin but Separated at Birth   1-1
Olivier Mousis, Sarah E Anderson, Adrienn Luspay-Kuti,
Kathleen E Mandt and Pierre Vernazza

1.1 Introduction   1-1
1.2 Composition of Pluto and Triton   1-2
    1.2.1 Atmospheric and Surface Composition   1-2
    1.2.2 Bulk Composition   1-4
    1.2.3 Time Evolution of the Volatile Budget in Pluto and Triton   1-5
1.3 Formation of Pluto and Triton's Building Blocks in the PSN   1-8
    1.3.1 Link with Atypical Comets   1-8
    1.3.2 Role of CO and N_2 Icelines   1-10
1.4 Dynamical Origin of Triton   1-13
1.5 Conclusion   1-14
    References   1-15

2 Pluto and Triton: Interior Structures, Lithospheres and Potential for Oceans   2-1
Francis Nimmo, Carver Bierson and William B McKinnon

2.1 Introduction   2-1
2.2 Bulk Structures   2-3
    2.2.1 Observations   2-3
    2.2.2 Differentiation State   2-6
2.3 Arguments for Subsurface Oceans   2-9
    2.3.1 Theory   2-9
    2.3.2 Observations   2-17
2.4 Lithospheric Structure and History   2-19
    2.4.1 Conduction   2-20
    2.4.2 Convection   2-21

2.4.3 Observations   2-21

2.5 Summary and Future Work   2-27

2.6 Future Work   2-28

    References   2-28

3 Triton and Pluto: Cryovolcanism in the Outer Reaches of Our   **3-1**
Solar System
Lynnae C Quick, Kelsi N Singer, Caitlin Ahrens, Candice J Hansen,
Jason D Hofgartner, Karl L Mitchell, Louise M Prockter,
Paul M Schenk and Orkan Umurhan

3.1 Cryovolcanism in the Solar System   3-1

3.2 Powering Cryovolcanism on Triton and Pluto: Internal Heating   3-3
    Mechanisms

3.3 Cryomagmatism: Melt Generation, Migration, and Composition   3-4

    3.3.1 Composition of Cryomagmatic Melts on Triton   3-5

    3.3.2 Composition of Cryomagmatic Melts on Pluto   3-6

3.4 Cryovolcanism on Triton and Pluto   3-7

3.5 Triton: A Cryovolcanic World?   3-8

    3.5.1 Triton's Plumes: A First Glimpse of Extraterrestrial Geysers   3-10

    3.5.2 Triton's Fans   3-15

3.6 Pluto: Evidence for Cryovolcanism at Both Large and Small Scales   3-16

    3.6.1 Virgil Fossae   3-20

    3.6.2 Change Detection   3-21

    3.6.3 Absence of Plumes and Limited Analogous Fans   3-22

3.7 Looking Ahead: Future Mission and Measurement Needs   3-23

3.8 Conclusions   3-24

    Acknowledgements   3-25

    References   3-25

4 Divergent: Triton and Pluto Morphology and Geology   **4-1**
Paul M Schenk

4.1 Introduction   4-1

4.2 Triton–Pluto Comparisons   4-4

    4.2.1 (Cryo)Volcanism   4-5

    4.2.2 Walled Planitia   4-10

    4.2.3 Nitrogen Ice Sheets   4-12

    4.2.4 Surface Modification by Lateral Flow: "Glaciation"   4-16

    4.2.5 Convective Overturn in the Interior   4-19

4.2.6 Erosive Processes: Etched Plains   4-20

4.3 Discussion and Conclusions   4-22

Acknowledgments   4-23

References   4-23

5 Volatile Cycles and Surface–Atmosphere Interactions   **5-1**
*Tanguy Bertrand, Will Grundy, Emmanuel Lellouch, François Forget
and Leslie Ann Young*

5.1 Introduction: Pluto and Triton as a Distinct Class of Planetary Objects   5-1

5.2 Basic Parameters and Observations of the Surface and Atmosphere of Pluto and Triton Relevant to Surface–Atmosphere Interactions   5-2

5.2.1 Orbits, Seasons and Climate Zones   5-2

5.2.2 The Frozen but Active Surfaces of Pluto and Triton: Two Fundamentally Different Landscapes   5-4

5.2.3 Characteristics of the Tenuous Condensable Atmospheres of Pluto and Triton   5-6

5.3 Fundamentals of Surface–Atmosphere Interactions on Objects Such as Pluto and Triton   5-10

5.3.1 General Properties of the Surface Ices: A Volatile Cocktail   5-10

5.3.2 The Important Role of Thermal Inertia on Distant Cold Bodies   5-11

5.3.3 Despite the Cold Temperatures, an Active Surface Exists Thanks to the Ice Volatility and Complex Feedbacks   5-11

5.3.4 The Dynamics of Tenuous N_2 Condensable Atmospheres   5-16

5.4 Building Models to Simulate the Volatile Cycles and the Climate System on Different Timescales   5-17

5.4.1 Energy Balance Models and Volatile Transport Models   5-18

5.4.2 Three-dimensional Global Climate Models   5-19

5.5 Volatile Transport on Seasonal and Astronomical Timescales on Pluto and Triton   5-19

5.5.1 How Obliquity Changes Everything, From the Surface to the Atmosphere   5-19

5.5.2 Formation of Glaciers and Thick Ice Deposits   5-23

5.5.3 The Seasonal Cycle of Volatile Ices   5-25

5.6 Conclusions and Remaining Questions   5-30

References   5-33

6 Clouds and Hazes in the Atmospheres of Triton and Pluto   **6-1**
Peter Gao and Kazumasa Ohno

6.1 Introduction   6-1

6.1.1 A Note on Nomenclature   6-1

6.2 Observations of Triton and Pluto Aerosols   6-2

    6.2.1 Voyager 2 Observations of Triton Aerosols   6-2

    6.2.2 Pre-New Horizons Observations of Pluto's Atmosphere   6-4

    6.2.3 New Horizons Observations of Pluto Aerosols   6-5

    6.2.4 Summary   6-7

6.3 Theory of Triton and Pluto Aerosols   6-8

    6.3.1 Photochemistry and Aerosol Formation on Triton and Pluto   6-8

    6.3.2 Formation, Evolution, and Dynamics of Triton and Pluto Aerosols   6-9

    6.3.3 Summary   6-14

6.4 Laboratory Investigations of Triton and Pluto Aerosols   6-15

6.5 Triton and Pluto Aerosols: Unknowns and Outlook   6-16

    References   6-18

7 Magnetospheric and Space Environment Interactions with the Upper Atmosphere and Ionosphere   7-1

Tom A Nordheim, Adrienn Luspay-Kuti, Lucas Liuzzo, Peter Gao and G Randy Gladstone

7.1 Introduction   7-1

7.2 Neutral Atmospheres   7-2

7.3 Ionospheres   7-3

7.4 Pluto's Solar Wind Environment and Atmospheric Ion Escape   7-8

7.5 Neptune's Magnetospheric Environment and Triton's Role as a Possible Plasma Source   7-11

7.6 Triton's Interaction with its Magnetospheric Plasma Environment   7-15

7.7 Conclusions   7-18

    References   7-20

8 On the Detection of Subsurface Oceans within Triton and Pluto   8-1

Corey J Cochrane, Julie C Castillo-Rogez, Steven D Vance and Benjamin P Weiss

8.1 Introduction   8-1

8.2 Neptune's Magnetic Environment   8-4

8.3 Modeling of Induced Magnetic Field   8-8

8.4 Ocean Detection and Characterization Methods   8-14

8.5 Trajectory Considerations   8-16

8.6	Instrument Selection	8-20
8.7	Summary	8-21
	References	8-22

9 Future Measurement Needs: Surface Processes 9-1

Alessandra Migliorini, Bryan J Holler, Will M Grundy, Tom S Stallard, Federico Tosi and Leslie A Young

9.1	Introduction	9-1
	9.1.1 Triton	9-2
	9.1.2 Pluto	9-7
9.2	Open Questions	9-10
9.3	Measurements in the Next Decade	9-12
	9.3.1 Ground-based Facilities	9-13
	9.3.2 Space-based Facilities	9-15
9.4	In Situ Exploration	9-16
	9.4.1 Spectroscopic Observation Needs	9-16
	9.4.2 Flyby Missions	9-19
	9.4.3 Orbiters	9-20
	9.4.4 Landers	9-22
9.5	Conclusions	9-23
	References	9-23

10 Future Measurement Needs for the Atmospheres of Pluto and Triton 10-1

Manuel Scherf, Audrey Vorburger, Peter Wurz and Helmut Lammer

10.1	Introduction	10-1
10.2	Brief Overview on Pluto's and Triton's Atmosphere	10-2
10.3	Open Questions and Related Measurement Needs	10-3
	10.3.1 Atmospheric Structure and the Potential Role of Haze Production	10-3
	10.3.2 The Origin of Their Atmospheres	10-5
	10.3.3 Atmospheric Pressure Evolution and Volatile Cycling	10-8
	10.3.4 Additional Uncertainties in Model Parameters	10-9
	10.3.5 Some Further Open Questions and Needs	10-10
10.4	Future Observations	10-12
	10.4.1 Ground-Based Observations	10-12
	10.4.2 *JWST* and Other Remote Space-Based Observations	10-14

10.4.3 In Situ Measurements   10-15

10.4.4 Laboratory Measurements and Experiments   10-18

10.4.5 Comparative Planetology   10-19

10.5 Conclusion   10-20

References   10-21

11   Planning for Long-Lived Missions   **11-1**
*Janet Vertesi, Matthew J Bietz, Marisa Cohn, Stephanie Jordan
and David Reinecke*

11.1 Introduction   11-1

11.1.1 Take a Sociotechnical Approach   11-2

11.2 Build a Multi-Generational Team   11-3

11.2.1 The Current Model Is Limited   11-3

11.2.2 Adopt a Long-Term Structure   11-4

11.2.3 Invest in Successful Succession   11-5

11.3 Transmit Knowledge Effectively to Newcomers   11-7

11.3.1 Embrace Tacit Knowledge   11-7

11.3.2 Structure Mentorship   11-8

11.3.3 Facilitate Culture   11-9

11.4 Anticipate Aging Systems   11-10

11.4.1 Lifetime Issues in Hardware and Software   11-10

11.4.2 Sustain Human Infrastructure   11-12

11.4.3 Accommodate Data (R)evolutions   11-13

11.5 Sustain Support   11-14

11.5.1 The Time Inconsistency Problem in Planetary Science   11-14

11.5.2 Sustainable Support   11-16

11.5.3 The People Equation   11-17

11.6 Invest in Meaningful Partnerships   11-18

11.6.1 Diversity and Inclusion   11-18

11.6.2 Encountering Differences   11-19

11.6.3 An Aging Workforce   11-20

11.7 Conclusion   11-21

References   11-21

12   Pluto and Triton: Open Questions and Decadal Linkages   **12-1**
*S Alan Stern, Candace J Hansen, William B McKinnon, Carol Paty,
Louise Prockter and Leslie A Young*

12.1 Introduction   12-1

12.2 Open Questions   12-1
    12.2.1 Origin   12-1
12.3 Interior/Ocean   12-4
12.4 Ocean World/Astrobiology   12-7
12.5 Surface/Geology   12-8
12.6 Atmosphere   12-10
12.7 Magnetospheric and Radiation Environments   12-12
12.8 Comparative Planetology of Dwarf Planets   12-14
12.9 Decadal Relevance and Response   12-15
    12.9.1 Decadal Relevance   12-15
12.10 Mapping of Open Questions to Decadal Science   12-17
12.11 Mission Needs for Pluto and Triton to Address Science   12-18
    and Open Questions
    References   12-19

Preface

Although Pluto and Triton have captured the imagination for decades as mysterious twins of the outer solar system, this is the first book devoted to comparing and contrasting these two fascinating worlds. The greatest advances in exploring Pluto and Triton have come from spacecraft flybys. The Voyager 2 spacecraft provided the first up close views of Triton in 1989, making amazing discoveries of its geology, atmosphere, and interaction with the magnetosphere. However, limitations in the spacecraft payload and in the time able to explore Triton left far more questions than the single flyby could answer. The New Horizons flyby of Pluto applied lessons learned from the Triton flyby, providing a more comprehensive dataset tailored to a twin of Triton. However, many questions still remain about Pluto because of the limitations of a single flyby. These flybys remain central to this book, as we review the state of knowledge and look ahead to the future.

This book is divided into two sections. Chapters 1-7 outline the current state of knowledge for Pluto and Triton in time from their formation to their current states, and in space from the deep interior to the interactions of their upper atmospheres with their space environments. The second section looks ahead to the future needs for exploring Pluto and Triton. Chapters 8, 9 & 10 outline future measurement needs for detecting interior oceans, understanding surface processes, and characterizing their atmospheres. In Chapter 11 we acknowledge the importance of human factors in space exploration. Because exploration of the outer solar system involves long-lived missions, this chapter outlines consideration in planning for them. We end this book with Chapter 12, which summarizes open questions and how they connect to the most recent Decadal Survey.

Acknowledgements

The results presented in this book would not have been possible without the support of NASA and ESA. Ground-based observations funded by NASA, ESA, and the ESA member states provided the earliest and ongoing observations of Pluto and Triton. The NASA Voyager, Cassini, and New Horizons missions as well as the ESA Rosetta mission provided data essential for these studies. Finally, much of the time that scientists required to analyze data, publish results, and summarize these results for this book was funded by NASA and ESA member state research grants. Scientific production of knowledge is only possible through such support and the Editors and Authors wish to express their gratitude.

The editors would like to thank each of the contributors to this book. Your combined expertise on these two amazing worlds, both what is currently known and the methods for future exploration, have made this work possible. Finally, no book is complete without the contributions of reviewers who ensure the high quality of each of the chapters. We thank the following reviewers:

- Erika Barth, Southwest Research Institute
- Chloe Beddingfield, Johns Hopkins Applied Physics Laboratory
- John Biersteker, Massachusetts Institute of Technology
- Andrew Coates, University College London
- Sarah Fagents, Hawai'i Institute of Geophysics and Planetology, University of Hawai'i at Mānoa
- Will Grundy, Lowell Observatory
- Candy Hansen, Planetary Science Institute
- Bryan Holler, Space Telescope Science Institute
- Perianne Johnson, Institute for Geophysics, University of Texas
- Stephan Loveless, University of Georgia
- Jeff Moore, NASA Ames
- Carol Raymond, Jet Propulsion Laboratory, California Institute of Technology
- Yasuhito Sekine, Earth-Life Science Institute, Tokyo Institute of Technology
- Christoph Sotin, University of Nantes

The editors wish to specifically acknowledge the Cassini and Rosetta projects and NASA grants 80NSSC18K1233, 80NSSC19K1306, 80NSSC23K1270, and 80NSSC18K1620. This series of funding opportunities made leadership and organization of this book possible.

Olivier Mousis specifically acknowledges funding from CNES. The project leading to this publication has received funding from the Excellence Initiative of Aix-Marseille Université-A*Midex, a French 'Investissements d'Avenir program' AMX-21-IET-018. This research holds as part of the project FACOM (ANR-22-CE49-0005-01_ACT) and has benefited from a funding provided by l'Agence Nationale de la Recherche (ANR) under the Generic Call for Proposals 2022.

William McKinnon would like to specifically thank NASA for supporting both the research and exploration of Triton and Pluto across the decades, through grants from the Planetary Geology and Geophysics, Outer Planets Research, and Origins of Solar Systems programs, and must fundamentally, with the Voyager Neptune Interstellar Mission and New Horizons Mission to Pluto and the Kuiper Belt. Additional thanks go to Clyde Tombaugh, Peter Goldreich, and Tom McCord for starting the fire.

Editor Biographies

Adrienn Luspay-Kuti

Dr. Adrienn Luspay-Kuti is a planetary scientist at the Johns Hopkins University Applied Physics Laboratory (JHU/APL) and the Principal Investigator for the Plasma Instrument for Magnetic Sounding (PIMS) on NASA's Flagship-class mission, Europa Clipper (launched on 14 October 2024). She received her MS degree in Astronomy at the Eötvös Loránd University of Sciences in 2009, and her PhD in Space and Planetary Sciences from the University of Arkansas in 2014. She held a postdoctoral fellowship from 2014 to 2016, and a Research Scientist position from 2016 to 2018 at the Southwest Research Institute. In 2018, she took up a Staff Scientist position at the Johns Hopkins University Applied Physics Laboratory in Maryland, where she currently resides and serves as a Section Supervisor for the Comets and Icy Small Bodies section. She has over a decade of experience in photochemical modeling of cold nitrogen-methane atmospheres such as Titan, Pluto and Triton, cometary science, volatile capture and release in the early solar system, and the thermodynamics of astrophysical ices and cryogenic liquids. She also has experience with data analysis of spacecraft-based instruments and flight instrument calibration. She was a member of the science team for the Rosetta Orbiter Spectrometer for Ion and Neutral Analysis from 2014 to 2018.

Kathleen Mandt

Kathleen E. Mandt is the Lab Chief for the NASA Goddard Space Flight Center (GSFC) Planetary Systems Laboratory. Her research includes the origin and evolution of volatiles throughout the solar system and the role of dynamics, chemistry and atmospheric evolution in understanding this. Dr. Mandt previously served as the Chief Scientist for Exoplanets, the Astrobiology Section Supervisor at the Johns Hopkins Applied Research Laboratory, an adjoint professor in the Department of Physics and Astronomy at the University of Texas at San Antonio, and as a Senior Research Scientist at the Southwest Research Institute. Dr. Mandt has served in several community and NASA mission leadership roles, including as the volatiles theme lead for the Lunar Reconnaissance Orbiter (LRO) mission; Project Scientist for the LRO Lyman Alpha Mapping Project instrument; Project Scientist for the Io Volcano Observer phase A study; Deputy Project Scientist for the Heliophysics Division-funded Interstellar Probe pre-decadal mission study; and is currently a member on the Europa Clipper Plasma Instrument for Magnetic Sounding science team. She served as a member of the steering committee of the Outer Planets Assessment Group and the Division for Planetary Science Professional Culture and Climate Subcommittee. Dr. Mandt earned a Ph.D. in environmental science and engineering from the University of Texas, San Antonio. Her National Academies service includes serving as Chair of the A Science Strategy for the Human Exploration of Mars: Astrobiology Panel, the Astro2020: Panel on Exoplanets, Astrobiology, and the Solar System and the Planetary Science and Astrobiology Decadal Survey 2023–2032: Panel on Giant Planet Systems.

List of Contributors

Caitlin Ahrens
NASA Goddard Space Flight Center, Greenbelt, MD, USA and The University of Maryland, College Park, College Park, MD 20742, USA

Sarah E. Anderson
Aix-Marseille Université, CNRS, CNES, Institut Origines, LAM, Marseille, France

Matthew J. Bietz
Department of Informatics, Donald Bren School of Information and Computer Sciences, University of California Irvine, Irvine, CA, USA

Tanguy Bertrand
LESIA, Observatoire de Paris, Paris, France

Carver Bierson
School of Earth and Space Exploration, Arizona State University, Tempe, AZ 85281, USA

Julie Castillo-Rogez
Jet Propulsion Laboratory, California Institute of Technology, Pasadena, CA 91109, USA

Corey J. Cochrane
Jet Propulsion Laboratory, California Institute of Technology, Pasadena, CA 91109, USA

Marisa Cohn
IT University of Copenhagen, Rued Langgaards Vej 7, 2300 København, Denmark

Francois Forget
Laboratoire de Météorologie Dynamique (LMD/IPSL), Sorbonne Université, Paris, France

Peter Gao
Earth & Planets Laboratory, Carnegie Institution for Science, 5241 Broad Branch Road NW, Washington, DC 20015, USA

G. Randy Gladstone
Southwest Research Institute, San Antonio, TX, USA

Will M. Grundy
Lowell Observatory, 1400 W. Mars Hill Rd., Flagstaff, AZ 86001, USA

Candice Hansen
The Planetary Science Institute, Tucson, AZ 85719, USA

Jason Hofgartner
Southwest Research Institute, 1301 Walnut St. #400, Boulder, CO 80302, USA

Bryan Holler
Space Telescope Science Institute, 3700 San Martin Dr, Baltimore, MD 21218, USA

Stephanie Jordan
Media + Information, Michigan State University, East Lansing, MI, USA

Helmut Lammer
IWF, Space Research Institute, Austrian Academy of Sciences, Graz, Austria

Emmanuel Lellouch
LESIA, Observatoire de Paris, Paris, France

Lucas Liuzzo
University of California, Berkeley, Berkeley, CA, USA

Adrienn Luspay-Kuti
Johns Hopkins Applied Physics Laboratory, Laurel, MD, USA

Kathleen Mandt
NASA Goddard Space Flight Center, Greenbelt, MD, USA

William McKinnon
Department of Earth, Environmental, and Planetary Sciences and McDonnell Center for the Space Sciences, Washington University in St. Louis, Saint Louis, MO 63130, USA

Alessandra Migliorini
Istituto Nazionale di Astrofisica—Istituto di Astrofisica e Planetologia Spaziali (INAF-IAPS), Via Fosso del Cavaliere, 100, 00133, Rome, Italy

Karl Mitchell
Jet Propulsion Laboratory, California Institute of Technology, Pasadena, CA 91109, USA

Olivier Mousis
Aix-Marseille Université, CNRS, CNES, Institut Origines, LAM, Marseille, France

Francis Nimmo
Department of Earth and Planetary Sciences, University of California Santa Cruz, CA 95064, USA

Tom Andre Nordheim
Johns Hopkins Applied Physics Laboratory, Laurel, MD, USA

Kazumasa Ohno
Division of Science, National Astronomical Observatory of Japan, 2-21-1 Osawa, Mitaka-shi, Tokyo, Japan

Carol Paty
Department of Earth Sciences University of Oregon, Eugene, OR 97403, USA

Louise Prockter
Johns Hopkins Applied Physics Laboratory, Laurel, MD, USA

Lynnae Quick
NASA Goddard Space Flight Center, Greenbelt, MD, USA

David Reinecke
Space Policy, Department of State, Washington, DC USA

Paul Schenk
Lunar and Planetary Institute/USRA, Houston TX 77058, USA

Manuel Scherf
IWF, Space Research Institute, Austrian Academy of Sciences, Graz, Austria

Kelsi Singer
Southwest Research Institute, 1301 Walnut St. #400, Boulder, CO 80302, USA

Thomas Stallard
Northumbria University, Newcastle upon Tyne, England, UK

S. Alan Stern
Southwest Research Institute, 1301 Walnut St. #400, Boulder, CO 80302, USA

Federico Tosi
Istituto Nazionale di Astrofisica—Istituto di Astrofisica e Planetologia Spaziali (INAF-IAPS), Via Fosso del Cavaliere, 100, 00133, Rome, Italy

Orkan Umurhan
NASA Ames Research Center, Moffett Field, CA 94035, USA and SETI Institute, Mountain View, CA 94043, USA

Steven D. Vance
Jet Propulsion Laboratory, California Institute of Technology, Pasadena, CA 91109, USA

Pierre Vernazza
Aix-Marseille Université, CNRS, CNES, Institut Origines, LAM, Marseille, France

Janet Vertesi
Sociology Department, Princeton University, Princeton, NJ, USA

Audrey Vorburger
Space Science and Planetology, Physics Institute, University of Bern, 3012 Bern, Switzerland

Benjamin Weiss
Department of Earth, Atmospheric and Planetary Sciences, Massachusetts Institute of Technology, Cambridge, MA 02139, USA

Peter Wurz
Space Science and Planetology, Physics Institute, University of Bern, 3012 Bern, Switzerland

Leslie Ann Young
Southwest Research Institute, 1301 Walnut St. #400, Boulder, CO 80302, USA

Introduction

Science is constantly evolving and is full of revelations we could not have thought of in our wildest imagination. In the era of space exploration, it is natural to assume that the biggest mysteries have already been unveiled by the numerous spacecraft we sent across the Solar System; but in reality, planetary *science* is no different. The view we have of our Solar System changes and expands with every mission for decades to come, especially with those exploring hard to reach, elusive worlds. This has been the case with Neptune's largest moon Triton since the Voyager 2 flyby of Neptune in 1989, and Pluto since New Horizons in 2015—just 9 years ago. The understanding, discoveries and connections we made from closely observing these planetary bodies for a few days only during a single flyby of each speak to the extent of how much there still is to learn. However, the majority of observations since has been limited to Earth-based telescopes with obvious limitations to the extent of obtainable information of these two bodies. Thus, Pluto and Triton remain two of the most mysterious icy worlds in the Solar System, with vast areas lacking observational coverage.

Despite all the unknowns, Pluto and Triton are unique laboratories for comparative planetology, as they are like 'twins' with nearly the same size, mass, bulk density and surface composition. They likely started out forming in the same region of the solar system, but then these twins were separated at birth, with Triton ending up as a moon of Neptune and Pluto remaining in the Kuiper belt. As such, a detailed comparison of Pluto and Triton may help establish the relative roles of 'nurture vs. nature' in their evolution.

Compositionally, the bulk of both Pluto and Triton have high abundances of N_2 (molecular nitrogen) and CO (carbon monoxide). If these volatiles were incorporated by forming in the protosolar nebula (PSN), then their building blocks were likely accreted from the outer regions of the PSN near the CO and N_2 snowlines. Water-poor, N_2-rich comets like C/2016 R2 (PanSTARRS) possibly point to a common region of origin of these icy worlds and comets. Today, Triton is composed of slightly more rock than Pluto, which may be indicative of early loss of volatiles on Triton due to intense tidal heating.

Both Pluto and Triton are thought to have formed subsurface oceans, which, despite being two of the most distant bodies in the solar system, are likely still present today. Nonetheless, the capture of Triton by Neptune also brought forth a number of differences between Triton and Pluto. Triton experienced intense heating during orbital circularization from its initially eccentric orbit to a tidally locked one, which suggests that its interior has fully differentiated. While Pluto's interior is also thought to be differentiated, full differentiation likely had not completed until after the Charon-forming impact. Obliquity tidal heating at Triton is thought to be at least an order of magnitude greater than experienced by other icy moons, and tidal dissipation may have led to the slowing of Triton's ice shell, preserving a subsurface ocean. Pluto's internal heat source comes from radiogenic decay in its interior, which may also allow a present-day subsurface ocean.

Thus, the very different thermal histories experienced by Triton and Pluto result in different internal evolution and geologic histories despite their very similar bulk compositions.

Perhaps the most apparent difference between the two bodies is how different their surfaces look. Pluto has both older, rugged terrains and younger, glaciated terrains, convecting, plains mainly composed of N_2 ice dominated by Sputnik Planitia, cryovolcanic terrains and extensional tectonics. In contrast, Triton's surface is much smoother thus much younger with few geological features, active plumes and diapirs consistent with present-day heat flow. This observed recent activity and the evidence for tidal heating are also clues for the possible existence of a present-day deep, subsurface ocean. Topographic features are roughly five times lower on Triton than on Pluto, also consistent with a prolonged history of elevated heat flow that may have led to a weaker ice shell unable to support topographic extremes. Pluto's Sputnik Planitia is the primary driver for Pluto's volatile redistribution, but there is no comparable impact basin found on Triton. On the other hand, Triton's cantaloupe terrain is indicative of solid-state crustal convection, but such terrain is not seen on Pluto. Patterns associated with convective overturn are only seen in Sputnik Planitia on Pluto, where the horizontal scale of the cell pattern in the N_2-CH_4 ice sheet is notably similar to the cell pattern in Triton's cantaloupe terrain. This difference in the location and topography of the cells may indicate a thicker and/or colder ice shell on Pluto than on Triton.

The known volatile surface ices are similar in composition on both Pluto and Triton; both dominated by N_2. However, while N_2 ice is primarily concentrated in Sputnik Planitia on Pluto and CH_4 (methane) ice is more evenly distributed on its surface, Triton's N_2 ice appears to be concentrated to an extended cap covering a band of $\sim 75°$ in latitude in the southern hemisphere. The lack of large impact basins on Triton is consistent with the lack of observable, deep N_2 ice sheets, as the low topography and large seasonal variations would likely prevent the accumulation of large N_2 deposits. Nevertheless, our knowledge of the inventory of volatile species on Triton, such as CO (carbon monoxide), HCN (hydrogen cyanide), C_2H_6 (ethane) and CH_4 (methane), and the distribution of volatile and non-volatile ices and complex hydrocarbons is strongly limited. Even on Pluto, where New Horizons provided observations for compositional mapping, we cannot be sure that seasonal condensates do not mask topographic compositions in some regions.

The volatile ices on both bodies participate in a seasonal cycle, and buffer the tenuous, N_2-dominated sublimation atmospheres of Triton and Pluto. These atmospheres evolve over the course of seasons, as evidenced by observed changes in their atmospheric pressures. At the time of the writing of this book, Pluto's northern hemisphere is experiencing spring and the observed pressure increase is due to the increased N_2 sublimation from Sputnik Planitia in response to the increased seasonal solar insolation. Both icy bodies' atmospheres also contain trace amounts of CH_4 and CO; however, Triton's atmosphere is significantly more CH_4-poor than Pluto's, the reason for which is still poorly understood. Both atmospheres are the venues of rich photochemistry initiated by the dissociation and ionization of N_2, CH_4, and CO molecules by solar ultraviolet (UV) photons. This photochemistry produces organic hazes composed of heavy hydrocarbons and nitriles. Even though Triton and Pluto have similar atmospheric compositions, surface pressure, and temperature, there are also notable differences that need to be resolved. Pluto's thermal profile is very different from what was originally expected, with much colder upper atmospheric temperatures than models predicted. Triton's atmosphere, on the other hand, does not seem to have the same thermal structure. The haze in Pluto's atmosphere may contribute to its cooling, as there are significant differences in the formation and distribution of the aerosols and hazes in their atmospheres. The differences in the CH_4 mole fractions between the two atmospheres lead to distinct aerosol distributions and probably composition. CH_4 photochemistry is thought to be controlled by the surface CH_4 mole fraction. Triton's surface CH_4 mole fraction only allows CH_4 photolysis within the lowest tens of kilometers of the atmosphere, which ultimately leads to the condensation of hydrocarbons into ice clouds. At the same time, Pluto's more CH_4-abundant atmosphere and higher surface mole fraction allows the photolysis of CH_4 at much higher altitudes, where it ultimately leads to the formation of complex haze particles. Thus, Triton possesses a global, low-altitude aerosol layer with distinct clouds below 10 km, whereas Pluto's detached aerosol layers likely composed of organic hazes extend up to hundreds of kilometers.

Regardless, the fundamental processes of surface–atmosphere interactions, atmospheric dynamics, and climate physics are very similar on both worlds. There is still a lot left to be learned on the interplay of the volatile cycles and complex feedback processes as related to internal activity and atmospheric hazes. There is little known about the atmospheric thermal balance in Pluto's and Triton's atmospheres, the exact reason for the colder-than-expected atmospheric temperatures in Pluto's atmosphere, and the role of orographic waves and sublimation tides, to mention a few. A synergy of *in situ* observations, laboratory measurements, and modeling will be crucial in furthering our understanding of surface–atmospheric processes.

The purpose of this book is to review the known similarities, differences, and the path forward in the exploration of these intriguing twin worlds separated at birth. The book is organized into two main parts: 1. Current knowledge and 2. Future needs. In Part 1, we review what we know about Triton and Pluto from the inside out: the formation of Triton and Pluto (Mousis et al., Chapter 1), the current understanding of their interiors and dynamics (Nimmo et al., Chapter 2), the cryovolcanic processes and features on their surfaces (Quick et al., Chapter 3), their geologies and morphologies (Schenk et al., Chapter 4), their surface–atmosphere interactions (Bertrand et al., Chapter 5), their atmospheric clouds and hazes (Gao & Ohno, Chapter 6), and their upper atmospheric interactions with their magnetic and space environments (Nordheim et al., Chapter 7) for a comprehensive overview of Pluto and Triton. In Part 2 we focus on the needs for future *in situ* and/or flyby exploration when it comes to subsurface ocean detection (Cochrane et al., Chapter 8), mapping of previously unilluminated or poorly-mapped surface regions (Migliorini et al., Chapter 9), and the atmospheric unknowns (Scherf et al., Chapter 10), including the need for new instrumentation and novel measurement methods. We discuss in detail the steps to planning this next generation of long-term missions based on lessons learned from previous missions in Chapter 11 (Vertesi et al.). As noted in this chapter, it is vital to pay attention to the need for a societal and sociotechnical approach that must be considered for distant solar system missions that by their nature span multiple generations. Chapter 11 leverages decades of studies by the authors who are experts in social sciences working with the planetary science community, and sheds light on the need for a shift in focus from purely technical to sociotechnical. We conclude this book with a discussion of the outstanding key science questions that span all aspects of planetology; from interiors to ocean words, to surfaces, atmospheres, magnetospheres, and origins to advance our knowledge for a well-rounded, comprehensive future understanding of our twins separated at birth (Chapter 12, Stern et al.). Perhaps this future wholesome view will allow us to isolate the relative roles of 'nurture' vs. 'nature', at least in the world of icy dwarf planets. In the meantime, we hope this book stimulates discussion and sparks the interest of the future planetary scientist to explore these mysterious and intriguing twin worlds.

Triton and Pluto
The long lost twins of active worlds
Adrienn Luspay-Kuti and Kathleen Mandt

Chapter 1

Triton and Pluto: Same Origin but Separated at Birth

**Olivier Mousis, Sarah E Anderson, Adrienn Luspay-Kuti,
Kathleen E Mandt and Pierre Vernazza**

Assessing the origin of Pluto and Triton has profound implications for the bigger picture of solar system formation and evolution. In such a context, this chapter reviews our current knowledge of the formation conditions of Pluto and Triton's constitutive building blocks in the protosolar nebula (PSN), which can be derived from their known or estimated volatile contents. Assuming that the ultravolatiles carbon monoxide and dinitrogen detected in Pluto and Triton are primordial, the presence of these molecules suggests that the two bodies accreted material originating from the vicinity of the carbon monoxide and dinitrogen icelines. Dinitrogen-rich and water-poor comets such as comet C/2016 R2 (PanSTARRS) obviously present a compositional link with Pluto and Triton, indicating that their building blocks formed in nearby regions of the PSN, despite the variation of the water abundance among those bodies. Also, the assumption of Triton's growth in Neptune's circumplanetary disk requires that its building blocks formed at earlier epochs in the PSN to remain consistent with its estimated composition.

1.1 Introduction

Our knowledge of the current composition of Pluto and Triton provides a starting point for determining where and when each of these bodies formed and how they ended up at their present locations. Assessing their origin has profound implications for the bigger picture of solar system formation and evolution, particularly the formation locations and the growth timescales of the four giant planets, the extent of their migration after formation, and how this latter factor shaped the present-day physical and orbital properties of smaller bodies.

Triton and Pluto have long been recognized as twin worlds with similar sizes, densities, and even compositions. While these uncanny similarities provide key

doi:10.1088/2514-3433/ad5278ch1
1-1

constraints on the formation conditions of their building blocks in the protosolar nebula (PSN), the nuances in their compositional differences carry just as much weight. Both bodies have predominantly dinitrogen (N_2)-dominated atmospheres with minor amounts of methane (CH_4) and carbon monoxide (CO), implying that their building blocks must have formed in nitrogen-rich regions of the PSN (Broadfoot et al. 1989; Cruikshank et al. 1993; Lellouch et al. 2017; Young et al. 2018). The atmospheres are buffered by sublimation equilibrium with some mixture of the same ices (predominantly N_2) on the surface (Yelle et al. 1995; Bertrand & Forget 2016). Solar ultraviolet flux initiates N_2–CH_4 photochemistry in both atmospheres, which leads to the production of more complex hydrocarbon species and haze (Wong et al. 2017; Benne et al. 2022). However, there are notable differences between Triton and Pluto when it comes to both the atmospheric abundances of CH_4 and CO, and their photochemistry (see Chapter 7).

Our knowledge about Triton's and Pluto's surface and atmospheric compositions is skewed toward Pluto, thanks to the close flyby by the *New Horizons* spacecraft of the Pluto system in 2015. In contrast, the only spacecraft to ever visit Triton so far was *Voyager 2* back in 1989. While a mission to Triton in the near future is absolutely necessary to further our understanding of this moon, the Neptune system, and potential ocean worlds in general, we are currently limited to the *Voyager 2* observations and applying lessons learned at Pluto to Triton.

This chapter reviews our current knowledge of the formation conditions of Pluto and Triton's constitutive building blocks, which can be derived from their known or estimated volatile contents. The known compositions of Pluto and Triton's atmospheric, surface, and bulk compositions are presented in Section 1.2. This section also describes the different mechanisms that may have been at play to shape Pluto and Triton's volatile budget over time. Section 1.3 establishes connections between the compositions of comets, in particular those of N_2-rich bodies such as comet C/2016 R2 (PanSTARRS) (R2), with those of Pluto and Triton. The formation conditions of Pluto and Triton's building blocks in the PSN are also discussed in light of recent works depicting the evolution of the radial abundance profiles of volatiles in the outer regions of the disk. The different scenarios depicting the dynamical origin of Triton are presented in Section 1.4. Section 1.5 is dedicated to discussion and conclusion.

1.2 Composition of Pluto and Triton

This section reviews the known or estimated abundances of the different volatile reservoirs in both Pluto and Triton. The processes that might have affected the volatile content of the two bodies over time are also discussed.

1.2.1 Atmospheric and Surface Composition

During its flyby, the *New Horizons* spacecraft found that Pluto's neutral atmosphere consists primarily of more than 99% N_2, ~0.30% CH_4 (Young et al. 2018), and ~0.05% CO (Lellouch et al. 2017). The hydrocarbons ethane (C_2H_6), acetylene

(C_2H_2), and ethylene (C_2H_4) were also detected below $\sim$500 km, with middle atmospheric mixing ratios of $\sim$0.001 for all three species, dropping to 2×10^{-5}, 5×10^{-6}, and 6×10^{-7} in the middle atmosphere at $\sim$100 km, respectively (Young et al. 2018). The near-surface CO/N_2 and the CO/CH_4 mixing ratios were found to be $\sim 4 \times 10^{-3}$ and $\sim 1.7 \times 10^{-3}$, respectively (Lellouch et al. 2017).

On the other hand, the *Voyager 2* flyby showed that Triton's atmospheric CH_4 is about one order of magnitude less abundant, compared with the *New Horizons* measurements made at Pluto. *Voyager* observations only provided an upper limit for CO in Triton's atmosphere (Broadfoot et al. 1989), but ground-based observations of its surface ice gave a CO/N_2 ratio of $\sim$0.1%, allowing to predict that the atmospheric mixing ratio would be 1.5×10^{-4} (Cruikshank et al. 1993). Later ground-based observations detected CO at abundances similar to the surface ice observations (Lellouch et al. 2010). The resulting CO/CH_4 ratio in Triton's atmosphere is then $\sim$3.5, roughly three orders of magnitude higher than in Pluto's atmosphere. The resulting CO/N_2 ratio in Triton's atmosphere is found to be $\sim$0.015%, namely a factor of $\sim$6 larger than in Pluto's atmosphere.

One of the biggest holes in our knowledge about Triton comes from the fact that no instrument on *Voyager 2* measured the composition of its surface. Hence, our information about Triton's surface composition is limited and solely comes from ground-based spectroscopic observations and modeling efforts. What we do know is that, similar to Pluto, Triton's surface is dominated by N_2 ice and to a lesser extent CH_4 and CO ices (Holler et al. 2016). These ices likely form solid solutions on the surface of Triton, but pure CH_4 ice may also form discreet patches. Considering the lack of spacecraft measurements, we have poor knowledge of the distribution of the ices on Triton's surface. Nevertheless, these ices are expected to migrate across the surface in response to the seasonally changing solar insolation (Cruikshank et al. 1984, 1993; Bauer et al. 2010; Buratti et al. 2011).

On Pluto, most of the N_2 ice seems to be concentrated in the basin called Sputnik Planitia, whereas the CH_4 ice is more widely distributed across the surface (Scipioni et al. 2021). The ice component of the bedrock on both Triton and Pluto includes H_2O ice. However, Triton's surface also has detectable amounts of CO_2 (Cruikshank et al. 1993), most likely in the form of exposed deposits, whereas this condensate appears to be absent from Pluto's surface. Small amounts of CO ice, H_2O ice, and NH_3 hydrates were also detected on Pluto's surface (Grundy et al. 2016; Dalle Ore et al. 2018; Cook et al. 2019). Heavier photochemistry products, such as methanol (CH_3OH) and hydrocarbon ices, are also present on the surface in trace amounts (Cook et al. 2019). A layer of heavier hydrocarbons and nitriles on the surface (Broadfoot et al. 1989) is also possible due to photochemistry, and hazes are observed at low altitudes as well (see Chapter 6). The noble gas argon (Ar) was not directly detected by *New Horizons* in Pluto's atmosphere, but an upper limit of 6% of the column density of CH_4 has been estimated (Steffl et al. 2020). Similarly, Ar has also been proposed to be present on Triton, with an upper limit of 10% of N_2's column density (McKinnon et al. 1995).

1.2.2 Bulk Composition

While surface ices and atmospheric compositions are two important reservoirs of volatiles, it is the bulk composition of a planetary body that needs to be assessed for understanding its origin and evolution. The estimates for the current bulk abundances should include volatiles and non-volatiles found on the surface and in the atmosphere, as well as what is predicted to be in the interior. Uncertainties on interior compositions can be large (McKinnon et al. 2019), and the full range of possibilities needs to be considered.

The bulk amount of H_2O can be estimated using the bulk density, which provides some potential information about the fractions of rock and water (ice and potential subsurface ocean) present. The average bulk density of Pluto is $\sim$1854 kg m^{-3} (Nimmo et al. 2017) and the bulk density of Triton is $\sim$2061 kg m^{-3} (McKinnon et al. 1995). In the case of Pluto, the water mass fraction in the bulk is between 0.28 and 0.36 (Glein & Waite 2018). A similar calculation for Triton results in a 0.21–0.28 water mass range in the bulk. Using the masses of Pluto and Triton and converting to moles, the abundance of H_2O in their bulks is estimated to be $(2–2.58) \times 10^{23}$ moles and $(1.5–2.03) \times 10^{23}$ moles, respectively.

The total amount of N_2 in Pluto and Triton is of special importance for constraining their region of formation within the early solar system, and understanding how closely related their formation was to each other. Whether present-day N_2 was originally accreted as N_2, NH_3, or N-bearing organic material has important implications on the conditions under which these twin worlds formed (see Section 1.3). Based on infrared absorption measurements, Pluto's surface N_2 reservoir is mainly concentrated to Sputnik Planitia. This reservoir is estimated to outweigh the atmospheric N_2 reservoir by orders of magnitude. Thus, taking the surface reservoir to represent the current amount of N_2 in the bulk is a reasonable first-order assumption (McKinnon et al. 2021; Glein & Waite 2018). The apparent amount of N_2 in Sputnik Planitia is $\sim(0.4–3) \times 10^{20}$ moles. Table 1.1 summarizes the estimated moles of N_2 and H_2O in the bulk composition of Pluto and Triton.

Unfortunately, no absorption measurement is available for Triton's surface, but the volatile abundances can be estimated from existing observations. Considering the *Voyager 2* atmospheric measurements, the energy-limited mass flux of N_2, and some limits to the thickness of N_2 polar frost deposits, an upper limit of 0.5–1 km thickness of N_2 averaged over the surface of Triton is obtained (McKinnon et al. 1995). Using these values, the amount of N_2 on Triton is estimated to be $(0.7–1.4) \times 10^{21}$ moles.

The second most abundant component, CH_4, has a mole fraction of $(3–3.6) \times 10^{-3}$ (Protopapa et al. 2017) on Pluto's surface diluted in N_2. CH_4 is predicted to be incorporated in the N_2 ice of the surface at a mole fraction of $\sim(1–5) \times 10^{-4}$ on Triton's surface, yielding a global equivalent layer of $\sim$1 m (Mandt et al. 2023). The CO mole fraction relative to N_2 is $\sim(2.5–5) \times 10^{-3}$ on Pluto's surface (Owen et al. 1993; Merlin 2015). The Pluto CO/N_2 ratio is about six orders of magnitude lower than the ratio measured in comet 67P/Churyumov-Gerasimenko ($\sim$35; (McKinnon et al. 2021)). Estimates for Triton's CO/N_2 surface mole fraction may vary between

Table 1.1. Estimated Moles of N_2 and H_2O in the Bulk Composition of Pluto and Triton

	Moles of N_2	Moles of H_2O
Pluto	$(0.4–3) \times 10^{20}$	$(2–2.58) \times 10^{23}$
Triton	$(0.7–1.4) \times 10^{21}$	$(1.5–2.03) \times 10^{23}$

The surface inventory is assumed to be representative of the inventory in the bulk.

Table 1.2. Mole Fractions of the Most Abundant Volatile Species, Relative to N_2, in the Total Inventory of Pluto and Triton

	N_2	
	Pluto	Triton
H_2O	$(0.7–6.5) \times 10^3$	$<107.2–291.5$
CH_4	$(3–3.6) \times 10^{-3}$	$\sim(1–5) \times 10^{-4}$
CO	$(2.5–5) \times 10^{-3}$	$<0.01–0.15$
CO_2	Not detected	$(1.5–100) \times 10^{-3}$

~ 0 and 0.15, depending on the atmospheric CO/N_2 estimate considered, and whether one assumes CO is mixed in the ice, or if it is physically separate from the N_2 ice. Table 1.2 summarizes the mole fractions of the most abundant volatile species, relative to N_2, in the total inventory of Pluto and Triton.

1.2.3 Time Evolution of the Volatile Budget in Pluto and Triton

The process for connecting the current volatiles of Pluto and Triton to their original building blocks begins with assessing the current bulk volatile composition of each body as described in Section 1.2.2. The measurements that are most useful for studying formation and evolution are noble gas abundances, noble gas isotopes, abundances of carbon, nitrogen, and oxygen, and their isotopes. Unfortunately, we currently only have limited information about carbon, nitrogen, and oxygen for Pluto and Triton, and only an upper limit on argon and on nitrogen isotopes at Pluto.

After determining the current composition, the next step is to evaluate what processes will influence volatiles in a way that changes the bulk composition of each body. Pluto and Triton have both experienced many processes that can change the composition of their volatiles, both during formation and following it. These processes will convert molecules from one form of volatile to another, like NH_3 to N_2, but will have no impact on the bulk elemental composition of volatiles in either body. The simplest approach, given such limited information, is to evaluate volatile abundances in terms of relative elemental abundances, such as N/C and O/C. This would enable a comparison with measurements of building block analogs like comets and chondrites, and models for icelines that take into consideration various forms of ices (see approach from Mandt et al. 2022 for lunar volatiles). We outline below the processes thought to have affected the volatile composition at Pluto and Triton and discuss whether they will have an impact on bulk elemental composition over time.

The earliest process would be accretional heating from impacts that take place during formation. Accretional heating could cause vaporization of hypervolatiles like N_2, CO, and CH_4 (McKinnon et al. 2019) leading to the formation of an early atmosphere. Any potential loss would be through atmospheric processes that are discussed later. The possible giant impact that formed the Pluto–Charon system (Canup et al. 2021), and any impact that took place as part of Triton's capture by Neptune (Rufu & Canup 2017) could also have caused heating (Canup et al. 2021) that could contribute to the formation of an early atmosphere and atmospheric loss processes. Once formation is complete, impacts would then add volatiles, including hypervolatiles, to the atmosphere (Simon et al. 2015). These processes would not only affect the bulk composition of the molecules present, but could also change the relative elemental composition by adding or removing nitrogen, carbon, oxygen, and hydrogen.

After formation, differentiation of the interior is thought to have led to the formation of a subsurface ocean where hydrothermal processes could have changed the composition of the molecular species that were originally accreted (Shock & McKinnon 1993; Glein & Waite 2018). These processes depend on the internal temperature and pressure, the pH in the water, the oxidation state of the system, and the bulk abundance of nitrogen and carbon. Of particular interest for Pluto and Triton, because of the presence of N_2, CH_4, and CO on the surface and in the atmosphere, are the conversions between N_2 and NH_3 and between CO_2 and CH_4 (Glein et al. 2008). Formation of N_2 and CO_2 tend to go together and are preferred at higher temperatures, lower pressures, and with systems that are more oxidized and have higher bulk abundances of nitrogen (Glein et al. 2008). Additionally, CO can also be converted to CO_2 or CH_4 through aqueous chemistry in the interior (Glein & Waite 2017). These processes change the molecular composition of the volatiles that are present, but will not affect the bulk elemental composition, such as N/C. In addition to the conversion reactions, hydrothermal reactions on early Pluto/Triton could have led to a loss of reactive volatiles (e.g., NH_3) at 100–300 °C, forming complex organic molecules (Sekine et al. 2017). When the rocky core temperature becomes high, e.g., $> \sim 500$ °C, organic matter contained in the rocky core could also have been thermally decomposed via a suite of irreversible reactions, possibly providing large amounts of volatiles, such as CO_2, CH_4, and N_2 (Okumura & Mimura 2011; Reynard & Sotin 2023). There are however aqueous reactions that can lead to permanent loss of CO (Shock & McKinnon 1993; Neveu et al. 2015; Glein & Waite 2017), which would change the bulk elemental ratios from what existed in the building blocks to values with reduced amounts of carbon and oxygen.

The surface compositions of Pluto and Triton potentially provide some constraints on the effectiveness of these hydrothermal processes (Mandt et al. 2023). Both Pluto and Triton have N_2, CH_4, and CO molecules on the surface and in the atmosphere. The presence of CH_4 suggests that production of N_2 and CO_2 was not complete in the interior unless all of the CH_4 currently on the surface was delivered through later impacts (Simon et al. 2015). Similarly, the presence of CO suggests either that any conversion of CO to CO_2 or CH_4 was also not complete unless all of the CO currently observed was also delivered through later impacts (Glein & Waite 2017). Triton has CO_2 on the surface, but none has been detected on Pluto (Ahrens et al. 2022), which might suggest that internal chemistry on Triton was more effective than at Pluto (Mandt et al. 2023). Additionally, NH_3 has been detected on the surface of Pluto further supporting the idea that any internal conversion to N_2 was not complete (Glein & Waite 2017; McKinnon et al. 2019; Mandt et al. 2023). Although NH_3 has not been detected on the surface of Triton, this may be due to the lack of high spatial and spectral resolution observations like the ones made by New Horizons at Pluto.

Two additional processes can remove molecules from the observed inventory without removing them from the bulk composition. The first is the sequestration of molecules in clathrates in the subsurface ocean. Clathrates are water-ice cages that trap other molecules and atoms. They are proposed to have formed at the base of Pluto's ice shell allowing the ice shell to harden while an ocean continues to exist, thanks to the heat retained by the insulating layer of clathrates (Kamata et al. 2019). Although N_2, CO, CO_2, and CH_4 can all be trapped in clathrates, CO clathrate is more stable than N_2 clathrate. Also, CH_4 and CO_2 clathrates are more stable than CO clathrate (McKinnon et al. 2019). This means that any of the carbon-bearing molecules produced in the interior through aqueous chemistry could be trapped in the interior long term in a layer of clathrates and appear to be missing from the bulk carbon inventory. The other process is the burial of CO ice under N_2 ice in locations like Sputnik Planitia on Pluto (Glein & Waite 2017). This would remove CO, and potentially CH_4, from the observed inventory reducing the apparent bulk carbon abundance.

Finally, atmospheric processes of photochemistry and escape will change the molecular composition and can lead to permanent loss of some volatiles through haze formation and by loss from the top of the atmosphere (Mandt et al. 2017). Photochemistry will remove N_2 and CH_4 by converting them to larger organic molecules that eventually form haze (Luspay-Kuti et al. 2017; Mandt et al. 2017; Wong et al. 2017), similar to processes on Titan (Mandt et al. 2009, 2014). This would reduce the observed N and C abundances over time, but is limited by the number of photons able to reach Pluto and Triton. Escape can occur through thermal processes where heating of the upper atmosphere gives molecules and atoms enough energy to escape the gravity of a planet. It can also occur through ionization and removal of ions by pickup processes in the solar wind or Triton's magnetosphere. In the case of the atmospheres of Pluto and Triton, both photochemistry and escape would preferentially remove CH_4 (Mandt et al. 2023).

A recent review of the elemental ratios found that the observed volatiles of Pluto and Triton are carbon-poor (Mandt et al. 2023), as shown in Figure 1.1. The only way to produce these observations is through the removal of carbon by preferentially removing CH_4 through atmospheric processes, a very reasonable possibility given that CH_4 would be the lightest main species in the atmosphere and the most easily removed (Mandt et al. 2023). This study also noted that the upper limit for Ar

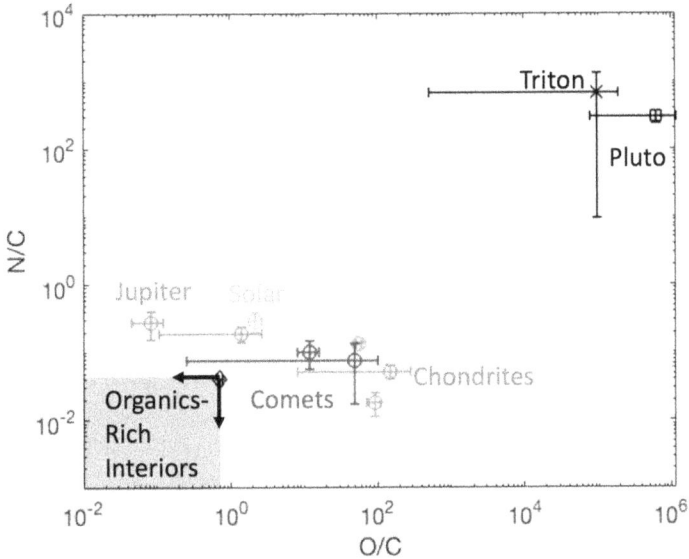

Figure 1.1. Elemental ratios providing information on the relative carbon abundance observed for Pluto and Triton. These ratios are compared to the solar abundances, abundances observed in Jupiter's atmosphere, and analogs for solid materials in the PSN. The two measurements for Jupiter were made by the Galileo Probe Mass Spectrometer and the Juno Microwave Radiometer (see Mandt et al. 2023 and references therein).

in Pluto's atmosphere based on New Horizons observations provides a lower limit for N/Ar that is much larger than the solar value. Because atmospheric loss processes would favor loss of nitrogen over argon, any primordial ratio is likely to be even larger than the lower limit suggesting that the nitrogen at Pluto originated as NH_3 or organics (Mandt et al. 2023).

1.3 Formation of Pluto and Triton's Building Blocks in the PSN

In this section, we draw connections between the compositions of comets, in particular those of N_2-rich bodies such as comet C/2016 R2 (PanSTARRS) (R2), with those of Pluto and Triton. The PSN conditions at which the building blocks of Pluto and Triton could have formed are also discussed.

1.3.1 Link with Atypical Comets

While Pluto and Triton exhibit N_2-rich surfaces (Cruikshank et al. 1993; Owen et al. 1993; Quirico et al. 1999; Merlin et al. 2018), bizarrely, comets are usually depleted in this molecule (Cochran et al. 2000). As cometesimals are thought to have formed from the materials available to them in the PSN at the location of their birth and are relegated to the outer solar system for the majority of their lifetimes, they preserve the composition of the disk at that precise area and moment in time: a core sample of the PSN, making them among the most pristine bodies currently populating our solar system. However, due to their inherent dynamical instability, it is impossible to trace a comet's dynamical history back to its formation location. Their ice-rich composition only shows that they formed in the outer parts of the solar system, beyond Jupiter.

For the most part, they have compositions similar to predicted protosolar abundances, with no depletion of carbon or oxygen, but a clear deficiency of nitrogen (Geiss 1987). For example, comet 1P/Halley has a N/O elemental abundance depleted by a factor of 3 with respect to the solar abundance (Jessberger 1991), and comet C/ 1995 O1 (Hale-Bopp) has an inferred N/O elemental depletion of 15 in the gas phase (Bockelée-Morvan et al. 2000). This low abundance ratio is attributed to a depletion of N_2, the least reactive of all N-bearing species and believed to be the main carrier of nitrogen in the PSN (Feldman et al. 2004). Laboratory studies show that ices incorporated into comets at around 50 K would have $N_2/CO \approx 0.06$ if N_2/CO is $\approx$ 1 in the solar nebula (Owen & Bar-Nun 1995). However, the typical N_2/CO ratio for most comets is $<10^{-3}$, which is much lower than expected (Cochran et al. 2000). Mousis et al. (2012) propose that the nitrogen deficiency in comets may be explained through two possible formation conditions: First, comets could have formed under colder conditions ($\leqslant 20$ K), thus incorporating N_2; the nitrogen would then disappear due to subsequent internal heating from the decay of radiogenic nuclides. Alternatively, comets could have formed under conditions warmer than 20 K, thereby circumventing the trapping of N_2. This dichotomy points toward the necessity of low temperatures (~ 20 K) for the formation of pure N_2 condensate and for the inclusion of N_2 in comets.

Recently, long-period comet R2 was revealed to be a CO-rich comet (Wierzchos & Womack 2018) and strongly depleted in water, with a H_2O/CO ratio of 0.0032 (McKay et al. 2019) with an upper limit of <0.1 (Biver et al. 2018). The spectrum was dominated by bands of CO^+ as well as N_2^+, the latter of which was rarely seen in such abundance in comets (Cochran & McKay 2018; Opitom et al. 2019). It was also found to be both CN-weak and dust-poor (Opitom et al. 2019). This CO-rich and water-poor composition, along with none of the usual neutrals seen in most cometary spectra, makes R2 a unique and intriguing specimen. The observed emission fluxes have been used to calculate ionic ratios of N_2^+/CO^+ in the coma of 0.09 (Anderson et al. 2022), the highest of such ratios observed for any comet so far with high-resolution spectroscopy and in line with the predictions of (Owen & Bar-Nun 1995). This is larger than the best measurement in comet 67P/Churyumov-Gerasimenko, with a N_2/CO ratio of $\sim 2.87 \times 10^{-2}$ (Rubin et al. 2020), though these measurements were obtained much closer to the nucleus using a mass spectrometer. To account for the high N_2/CO and CO/H_2O ratios measured in R2, it has been proposed that its precursor grains condensed in the vicinity of the CO and N_2 icelines (Mousis et al. 2021). This would indicate that R2-like comets formed in a colder environment than the other comets sharing more usual compositions.

Despite its H_2O deficiency, the CO- and N_2-rich composition of comet R2 bears a closer resemblance to the compositions of Pluto and Triton than to those of any other comet observed thus far, hinting at the possible formation from similar building blocks (see Section 1.3.2). Nevertheless, there are discrepancies when comparing R2's relative abundances of CO, CH_4, and N_2 to the surface spectra of Pluto. These inconsistencies may be attributed to variations in the surface compositions of large KBOs, as their surfaces are likely modified by the processes mentioned in Section 1.2.3. Additionally, (Biver et al. 2018) suggested that R2 could be a fragment of a differentiated KBO object, also supported by its dust-poor composition. Desch & Jackson (2021) proposed that R2 could be a nitrogen iceberg, a fragment of a differentiated KBO surface formed during periods of dynamical instability. While this hypothesis presents an intriguing explanation for R2's composition, it is challenged by the difficulty of preserving N_2 content during the energetic collisions necessary to create such fragments (Levine et al. 2021). Consequently, although R2's N_2-rich composition provides compelling insights, the mechanisms of its N_2 retention, as well as its precise formation conditions, warrant further exploration in future studies.

1.3.2 Role of CO and N_2 Icelines

The solar system began as a cloud of interstellar gas and micrometric dust particles, which collapsed to form a protoplanetary disk (PPD), commonly named the PSN. Within such a gas-dominated PPD, dust grains stuck together to form pebbles in the mm–cm range. The difference in the motions of the gas and solid particles in the PPD led to streaming instabilities that locally eased the formation of planetesimals (Youdin & Goodman 2005; Johansen et al. 2012, 2015). These kilometer-sized objects were the building blocks of the solar system, some growing large enough for their gravity to shape them into spheres, becoming planetary embryos. While gas was still in the disk, these planetary embryos became cores of the giant planets, which were embedded in the disk and could migrate, leading to dynamical instability. Once the gas in the disk was depleted, these went on to become the basis of the formation of the rocky planets, as well as dwarf planets, and large moons.

The composition of available planetesimals was determined by their proximity to specific icelines in the disk. An iceline (also ice line, snowline, frostline, or simply condensation line) is defined as the distance where the surface density of vapor of a given species is equal to that of its solid form in PPDs, the PSN, and circumplanetary disks where satellites form. Inside the snowline, water ice evaporates into water vapor. Outside the snowline, ice is present due to the condensation of vapor, though the motion of particles within the disk allows for solids to exist in front of this line as well as some vapors to exist beyond, due to the kinetics of condensation/sublimation.

The location of early icelines determined the distribution of volatile species in the disk, which in turn affected the formation of planetesimals and the building blocks of the planets. Factors such as the luminosity of the Sun, the size of dust grains, and the turbulence of the disk also affected the location of icelines in the PSN. The luminosity of the Sun played a critical role because it determined the temperature of the disk at different distances. As the disk evolved over time, so too have the positions of the icelines in our solar system. This evolution can be caused by various factors such as disk dissipation (Mousis et al. 2020), the presence of an inner gap (Liu 2021), and interplay between planet formation, pebble accretion, and disk evolution (Bitsch et al. 2019). The size of dust grains also played a role because smaller grains could be more easily lifted by the turbulence of the disk, affecting the transport of volatile species and altering the location of icelines. Studies of the asteroid belt show that the location of the water iceline played a crucial role in determining the distribution of hydrated minerals within the belt (Rivkin et al. 2003). Similarly, the location of the CH_4, CO, and N_2 icelines had significant impacts on the formation and composition of the outer planets and their moons (Aguichine et al. 2022; Anderson et al. 2021). The interaction of icy pebbles and their vapors around icelines can lead to significant changes in the composition of the PSN, creating a compositional gradient that may be responsible for the volatile enrichment observed in the giant planets of our solar system (Monga & Desch 2015; Booth et al. 2017; Desch et al. 2017; Mousis et al. 2019; Aguichine et al. 2022).

Figure 1.2 represents the time evolution of the radial profiles in the PSN of the N_2/CO and CO/H_2O ratios relatives to their initial abundances, defined by the enrichment factor f, in both solid and gas phases, and a value of the viscosity parameter α set to 10^{-3}, based on the disk and transport model of Mousis et al. (2021) to which the reader is referred for full details. The adopted α value is well within the range of those typically used in models of PPDs (Hersant et al. 2001; Nelson et al. 2013; Simon et al. 2015). The figure shows that f is flat almost everywhere in the PSN, except in the vicinity of the CO and N_2 icelines, which are located at very close distances from each other, i.e., less than a few tenths of AU, in the 10–15 AU region of the PSN, with the CO iceline situated a bit closer in the disk.

The figure shows that the N_2/CO ratio increases to more than $\sim$10 in the gas phase in the area of the two icelines. This peak corresponds to the supply of N_2 vapor when N_2-rich dust drifts inward the N_2 iceline, which is in excess compared to the CO vapor supplied via backward diffusion beyond the CO iceline. When significant, this peak is preceded by a decrease of the N_2/CO ratio, which corresponds to the supply of CO vapor when CO-rich dust drifts inward the CO iceline. The N_2/CO ratio in the solid phase also experiences important depletions that can reach several orders of magnitude, which correspond to the location where N_2 essentially forms vapor while CO remains in solid phase. This calculation shows it is possible to form dust in a narrow region of the PSN, i.e., within 10–15 AU, with N_2/CO ratios that not only can match the value estimated in comet R2 but also those estimated in the bulk inventory of Pluto and Triton (see Table 1.2). On the other hand, when approaching the CO iceline, the CO/H_2O ratio varies over several orders of magnitude. It forms a peak at the location of the iceline whose magnitude depends on the adopted α–value in the disk (Mousis et al. 2021), as a result of the backyard diffusion and condensation of the CO vapor. If the building blocks of R2 were assembled from such grains, they should present a CO-rich and H_2O-poor composition, in agreement with the observations. However the CO/H_2O ratio is deeply depleted in the solids at distances inward of the CO iceline because this species is in gaseous form. Solids present at very close locations inward the CO iceline would then present CO/H_2O ratios compatible with those estimated for Pluto and Triton.

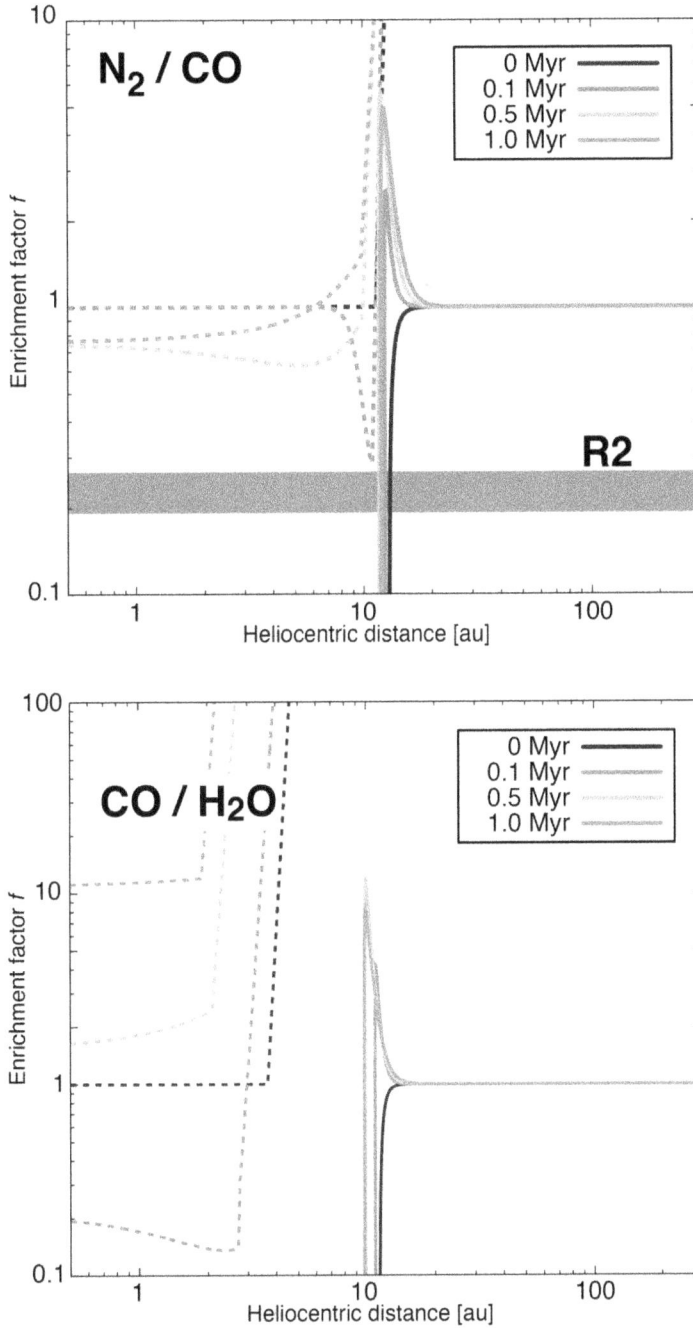

Figure 1.2. Radial profiles of the N_2/CO and CO/H_2O ratios relative to their initial abundances (defined by the enrichment factor f) calculated as a function of time in the PSN for a viscosity parameter $\alpha = \times 10^{-3}$. Dashed and solid lines correspond to vapor and solid phases, respectively. The purple bar corresponds to the N_2/CO ratio measured in comet R2 (see Mousis et al. (2021) for details).

As mentioned above, the heliocentric distances of the iceline locations are purely indicative as they remain strongly model-dependent. Also, the thermodynamic data for pure condensates should be considered with caution at low-pressure conditions, as they have been extrapolated from laboratory data that are several orders of magnitude higher (Fray & Schmitt 2009).

1.4 Dynamical Origin of Triton

Triton is unique among the large moons of the solar system in that it moves in a retrograde and highly inclined—albeit circular—orbit around Neptune.

The origin of Triton is still an open question, with the two competing hypotheses being 1) the capture of Triton from a heliocentric orbit (McKinnon 1984; McKinnon & Leith 1995; Agnor & Hamilton 2006; Vokrouhlický et al. 2008; Nogueira et al. 2011) and 2) an *in situ* formation in the circumplanetary disk (CPD) of Neptune (Harrington & van Flandern 1979; Szulágyi et al. 2018; Li & Christou 2020). In either case, the extraordinary orbital properties of Triton may be a natural consequence of the formation and early dynamical evolution of the outer solar system.

Even if the ultimate proof is still missing, there is some level of consensus in the planetary science community regarding the fact that the giant planets may have formed at different heliocentric distances from those where they are presently observed. The idea of a radial migration of the giant planets—and in particular that of Uranus and Neptune—in the early solar system (Fernandez & Ip 1984; Malhotra 1993, 1995) served as a foundation for many subsequent works including the Nice model (Tsiganis et al. 2005). In this model and its subsequent evolutions (Nesvorný & Morbidelli 2012; Nesvorný 2018), the giant planets are assumed to have formed in a compact configuration (all were located between $\sim$5 and $\sim$ 15 AU from the Sun), to be possibly more numerous (five instead of today's four; Nesvorný 2011; Nesvorný & Morbidelli 2012) and to be surrounded by a planetesimal disk. Eventually, the orbits of the giant planets became unstable. Uranus and Neptune were gravitationally scattered outwards, thereby penetrating the outer disk and scattering its constituents throughout the solar system. In this scenario, Triton and Pluto may have a similar origin in the primordial outer disk with Triton being captured by Neptune via gas drag (McKinnon 1984; McKinnon & Leith 1995) or three-body gravitational interaction (Agnor & Hamilton 2006; Vokrouhlický et al. 2008; Nogueira et al. 2011). In this latter case, Triton's capture efficiency has been estimated to be between 2% (Vokrouhlický et al. 2008) and 50% (Nogueira et al. 2011).

The Nice model scenario may also be compatible with an *in situ* formation of Triton, but requires the existence of five giant planets in the early solar system as advocated by some versions of the Nice model (see above). The fifth giant planet, before its ejection from the solar system, probably encountered other planets as well. This might have been the case for Neptune, with the close encounter causing a drastic flip to Triton's initial prograde, circular and equatorial orbit (Li & Christou 2020). Li & Christou (2020) found that the encounter with the fifth giant planet injects a large moon onto a retrograde orbit with specific angular momentum similar to Triton's in 0.3%–3% of their simulations, which is an order of magnitude less efficient than the above capture scenario.

One should note that, if Triton formed in Neptune's CPD instead of the outer PSN, the presence of important amounts of N_2 and CO in its interior indicates that accretion from building blocks originating from the PSN instead of having condensed in the CPD itself, assuming those two species are primordial. This would imply that the moons formed in a cool and late CPD around Neptune. In contrast, if CH_4 and NH_3 were the dominating N- and C-bearing species in Triton, this would have indicated that their building blocks condensed in a warm and dense CPD (Prinn & Fegley 1981). This implies that, whatever the considered formation scenario is, accretion of Triton in the outer PSN or in Neptune's CPD, its building blocks always originated from the PSN.

1.5 Conclusion

Assuming that CO and N_2 have been captured by Pluto and Triton during their formation, the presence of large amounts of these molecules suggest that the two bodies accreted material originating from the outer regions of the PSN. Specifically, volatile transport/condensation models of the PSN suggest that the formation of Pluto and Triton's building blocks in the vicinity of the N_2 and CO icelines would be compatible with their present-day compositions. With its CO- and N_2-rich composition, and despite a significant H_2O deficiency compared with Pluto and Triton, the unusual comet R2 obviously presents a compositional link with the two bodies, indicating that their building blocks formed in nearby regions of the PSN. Those regions, located around the N_2 and CO icelines present substantial variations of the N_2/H_2O and CO/H_2O ratios in the condensed solids and could encompass the peculiar compositions attributed to R2, Pluto, and Triton. Also, the possible formation of Triton in Neptune's CPD remains consistent with its current composition, provided that its building blocks formed before in the PSN.

On the other hand, nothing warrants that Pluto and Triton's current volatile budgets are primordial, as many post-formation processes could have shaped their current compositions. This implies that any conclusion regarding the origin of the two bodies, and based on their current compositions, must be taken with caution.

To assess the origin of Pluto and Triton's building blocks, a measurement of the deuterium-to-hydrogen ratio in their H-bearing volatiles would be needed. The comparison of such a value with those measured in comets and the ice giants Uranus and Neptune would provide a useful context to the formation of Pluto and Triton's building blocks in the PSN. Measurements of the $^{14}N/^{15}N$ ratio in the N_2 present in Triton and Pluto would be useful as well. Comparing this value with those measured in comets (~127; Rousselot et al. 2014), solar wind (441; Marty et al. 2011), or Titan (143; Mandt et al. 2009) would provide hints about the primordial form under which nitrogen was accreted by the two bodies, as the $^{14}N/^{15}N$ ratio in primordial NH_3 is expected to strongly depart from the one in primordial N_2 (Rousselot et al. 2014).

References

Agnor, C. B., & Hamilton, D. P. 2006, Natur, 441, 192

Aguichine, A., Mousis, O., & Lunine, J. I. 2022, PSJ, 3, 141

Ahrens, C., Meraviglia, H., & Bennett, C. 2022, Geosc, 12, 51

Anderson, S. E., Mousis, O., & Ronnet, T. 2021, PSJ, 2, 50

Anderson, S. E., Rousselot, P., Noyelles, B., et al. 2022, MNRAS, 515, 5869

Bauer, J. M., Buratti, B. J., Li, J.-Y., et al. 2010, ApL, 723, L49

Benne, B., Dobrijevic, M., Cavalié, T., Loison, J. C., & Hickson, K. M. 2022, A&A, 667, A169

Bertrand, T., & Forget, F. 2016, Natur, 540, 86

Bitsch, B., Raymond, S. N., & Izidoro, A. 2019, A&A, 624, A109

Biver, N., Bockelée-Morvan, D., Paubert, G., et al. 2018, A&A, 619, A127

Bockelée-Morvan, D., Lis, D. C., Wink, J. E., et al. 2000, A&A, 353, 1101

Booth, R. A., Clarke, C. J., Madhusudhan, N., & Ilee, J. D. 2017, MNRAS, 469, 3994

Broadfoot, A. L., Atreya, S. K., Bertaux, J. L., et al. 1989, Sci, 246, 1459

Buratti, B. J., Bauer, J. M., Hicks, M. D., et al. 2011, Icar, 212, 835

Canup, R. M., Kratter, K. M., & Neveu, M. 2021, The Pluto System After New Horizons, ed. S. A. Stern, J. M. Moore, W. M. Grundy, L. A. Young, & R. P. Binzel, (Tuscon, AZ: Univ. Arizona Press) 475

Cochran, A. L., Cochran, W. D., & Barker, E. S. 2000, Icar, 146, 583

Cochran, A. L., & McKay, A. J. 2018, ApL, 854, L10

Cook, J. C., Dalle Ore, C. M., Protopapa, S., et al. 2019, Icar, 331, 148

Cruikshank, D. P., Brown, R. H., & Clark, R. N. 1984, Icar, 58, 293

Cruikshank, D. P., Roush, T. L., Owen, T. C., et al. 1993, Sci, 261, 742

Dalle Ore, C. M., Protopapa, S., Cook, J. C., et al. 2018, Icar, 300, 21

Desch, S. J., Estrada, P. R., Kalyaan, A., & Cuzzi, J. N. 2017, ApJ, 840, 86

Desch, S. J., & Jackson, A. P. 2021, JGRE, 126, e06807

Feldman, P. D., Cochran, A. L., & Combi, M. R. 2004, Comets II, ed. M. C. Festou, H. U. Keller, & H. A. Weaver, (Tuscon, AZ: Univ. Arizona Press) 425

Fernandez, J. A., & Ip, W. H. 1984, Icar, 58, 109

Fray, N., & Schmitt, B. 2009, P&SS, 57, 2053

Geiss, J. 1987, A&A, 187, 859

Glein, C. R., & Waite, J. H. 2018, Icar, 313, 79

Glein, C. R., & Waite, J. H. 2017, AGU Fall Meeting Abstracts Vol 2017, P13F

Glein, C. R., Zolotov, M. Y., & Shock, E. L. 2008, Icar, 197, 157

Grundy, W. M., Binzel, R. P., Buratti, B. J., et al. 2016, Sci, 351, aad9189

Harrington, R. S., & van Flandern, T. C. 1979, Icar, 39, 131

Hersant, F., Gautier, D., & Huré, J.-M. 2001, ApJ, 554, 391

Holler, B.J., Young, L.A., Grundy, W.M., & Olkin, C.B. 2016, Icar, 267, 255

Jessberger, E. K. 1991, SSRv, 56, 227

Johansen, A., Mac Low, M.-M., Lacerda, P., & Bizzarro, M. 2015, SciA, 1, 1500109

Johansen, A., Youdin, A. N., & Lithwick, Y. 2012, A&A, 537, A125

Kamata, S., Nimmo, F., Sekine, Y., et al. 2019, NatGeo, 12, 407

Lellouch, E., de Bergh, C., Sicardy, B., Ferron, S., & Käufl, H.-U. 2010, A&A, 512, L8

Lellouch, E., Gurwell, M., Butler, B., et al. 2017, Icar, 286, 289

Levine, W. G., Cabot, S. H. C., Seligman, D., & Laughlin, G. 2021, ApJ, 922, 39

Li, D., & Christou, A. A. 2020, AJ, 159, 184

Liu, Y. 2021, RAA, 21, 164

Luspay-Kuti, A., Mandt, K., Jessup, K.-L., et al. 2017, MNRAS, 472, 104

Malhotra, R. 1993, Natur, 365, 819

Malhotra, R. 1995, AJ, 110, 420

Mandt, K. E., Luspay-Kuti, A., Mousis, O., & Anderson, S. E. 2023, ApJ, 959, 57

Mandt, K. E., Mousis, O., Hurley, D., et al. 2022, NatCo, 13, 642

Mandt, K. E., Mousis, O., Lunine, J., & Gautier, D. 2014, ApL, 788, L24

Mandt, K. E., Waite, J. H., Lewis, W., et al. 2009, P&SS, 57, 1917

Mandt, K., Luspay-Kuti, A., Hamel, M., et al. 2017, MNRAS, 472, 118

Marty, B., Chaussidon, M., Wiens, R. C., Jurewicz, A. J. G., & Burnett, D. S. 2011, Sci, 332, 1533

McKay, A. J., DiSanti, M. A., Kelley, M. S. P., et al. 2019, AJ, 158, 128

McKinnon, W. B. 1984, Natur, 311, 355

McKinnon, W. B., Glein, C. R., Bertrand, T., & Rhoden, A. R. 2021, The Pluto System After New Horizons, ed. S. A. Stern, J. M. Moore, W. M. Grundy, L. A. Young, & R. P. Binzel, (Tuscon, AZ: Univ. Arizona Press) 507

McKinnon, W. B., Glein, C. R., & Rhoden, A. R. 2019, Pluto System After New Horizons (Houston, TX: LPI) 7067

McKinnon, W. B., & Leith, A. C. 1995, Icar, 118, 392

McKinnon, W. B., Lunine, J. I., & Banfield, D. 1995, Neptune and Triton (Tuscon, AZ: Univ. Arizona Press) 807

Merlin, F. 2015, A&A, 582, A39

Merlin, F., Lellouch, E., Quirico, E., & Schmitt, B. 2018, Icar, 314, 274

Monga, N., & Desch, S. 2015, ApJ, 798, 9

Mousis, O., Aguichine, A., Bouquet, A., et al. 2021, PSJ, 2, 72

Mousis, O., Aguichine, A., Helled, R., Irwin, P. G. J., & Irwin, J. I. 2020, RSPTA, 378, 20200107

Mousis, O., Guilbert-Lepoutre, A., Lunine, J. I., et al. 2012, ApJ, 757, 146

Mousis, O., Ronnet, T., & Lunine, J. I. 2019, ApJ, 875, 9

Nelson, R. P., Gressel, O., & Umurhan, O. M. 2013, MNRAS, 435, 2610

Nesvorný, D. 2011, ApL, 742, L22

Nesvorný, D. 2018, ARA&A, 56, 137

Nesvorný, D., & Morbidelli, A. 2012, AJ, 144, 117

Neveu, M., Desch, S. J., & Castillo-Rogez, J. C. 2015, JGRE, 120, 123

Nimmo, F., Umurhan, O., Lisse, C. M., et al. 2017, Icar, 287, 12

Nogueira, E., Brasser, R., & Gomes, R. 2011, Icar, 214, 113

Okumura, F., & Mimura, K. 2011, GeCoA, 75, 7063

Opitom, C., Hutsemékers, D., Jehin, E., et al. 2019, A&A, 624, A64

Owen, T., & Bar-Nun, A. 1995, Icar, 116, 215

Owen, T. C., Roush, T. L., Cruikshank, D. P., et al. 1993, Sci, 261, 745

Prinn, R. G., & Fegley, B. J. 1981, ApJ, 249, 308

Protopapa, S., Grundy, W. M., Reuter, D. C., et al. 2017, Icar, 287, 218

Quirico, E., Douté, S., & Schmitt, B. 1999, Icar, 139, 159

Reynard, B., & Sotin, C. 2023, Earth Planet. Sci. Lett., 612, 118172

Rivkin, A. S., Davies, J. K., Johnson, J. R., et al. 2003, M&PS, 38, 1383

Rousselot, P., Pirali, O., Jehin, E., et al. 2014, ApL, 780, L17

Rubin, M., Engrand, C., Snodgrass, C., et al. 2020, SSRv, 216, 102

Rufu, R., & Canup, R. M. 2017, AJ, 154, 208

Scipioni, F., White, O., Cook, J. C., et al. 2021, Icar, 359, 114303

Sekine, Y., Genda, H., Kamata, S., & Funatsu, T. 2017, NatAs, 1, 0031

Shock, E. L., & McKinnon, W. B. 1993, Icar, 106, 464

Simon, J. B., Lesur, G., Kunz, M. W., & Armitage, P. J. 2015, MNRAS, 454, 1117

Steffl, A. J., Young, L. A., Strobel, D. F., et al. 2020, AJ, 159, 274

Szulágyi, J., Cilibrasi, M., & Mayer, L. 2018, ApL, 868, L13

Tsiganis, K., Gomes, R., Morbidelli, A., & Levison, H. F. 2005, Natur, 435, 459

Vokrouhlický, D., Nesvorný, D., & Levison, H. F. 2008, AJ, 136, 1463

Wierzchos, K., & Womack, M. 2018, AJ, 156, 34

Wong, M. L., Fan, S., Gao, P., et al. 2017, Icar, 287, 110

Yelle, R. V., Lunine, J. I., Pollack, J. B., & Brown, R. H. 1995, Neptune and Triton (Tuscon, AZ: Univ. Arizona Press) 1031

Youdin, A. N., & Goodman, J. 2005, ApJ, 620, 459

Young, L. A., Kammer, J. A., Steffl, A. J., et al. 2018, Icar, 300, 174

Chapter 2

Pluto and Triton: Interior Structures, Lithospheres and Potential for Oceans

Francis Nimmo, Carver Bierson and William B McKinnon

We review our current understanding of the internal structures and dynamics of Pluto and Triton. Pluto has ancient surfaces, rugged topography, extensional tectonics, cryovolcanism, and actively convecting nitrogen plains; Triton exhibits low topographic relief, few tectonic features apart from apparent diapirs and possible cryovolcanic flows, and has a mysteriously young surface age ($\sim$10 Myr). The two bodies have similar radii and bulk densities, suggesting a roughly 2:1 rock: ice ratio by mass, together with an uncertain (but possibly large) fraction of carbonaceous material. Triton is almost certainly differentiated into an ice shell, rocky mantle, and metallic core; Pluto is probably also differentiated, though this process may have only been completed after the Charon-forming impact. Both bodies likely started life with subsurface oceans, and at least for Pluto several lines of evidence suggest an ocean exists at present, beneath a 100–200 km thick conductive ice shell. Little direct evidence exists for an ocean in Triton, but it is energetically plausible and would be a site of strong tidal heating via obliquity tides. Although obliquity tides represent an important difference between Pluto and Triton, Triton's young surface age remains a puzzle. Possible explanations include near-surface convection of weak ices (e.g., CO_2) or cryovolcanic resurfacing, perhaps driven by a slowly thickening conductive ice shell.

2.1 Introduction

Triton and Pluto represent two of the most mysterious worlds of the outer solar system. Part of the mystery arises from a paucity of data: Each body was imaged during a single, distant spacecraft flyby (*Voyager 2* in 1989 and *New Horizons* in 2015), so that in each case roughly half the body remains unimaged or poorly imaged. But the deeper mystery is why these bodies look so different, and what that is telling us. Both bodies are similar in mass, size (Table 2.1) and surface

doi:10.1088/2514-3433/ad5278ch2
2-1

Table 2.1. Characteristics of Triton and Pluto

Quantity	Units	Triton	Pluto	Charon	Notes
$\bar{R}$	km	1353.4 ± 0.9	1188.3 ± 1.6	606 ± 1.0	
ρ_b	kg m^{-3}	2060 ± 10	1854 ± 11	1701 ± 33	
f_s	—	0.75	0.68	0.62	$\rho_s = 3.5$ g cc^{-1}
					and $\rho_i = 0.95$ g cc^{-1}
R_c	km	1028	840	412	
P	days	5.877	6.387	6.387	Synchronous
q	—	2.67×10^{-4}	2.51×10^{-4}	2.73×10^{-4}	
$(a - c)_{obs}$	km	2.2 ± 0.8	<7	<3	Thomas (2000) and
					(Nimmo et al. 2017)
$(a - c)_{pred}$	km	1.8	0.5	0.8	Assuming $h_2 = 2.5$
g	m s^{-2}	0.78	0.62	0.29	
F_s	mW m^{-2}	3.1	2.1		Conductive (Equation (2.6))
F_{rad}	mW m^{-2}	3.1	2.2	1.0	Radiogenic, present-day

Charon is also shown for comparison.

composition, and Triton likely originated in the transneptunian Kuiper Belt, or its precursor, where Pluto resides today. Yet Triton has a young, smooth surface and active geysers, while Pluto has ancient, faulted terrains, high mountain ranges, nitrogen ice glaciers, and possible cryovolcanoes. These observations provide clues to the interior structure and evolution of these bodies, which is the focus of this chapter.

An important reason for considering this topic is that Pluto and Triton may both possess subsurface oceans (see below). This makes them potentially habitable environments, a major focus of the most recent Planetary Science & Astrobiology Decadal Survey, *Origins, Life, and Worlds* (National Academies of Sciences, Engineering, and Medicine, 2023). As such, the information and hypotheses presented here could in principle form the basis for future spacecraft mission proposal efforts.

The interior structures of these bodies have been reviewed by various authors: McKinnon et al. (1995) (Triton), McKinnon et al. (1997) (Pluto), Nimmo & McKinnon (2021) (Pluto) and Hussmann et al. (2010). Good general overviews of each body include McKinnon & Kirk (2014) (Triton), Stern et al. (2018) and Moore & McKinnon (2021) (Pluto), while Schenk et al. (2021) provide a modern view of Triton's topography and geological character. However, these works (with the notable exception of Schenk et al. 2021) did not in general perform a comparative analysis of the similarities and differences between Pluto and Triton, which we attempt here. Other chapters in this volume relevant to our chapter include those on Pluto and Triton's formation (Chapter 1), surface (Chapter 4), cryovolcanism (Chapter 3), and open questions (Chapter 12).

There are two important differences between Pluto and Triton. The first is scientific: the difference in surface age of the two bodies. Pluto has heavily cratered terrains that most likely date to shortly after the formation of the solar system (Moore et al. 2016; Moore & McKinnon 2021; Singer et al. 2021), as well as actively convecting nitrogen deposits that have negligible surface ages (McKinnon et al. 2016). Triton is very different: Its surface age is young (possibly <10 Myr) everywhere imaged. The cratering arguably appears to be dominated by planet-ocentric impactors (Schenk & Zahnle 2007 but see Stern & McKinnon 2000; Mah & Brasser 2019), unlike most outer solar system bodies where heliocentric impactors dominate. Why Triton has such a young surface presents a major unsolved problem.

The second difference is practical. The *New Horizons* spacecraft carried spectrometers which allowed high-resolution compositional maps of Pluto's surface to be produced (e.g., Grundy et al. 2016). At Triton, however, the only compositional information is from Earth-based instruments, with correspondingly poor (i.e., hemispheric) spatial resolution (e.g., Grundy et al. 2002). Interpretation of Triton surface features is significantly handicapped as a result.

The rest of this chapter is organized as follows. We begin with a discussion of the constraints on the bulk structures of these bodies (Section 2.2). We then discuss the theoretical and observational arguments for and against the presence of subsurface oceans (Section 2.3), and then look at the evidence concerning the near-surface structure (i.e., the lithosphere; Section 2.4). We conclude with a summary and suggestions for future work (Section 2.5). We note here, and will repeat this point periodically below, that many of our conclusions are tentative and based on scant evidence. Readers are advised to retain an appropriate level of scepticism.

2.2 Bulk Structures

In considering the bulk structure of a planetary body, there are at least two fundamental questions: What is it made of? and is it homogeneous, or have its constituent components separated (differentiation)? Below we review the limited information available in an attempt to answer these questions.

2.2.1 Observations

2.2.1.1 Bulk Density
The most basic constraint on a body's internal structure is its bulk density. For Pluto and Triton, the densities are intermediate between those of water ice (≈ 1 g cc^{-1}) and (ordinary) chondritic rock (≈ 3.5 g cc^{-1}) (see Table 2.1). If we assume a simple two-layer model with a rocky core overlain by an ice/water layer (the hydrosphere), then the inferred rock mass fraction can be derived assuming end-member densities and the following equation:

$$\frac{1}{\rho_b} = \frac{f_s}{\rho_s} + \frac{(1 - f_s)}{\rho_i} \tag{2.1}$$

Here, ρ_b is the bulk density, ρ_s and ρ_i are the silicate and ice end-member densities, and f_s is the rock mass fraction. If we take $\rho_s = 3.5$ g cc^{-1}, similar to that of ice-free Io, and $\rho_i = 0.95$ g cc^{-1}, approximating the density of the hydrosphere, then we find that both Pluto and Triton have rock mass fractions of roughly 70% (Table 2.1).

At face value, Triton is more rock-rich, but if self-compression of Triton's hydrosphere is taken into account the rock mass fractions of the two are closer (McKinnon et al. 1997). The corresponding radius of the rocky core R_c is also about 70% of the full radius for both bodies, while the nominal hydrosphere thicknesses are 345 and 325 km, respectively. These values could be significantly modified if porosity, high-pressure ices, hydrated silicates or carbon compounds play an important role (see below). Taken at face value, the rock/ice mass ratio of about 2:1 or 3:1 is higher than the expected initial solar nebula ratio of about 2:3 to 1:2, if all volatile ice formers (e.g., CH_4) accrete as solids. If, however, oxygen is sequestered as CO in the outer protosolar nebula, a highly volatile and difficult to condense compound, and thus unavailable to form silicates or oxides, then the high rock/ice ratios of Pluto and Triton can be rationalized (McKinnon & Mueller 1988; Wong et al. 2008).

Porosity, if present, would reduce the effective density of the different layers. Porosity is typically removed by ductile flow once temperatures get high enough. In Pluto, porosity in the ice shell (if present initially) could survive down to depths of 50–170 km, depending on its heat flow history (Bierson et al. 2020). This would reduce the shell density by of order 10%, reducing the hydrosphere thickness by a few tens of kilometers. For Triton, where heat fluxes were very high in the past and the surface is young, ice shell porosity is much less likely to be an issue.

The silicate core of the Moon Enceladus (radius 252 km) has a density of about 2.5 g cc^{-1} indicating either high porosity or hydrated silicates or both (Hemingway et al. 2018). Since pressures within Pluto and Triton are higher than on Enceladus, silicate porosity is less likely to survive, except perhaps in the shallowest rocky layers. Furthermore, Triton's early heating episode (see Section 2.2.2.2) probably efficiently annealed any initial silicate porosity. At present, there is only marginal evidence for the existence of hydrated silicates on either Pluto or Triton (see Denton et al. 2021). Such minerals are certainly theoretically possible (Cioria & Mitri 2022). However, the inner portion of Pluto's core (perhaps 50% by volume) will likely have reached the $\approx$ 800 K dehydration temperature of hydrated minerals, based on radiogenic heating models (see McKinnon et al. 1997; Kamata et al. 2019; Bierson et al. 2020 and Figure 2.2), thus driving off water and increasing the local density. If ρ_s were set to 2.5 g cc^{-1}, the values of f_s would increase and the hydrosphere thicknesses decrease, to about 190 and 150 km for Pluto and Triton, respectively.

So far, we have implicitly assumed that rock and water ice are the main constituents of Pluto and Triton. However, a third important contributor may be organic carbon compounds (broadly speaking, carbonaceous, including graphite). As reviewed in McKinnon et al. 2021, models based on cometary and especially on *Rosetta* data have proposed bulk compositions consisting of up to about 30% by mass organics/hydrocarbons/graphite. While there is no direct evidence for either soluble or insoluble organic material (IOM) in substantial quantities in the interior of either Pluto or Triton, their presence (at least) at depth cannot be excluded, and may even be indicated, based on observations of reddish, apparent cryovolcanic deposits on Pluto (Cruikshank et al. 2019). If so, this could have significant effects on the bulk chemistry and inferred structure of Pluto and Triton (organic matter being low density). In particular, a subsurface carbon reservoir would also imply the presence of nitrogen (based on a cometary N/C ratio of a few percent (Fray et al. 2017)). Heating of this reservoir (by either radioactive decay or tides) would produce N_2 among other compounds (McKinnon et al. 2021) and could explain the low CO/N_2 ratio at Pluto's surface (Glein & Waite 2018).

Finally, the role of ammonia should be mentioned. Based on typical cometary ammonia/water ratios of about 1% (Mumma & Charnley 2011), NH_3 is not likely to be volumetrically significant, unless it too was produced thermochemically (Shock & McKinnon 1993; see discussion in McKinnon et al. 2021). Ammoniated compounds have been observed in some regions of Pluto and have a relatively short lifetime at the surface, perhaps indicating eruption from a subsurface ocean (Cruikshank et al. 2019, 2021). If present, ammonia can greatly reduce the freezing point of any water (Kargel 1992; Leliwa-Kopystyński et al. 2002) and therefore has an important effect on the thermal evolution of water-rich bodies. Methanol (CH_3OH) has similar antifreeze properties and is several times more abundant in comets than NH_3, so may be equally or more important on Pluto and Triton. The potential role of ammonia is discussed in more detail below (Section 2.3.1.1 and 2.3.2).

2.2.1.2 Moment of Inertia

The next most basic constraint is the body's moment of inertia (MoI), which indicates the extent to which a body is differentiated. If the body acquired its shape while behaving like a fluid (i.e., satisfying hydrostatic equilibrium), then its shape can be used to derive its moment of inertia via the so-called Darwin–Radau relation (e.g., Gao & Stevenson 2013). Unfortunately, current shape measurements are insufficiently precise to provide strong constraints.

Hydrostatic planets and satellites adopt oblate or ellipsoidal shapes, respectively, with short (c) and long (a) axes. For a slowly spinning hydrostatic body, the predicted flattening is given to first order by

$$(a - c)_{\mathrm{pred}} = afqh_2 \tag{2.2}$$

where q is a factor comparing the importance of spin (centrifugal acceleration) and gravity, h_2 is a so-called Love number, which depends on the moment of inertia of the body, and f is a factor that is $\frac{1}{2}$ for a planet and 2 for a synchronously spinning satellite.[1] The quantity q is given by

$$q = \frac{3\omega^2}{4\pi G \rho_b} \tag{2.3}$$

where ω is the spin angular frequency. The Love number h_2 describes the ability of a body to resist tidal or rotational deformation. The h_2 of a uniform, fluid body is 2.5, and this value is reduced if the body's mass is more centrally concentrated (i.e., it has a lower MoI) or if the body is more rigid. Given the Love number (derived from the observed flattening via Equation (2.2)), and the assumption of hydrostatic equilibrium, the MoI can be determined from the Darwin–Radau relation.

---

[1] For Pluto, $f = 0.72$ because of the tidal distortion caused by Charon; see Nimmo & McKinnon 2021.

As shown in Table 2.1, only upper bounds on the flattening of Pluto and Charon have been determined, so no constraint on MoI can be derived. For Triton, *Voyager* measured a flattening of 2.4 ± 0.8 km (Thomas 2000), which yields a lower bound of 1.6 km. This value indicates that h_2 exceeds 2.2, which via the Darwin–Radau relationship implies a moment of inertia larger than 0.36. This value is comparable to the MoI derived for Callisto (Anderson et al. 2001), which might indicate a partially differentiated interior. However, given the uncertainties in the shape and the expectation of complete differentiation, we do not view this constraint as a strong one.

In principle, measurement of degree-2 gravity moments can be used instead of shape to determine the moment of inertia, again via the Darwin–Radau relationship. However, because both the *Voyager 2* and *New Horizons* flybys were distant from their respective targets (40,000 and 13,700 km, respectively), no such gravity moments were measured.

2.2.2 Differentiation State

We concluded above that Pluto and Triton both consist of a roughly 2:1 rock–water ice mixture, possibly with carbon compounds mixed in. Here, we consider whether these components have separated (differentiated) or not.

2.2.2.1 Pluto

One important observation is the dominance of extension seen in surface tectonics (Moore et al. 2016). If Pluto were an undifferentiated, solid ice–rock mixture, the deeper ice would undergo a phase change to higher-density ices as the body cooled over time, the ice II stability field being especially temperature sensitive (McKinnon et al. 1997). Such phase changes would lead to surface contraction, for which there is essentially no geological evidence. This is currently the strongest argument that Pluto has undergone differentiation in bulk. Crustal blocks (mountains) exposed at the edge of Sputnik basin are also water–ice rich spectrally (e.g., Protopapa et al. 2017), inconsistent with a dark, primordial ice/carbon/rock KBO composition, implying at least some degree of differentiation.

However, at its formation, Pluto may have been undifferentiated or only partially differentiated. If Charon is the product of a giant impact with Pluto, the Charon-forming material would have been derived preferentially from the near-surface of both proto-Pluto and its comparably sized impactor, if an impact-generated disk forms (Canup 2005). If the Pluto progenitors had already differentiated at this point, Charon would have ended up much more ice rich than is the case (Table 2.1). Moreover, it is much easier to explain the large Charon/Pluto mass ratio if Charon is the product of a grazing, intact capture, which is facilitated dynamically if both impacting bodies are undifferentiated (no central mass concentrations). Subsequent work (Canup 2011) showed that slightly (10%) differentiated progenitors favor both the formation of a large rock-rich Charon and the ejection of icy debris into a disk that could give rise to Pluto–Charon's small, exterior icy satellites. Taken together, these arguments suggest that Pluto differentiated, or mostly differentiated, after the

Charon-forming impact, either as a result of the impact itself or because of long-lived radioactive decay (see Section 2.3.1.3).

A further argument is based on the identification of lineated features antipodal to Sputnik Planitia, a presumed impact basin. These features are only seen in very-low-resolution images, so their origin is uncertain. Nonetheless, if they formed by focusing of seismic waves antipodal to the impact site (as has been suggested for other bodies such as Mercury; Schultz & Gault 1975), models show that such an origin is consistent with the existence of a hydrated silicate core with an ocean on top (Denton et al. 2021). Unfortunately, models without a silicate core were not run, so it is not yet clear whether this argument requires a core. A key parameter in these calculations is the ocean thickness, which would also be sensitive to ice shell thickness and the presence of any carbon-rich layer at the time of the impact.

If Pluto were fully differentiated, then one might expect a central iron core to have formed provided FeS and iron metal were present, as even long-lived radiogenic heating reaches the Fe–FeS eutectic (lowest) melting temperature of about 1270 K (McKinnon et al. 1997). There is no evidence for a core-generated magnetic field (McComas et al. 2016), indicating that either no such core formed, that an iron core exists but is solid, or that a liquid iron core has formed but is not currently convecting.

Finally, the existence of a subsurface ocean on Pluto is suggested by a variety of arguments (see Section 2.3). If these are correct, then Pluto must be at least partially differentiated. Further, any initial differentiation, i.e., separation of rock from ice under gravity due to ice melting, brings with it the possibility of water–rock chemical interactions. Specifically, the formation of hydrated minerals from the alteration of common cometary (and presumed KBO) minerals such as olivine and pyroxene (serpentization) is generally highly exothermic (e.g., Grimm & McSween 1989; Loveless et al. 2022). The heat released can melt more ice and thus accelerate the differentiation process.

2.2.2.2 Triton

For Triton, there are almost no observational constraints on its differentiation state, other than the obvious: its surface composition and muted topography indicate a near-surface dominated by ices. Its shape permits a wide range of internal structures (see Section 2.2.1.2). Although there are a few apparently extensional features (Chapter 4), extensional tectonics does not dominate as on Pluto, and its very young surface age (as low as ~10 Myr) means that no evidence of early evolution is retained.

However, there are strong theoretical arguments in favor of Triton being fully differentiated. Its retrograde orbit is most easily explained by capture, presumably from a body originally in the Kuiper Belt (McKinnon 1984; Goldreich et al. 1989; Nogueira et al. 2011). The currently favoured mechanism for capture is binary exchange, in which a binary KBO lost one member of its pair of bodies during a close encounter with Neptune (Agnor & Hamilton 2006). But the key point is that Triton's initial eccentricity will have been of order unity, while at the present day its orbital is almost perfectly circular. The circularization of the orbit was achieved by tidal dissipation and interaction with any original Neptune satellites (Ćuk & Gladman 2005; Rufu & Canup 2017), which potentially released very large amounts of heat. Expressing the heat released as an equivalent change in mean temperature ΔT_{circ} yields

$$\Delta T_{\text{circ}} = \frac{GM_p}{2aC_p} \approx 10{,}000 \ K \qquad (2.4)$$

where M_p is the mass of Neptune, a is the semimajor axis, and C_p is the mean specific heat capacity of Triton, taken to be 10^3 J kg$^{-1}K^{-1}$. Such a large temperature increase strongly suggests that complete differentiation, and prompt formation of an ocean, occurred.

In reality, of course, what really matters is the rate at which heat was generated: a sufficiently slow rate of heat generation would avoid the total melting and volatile loss implied by Equation (2.4). The heating rate in turn depends on the dissipative properties of Triton's interior. However, models which take these effects into account (e.g., Ross & Schubert 1990) do not change the basic conclusion that Triton experienced complete melting and differentiation as its orbit circularized. We do note that late capture or circularization, such as advocated by Ross & Schubert (1990) is ruled out by the presence/survival of Neptune's irregular satellite system. These bodies were unknown in the *Voyager* era and are thought to be a relict of early giant planet instability and migration (e.g., Nesvorný 2018); they would have been scattered or destroyed if Triton's circularization had occurred subsequent to their emplacement.

A potentially important consequence of this massive tidal heating is that Triton may have lost some of its more volatile constituents (Lunine & Nolan 1992). The fact that Triton is slightly more rock-rich than either Pluto or Charon is possible evidence for this hypothesis, but many volatiles (N_2, CO, CO_2, CH_4) have evidently been retained, at least in part (Cruikshank et al. 1998). The heating would also surely have driven hydrothermal reactions in the silicate core (Shock & McKinnon 1993). Evolution of ocean chemistry and composition of Triton's atmosphere was likely as a result. For a massive atmosphere raised by tidal heating, and possibly greenhouse sustained, further processing and/or loss of volatile components such as CH_4 and NH_3 by impacts, photochemistry, and EUV heating is conceivable (Lunine & Nolan 1992; see discussion in McKinnon et al. 1995). Formation of clathrates—gas molecules encased in an ice scaffold—at the base of Triton's growing ice shell, such as proposed by Kamata et al. (2019) for Pluto, may also have been possible.

The immense energy release also strongly suggests that Triton should possess a central iron core. No magnetic field measurements are available, and no theoretical models of the evolution of such a core appear to have been published. But there is at least the possibility that Triton, like Ganymede, could be generating an active magnetic field at present. The size of any such core is unknown as it depends on the availability of core-forming elements, not just Fe–Ni, but light alloying elements such as S, O, and C as well.

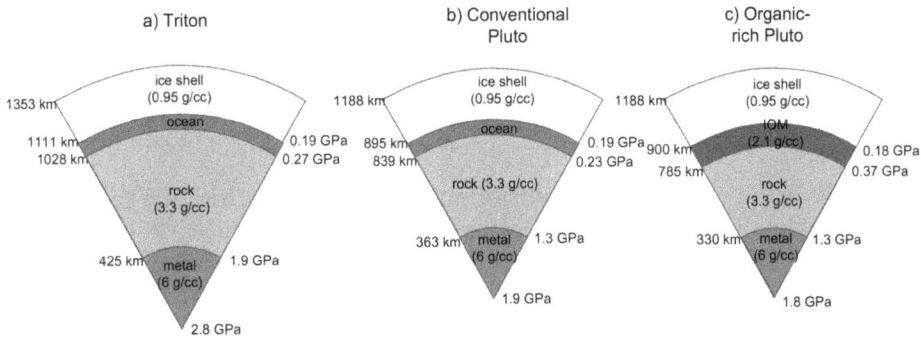

Figure 2.1. Schematic structures of Pluto and Triton, shown at the correct relative scale. Two hypothetical Plutos are shown, one organic-poor and the other organic-rich (see text). IOM is the insoluble organic material. There is no direct evidence for a metallic core on either body although they are expected on theoretical grounds. In principle, an ocean and an IOM-like layer could both be present.

2.2.2.3 Summary

Pluto and Triton both probably consist of a ~2:1 rock–ice mixture, perhaps including a significant fraction of carbon compounds. Triton's volatile inventory may have been reduced during early tidal heating. The cores of these bodies may have undergone partial hydrothermal alteration, but are unlikely to retain significant porosity today. For Pluto, several observations point toward a differentiated structure—but full differentiation most likely occurred after the Charon-forming impact. For Triton, few observations are available, but theoretical calculations of the heating experienced during orbital circularization strongly point toward a fully differentiated structure.

Figure 2.1 shows hypothetical internal structures for the two bodies, at the correct relative scale. Both organic-rich and organic-poor Plutos are shown. Iron cores are depicted although (as discussed above) there are only theoretical arguments, and no direct evidence, of their existence. The existence (or not) of subsurface oceans is discussed below.

2.3 Arguments for Subsurface Oceans

Whether or not Pluto and Triton possess subsurface oceans is a question of both geophysical and astrobiological significance. Here we are particularly interested in present-day oceans: smaller bodies like Charon may have possessed oceans in the past, but are unlikely to do so at the present day.

2.3.1 Theory

A basic requirement for the long-term maintenance of an ocean is that the heat transferred to the bottom of the ocean equals or exceeds the heat transported out of the ocean and across the ice shell. The heat input depends on radioactive decay and any primordial heat left over from accretion or tidal heating; the heat loss is determined by the ice shell characteristics and in particular whether it is conductive or convective. We explore both possibilities below.

2.3.1.1 Steady-state Conduction

If the ice shell is conductive, then the steady-state temperature structure is given by

$$\ln(T/T_s) = \ln(T_i/T_s)\frac{R_i}{R - R_i}\left(\frac{R}{r} - 1\right) \tag{2.5}$$

where here we are accounting both for spherical geometry and the fact that the thermal conductivity of ice is temperature-dependent, going as k_0/T where k_0 is a constant. In this equation, T_s is the surface temperature, T_i and R_i are the temperature and radius at the base of the ice shell, and r is the radial coordinate. The corresponding steady-state heat flux, expressed at the surface, is given by

$$F_s = k_0\frac{\ln(T_i/T_s)}{(R - R_i)}\frac{R_i}{R} \tag{2.6}$$

The minimum heat flux required is obtained in the case of a vanishingly thin ocean, in which case $R_i = R_c$, the location of the top of the rocky core. Taking $k_0 = 567$ W m^{-1} (Klinger 1980), $T_i = 251$ K (the ice I–water–ice III triple-point temperature), $T_s = 40$ K and using the values of R_c from Table 2.1, we find a minimum surface heat flux of 2.1 mW m^{-2} would be required to maintain a present-day thin ocean on Pluto. On Pluto, the triple-point pressure (0.21 GPa) is just exceeded at the bottom of the hydrosphere (Figure 2.1). For Triton, its greater gravity implies that the triple-point pressure is reached at a correspondingly shallower depth, about 260 km, implying a minimum heat flux of 3.1 mW m^{-2} is necessary to sustain a present-day thin ocean (in this case a perched ocean, sandwiched between ice I and ice III). Porosity, if present on either body, would reduce the effective thermal conductivity and the heat required to maintain an ocean.

One such source of heat is radioactive heating in the silicate core. Different authors have assumed different present-day chondritic heating rates in the silicates: 3.4×10^{-12} W kg^{-1} (Robuchon & Nimmo 2011), 4.5×10^{-12} W kg^{-1} (Hussmann et al. 2006) and 5.5×10^{-12} W kg^{-1} (Brown et al. 1991). Using the intermediate value, the resulting present-day surface heat fluxes are 3.1 and 2.2 mW m^{-2}, respectively, for Triton and Pluto. These values would be reduced if carbon compounds, rather than rock, made up a significant fraction of the interior. Comparison of these heat production rates with the minimum conductive rates required to sustain an ocean suggests that oceans on both Triton and Pluto would be marginal.

However, this simple estimate neglects several important factors. Heat produced at earlier times (e.g., due to accretion and early radiogenic decay) and then stored will also contribute to the present-day heat flux. Lower-conductivity materials (porous ice, clathrates) would reduce the conductive heat flux and make oceans more viable. The presence of NH$_3$ in the ocean would reduce the temperature at the base of the ice shell (see Section 2.3.1.2), allowing an ocean to remain viable at lower heat fluxes. Hussmann et al. (2006) concluded that 5 wt% NH$_3$ was sufficient to permit present-day oceans at both Triton and Pluto; progressive enrichment of NH$_3$ in the remaining liquid as freezing proceeds aids this process (Hammond et al. 2018).

At least for Triton, tidal heating may provide an additional source of energy (see Section 2.3.1.3).

These factors will all contribute to the survival of an ocean. The one factor that would rule out an ocean is if the ice shell is convecting. In this case, heat removal is more rapid and an ocean is unlikely to develop (or survive). Robuchon & Nimmo (2011) found that for Pluto ice shell convection meant an ocean never developed. Whether or not convection occurs depends mostly on the temperature at the base of the ice shell, and is discussed in more detail below.

2.3.1.2 Convection

Whether convection occurs in an ice shell heated from below depends on the Rayleigh number, which is defined as

$$Ra = \frac{\rho_i g \alpha (T_b - T_s) d^3}{\kappa \eta_b} \tag{2.7}$$

where g is the surface acceleration due to gravity, α is the thermal expansivity, T_b and T_s are the temperatures at the surface and the base of the ice shell, d is the shell thickness, κ is the thermal diffusivity of ice, and η_b is the viscosity of ice at the base of the shell. For strongly temperature-dependent materials like ice, convection occurs when the Rayleigh number exceeds some critical value Ra_c. Convection in spherical shells operates slightly differently to convection in a Cartesian box (Yao et al. 2014), so for Newtonian fluids we write (Robuchon & Nimmo 2011):

$$Ra_c = 20.9[x\gamma(T_b - T_s)]^4 \tag{2.8}$$

where x is a dimensionless quantity that takes into account the spherical geometry of the shell and γ defines the sensitivity of viscosity to temperature: $\eta(T) = \eta_b \exp(-\gamma[T - T_b])$. Here, $\gamma = Q/RT_b^2$, where Q is the activation energy (≈ 50 kJ mol$^-$; Goldsby & Kohlstedt 2001) and R is the gas constant. The quantity x depends on the ratio R/R_c, which for both Pluto and Triton is around 0.75. The corresponding value of x is 1.65 (Robuchon & Nimmo 2011), while $\gamma(T_b - T_s) \approx 20$, $T_b = 270$ K, $T_s = 40$ K and the γ is appropriate for the grain-boundary sliding rheology (Goldsby & Kohlstedt 2001). So for both Pluto and Triton, $Ra_c \approx 2 \times 10^7$.

Taking d to be the entire ice shell thickness, the actual Rayleigh numbers for Pluto and Triton are then

$$Ra_P = 5.6 \times 10^9 \left(\frac{d}{345 \text{ km}}\right)^3 \left(\frac{10^{14} \text{ Pa s}}{\eta_b}\right),$$

$$Ra_T = 5.7 \times 10^9 \left(\frac{d}{325 \text{ km}}\right)^3 \left(\frac{10^{14} \text{ Pa s}}{\eta_b}\right) \tag{2.9}$$

Here, we are taking $\alpha = 10^{-4}$ K^{-1} and $\kappa = 10^{-6}$ m^2 s^{-1}; ice near its melting point typically has a viscosity of around 10^{14} Pa s (Durham & Stern 2001). What Equation

(2.9) implies is that convection will occur unless the shell is less than roughly 50 km thick (for our choice of $T_b = 270$ K), or the basal viscosity of the shell exceeds roughly 2×10^{16} Pa s. A basal viscosity high enough to prevent convection implies a basal temperature lower than roughly 210 K.

There are at least two ways to achieve such low temperatures and prevent convection. The first is to appeal to a cold ocean. While salts (e.g., NaCl) and pressure changes would only lower the freezing point by a few tens of K at most, an ocean containing between 20 and 25 wt% NH_3 will have a freezing temperature of about 210 K, depending on the pressure (Hogenboom et al. 1997). Although ammoniated compounds have been observed at the surface (Cruikshank et al. 2019), the very high NH_3 concentrations required are hard to reconcile with relatively low cometary NH_3 concentrations (Le Roy et al. 2015), unless there were other (thermochemical) sources of NH_3 (say, derived from ammonium salts; Altwegg et al. 2020), or speculatively, NH_3 became concentrated in the ocean as it froze and the ice shell thickened over time (Hammond et al. 2018).

The second option is slightly more subtle and appeals to a methane clathrate layer at the base of the ice shell (Kamata et al. 2019). Because clathrates have a thermal conductivity an order of magnitude less than that of water ice (Ross & Kargel 1998), there will be a large temperature drop across the clathrate layer. This in turn means that the water ice shell will be thinner and cold and unlikely to convect.

If the ice shell is convecting, an estimate of the effect on the cooling rate can be obtained by calculating the Nusselt number, Nu, which is the ratio of the convective heat flux to the equivalent conductive heat flux across the layer (e.g., Gaeman et al. 2012). For stagnant-lid convection, we have (Solomatov 1995)

$$Nu \approx \frac{1}{2} Ra^{1/3} [\gamma (T_b - T_s)]^{-4/3} \approx 16 \left(\frac{Ra}{5.4 \times 10^9} \right)^{1/3} \tag{2.10}$$

when Ra is not too close to Ra_c.

Thus, a warm convecting ice shell would lose heat rapidly. However, as discussed in Section 2.4.3.1, at least for Pluto there are arguments that the shell is cold and conductive. We also note that Ra_c provides a minimum Nu for Triton of ≈ 2.5, which even for a maximally thick ice shell could exceed the heat flow available from the core and ocean below. In this case, a convective steady state is not possible today and a conductive solution is preferred.

Summary A pure water-ice shell on Pluto or Triton would convect unless either the base of the ice shell were cold (<210 K) or the ice shell were thin (<50 km), subject to the usual rheological uncertainties. Such a thin shell is unlikely, but a cold shell could arise because of an NH_3-rich ocean or a layer of clathrates at the base of the ice shell.

2.3.1.3 Thermal Evolution

A little more insight into the origin and survival of subsurface oceans can be obtained by looking at their evolution over time. However, because Pluto and Triton are relatively small, they only retain a limited "memory" of conditions at early times. A conductive sphere 1000 km in radius has a heat diffusion timescale of $R^2/\pi^2\kappa \approx 3$ Gyr (Carslaw & Jaeger 1986), so that any primordial heat locked up in the silicate core will have mostly leaked out by the present day. Before examining thermal evolution in detail, we will consider the most likely energy sources.

Accretion

Accretion is the process by which material accumulates to form a planetary body. The maximum temperature change resulting from accretion (assuming no heat loss via surface radiation and zero velocity at infinity) is given by

$$\Delta T_{acc} = \frac{4}{5}\frac{\pi G R^2 \rho_b}{C_p} \tag{2.11}$$

For Pluto and Triton, the corresponding values are 440 and 630 K, respectively.

If the majority of accretion energy is retained, these values suggest that ice should have easily reached its melting point (250-to-270 K, depending on pressure) and generated an early ocean. However, if the accreting objects are small and the interval between impacts is sufficiently long, heat will not be buried in the interior of the growing body but will rather be radiated to space (e.g., McKinnon et al. 2017). The size distribution of impactors and the duration of accretion is thus important; Bierson et al. (2020) concluded that an accretion time <30 kyr was sufficient to guarantee formation of an early ocean, for small impactors. Whether such accretion timescales are realistic is unclear. So-called "pebble accretion" in the primordial Kuiper Belt gas nebula is expected to be slow and inefficient initially (Johansen et al. 2015), whereas the early "streaming instability" produces very rapid growth, but only for initial planetesimal sizes circa 100 km (Morbidelli & Nesvorný 2020). Oligarchic collisions (Ormel et al. 2010) and growth among these planetesimals, to the degree that these occur, should deeply bury impact heat irrespective of accretion timescale. Some degree of melting was thus likely as KBOs grew to proto-Pluto scale (pre giant impact; McKinnon et al. 2021). But given the overall uncertainties in KBO growth, we conclude that an accretion-driven early ocean is permitted but not assured.

Radiogenic Heating

Long-lived radioactive elements (^{40}K,U,Th) can potentially generate and maintain an ocean. The equivalent temperature change due to long-lived radiogenic heating ΔT_{rad} depends on the silicate mass fraction of the body f_s and is calculated making the (unrealistic) assumption that all heat is retained (i.e., none is lost via conduction). Robuchon & Nimmo (2011) found a value of 780 K for Pluto. Carbon compounds, if present in significant quantities, would reduce this value. Nonetheless, the calculations shows that the total radiogenic energy generated is comparable to that produced by accretion, but is of course released over a much longer timescale.

An alternative method of generating an early ocean is heating via ^{26}Al decay (Merk & Prialnik 2006). This isotope is extremely energetic, but has a half-life of only 0.7 Myr and is thus effectively exhausted by 4 Myr after solar system formation. On the basis of the apparent survival of porosity in smaller Kuiper

Belt Objects, Bierson & Nimmo (2019) argued that [26]Al heating was not important. If it had been important, Pluto would have differentiated very early, which is hard to reconcile with the similar bulk densities of Pluto and Charon (see Section 2.2.1.1 and McKinnon et al. 2021), though it may not be impossible (Arakawa et al. 2019). We thus conclude that [26]Al is unlikely to have mattered on the timescale of these bodies' accretion, or perhaps contributed just enough heat to provide the 10%–15% differentiation required in the Charon-forming collisional models of Canup (2011).

Serpentinization

On Earth, hot ultramafic rocks react with sea-water to produce serpentinite, in a reaction that liberates heat. The energy release of this exothermic reaction is 233 kJ kg^{-1} (Grimm & McSween 1989), comparable to the latent heat of fusion of ice. If such hydrothermal reactions take place within the silicate cores of Pluto or Triton, they could constitute an additional heat source (e.g., Loveless et al. 2022). The mean resulting temperature increase if the entire core underwent serpentinization would be about 130 K. In reality, such reactions will be limited to areas of the core with both sufficient permeability to permit flow of water (if needed) and sufficiently high temperatures (Section 2.2.1.1); in practice, these two requirements make it uncertain that serpentinization will be an important energy source or on what timescale. But as long as primordial anhydrous condensates remain in contact with liquid water, the possibility of reactions and heat release will remain.

Tidal heating

A body in orbit around a more massive body becomes tidally distorted. Viewed from a point on the orbiting body, the central body remains fixed in the sky if the orbit is synchronous and circular, and the orbiting body has no axial tilt (obliquity). If any of these conditions is violated, the central body will display apparent motion in the sky and the size and/or orientation of the tidal bulge will change with time. If there is any friction, this time-varying behavior will cause dissipation: tidal heating. For a synchronous body the rate at which tidal heat is produced is given by (e.g., Wisdom 2008)

$$\dot{E} = \frac{3}{2} \frac{n^5 R^5}{G} \frac{k_2}{Q} (7e^2 + \sin^2 \theta) \tag{2.12}$$

where n is the mean motion, e is the eccentricity, θ the obliquity, k_2 the Love number at tidal frequencies, and Q the so-called dissipation factor (a low Q indicates a high dissipation rate). This expression assumes e and θ are both small. The quantity k_2/Q depends on the mechanical properties of the interior, for instance whether the body has an ocean or not, and is usually very uncertain.

For Pluto, there is no nearby massive body to cause tidal heating. Charon is too small to have caused truly substantial tidal heating in the past, though it could have caused some, and the spin-down of Pluto to its present-day synchronous state (due to Charon tides) will have contributed a modest amount of heat (about 50 K; Robuchon & Nimmo 2011; Bagheri et al. 2022).

While Triton was extensively tidally heated during circularization of its orbit (see Section 2.2.2.2), its present-day eccentricity is so low that present-day eccentricity tidal heating is very small. However, Triton may be being heated by obliquity tides at present, because θ is predicted to be non-zero. Although there is no measurement of Triton's obliquity, its high inclination (157°) has been used to predict an obliquity value of 0.35° (Chen et al. 2014). This value is relevant to a solid body; if a decoupling ocean exists, the obliquity is likely to be higher, as at Titan (Bills & Nimmo 2011; because tidal torques are being applied to the ice shell rather than the entire body). Application of Equation (2.12) with $k_2/Q = 0.01$ yields a solid-body heat flow of 11 GW, or a heat flux of 0.5 mW m^{-2}, which is geophysically trivial. If Triton possesses a subsurface ocean, tidal heating may be larger because ocean tides are preferentially excited by obliquity forcing (Tyler 2008; Chen et al. 2014) and because the obliquity is expected to be larger. In this case, tidal heat fluxes could be comparable to or exceed the present-day radiogenic value, with a total present-day heat flux of up to 18 mW m^{-2} (Nimmo & Spencer 2015).

This ocean tidal heating is expected to be independent of ocean thickness, to first order (Nimmo & Spencer 2015), but more complicated scenarios, for instance the case of a stratified ocean, could have different heating characteristics (Rovira-Navarro et al. 2023) and have not been investigated for Triton.

Long-term evolution models

Assuming Triton's capture occurred shortly after its formation, its similar size to Pluto leads to broad similarities in the predicted long-term thermal evolution of the two bodies, as well as some differences.

For Pluto two possible initial conditions have been discussed. The most common initial condition is the "cold start" which assumes Pluto formed slowly enough as to not trap significant accretional energy (Robuchon & Nimmo 2011; Hammond et al. 2016; Bierson et al. 2018; Kimura & Kamata 2020; Loveless et al. 2022). Under this condition, the release of heat from long-lived radioactive elements melts an ocean which reaches its maximum thickness between 1 and 2 Gyr after formation. This corresponds to roughly one half-life of the major heat producing elements plus some time for the heat to be conductively transported to the base of the ice shell. Most models assume a conductive ice shell, which is consistent with other arguments (see Section 2.4.3.1). However, Robuchon & Nimmo (2011) also examined convecting shells and found that, in this case, heat transport was sufficiently effective that an ocean never formed.

The alternative initial condition assumes that Pluto's accretional heat was efficiently retained causing the presence of a subsurface ocean at the end of formation (Bierson et al. 2020), i.e., after the giant impact. As discussed above, this requires that Pluto formed quickly enough or from large enough bodies (oligarchic growth) to not efficiently radiate its accretional energy during formation. In this "hot start," the initial ocean thins rapidly within the first 1–2 Gyr until an equilibrium is reached between radioactive decay and the overlying heat loss through the ice shell (Bierson et al. 2020; Kimura & Kamata 2020). This early freezing is expected to cause large extensional stresses which have been invoked to explain large ancient extensional features (the so-called "Ridge-Trough System") on Pluto's surface (Schenk et al. 2018; Bierson et al. 2020).

Figures 2.2(a) and (c) show, respectively, a "cold start" and "hot start" Pluto. It can be seen that, after the initial 1–2 Gyr period Pluto's subsequent evolution consists of a slowly freezing ocean and the present-day state is almost independent of the initial conditions (as expected). The rate of heat loss through the ice shell depends critically on whether the ice shell is conductive or convective (discussed in Section 2.3.1.2). The models shown here assume a conductive shell and core and yield present-day shell and ocean thicknesses of roughly 200 and 100 km, respectively, with a peak surface heat flux of about 6 mW m^{-2} in the cold start case. If Pluto's ice shell is convective, heat loss is fast enough that the entire subsurface ocean is expected to have frozen by the present day (or never formed initially). Convection, however, is not favoured based on arguments presented in Section 2.4.3.1. Conductive models predict an ocean at present that is between 20 and 200 km thick depending on the other assumptions made (Robuchon & Nimmo 2011; Hammond et al. 2016; Bierson et al. 19, 2018; Kimura & Kamata 2020).

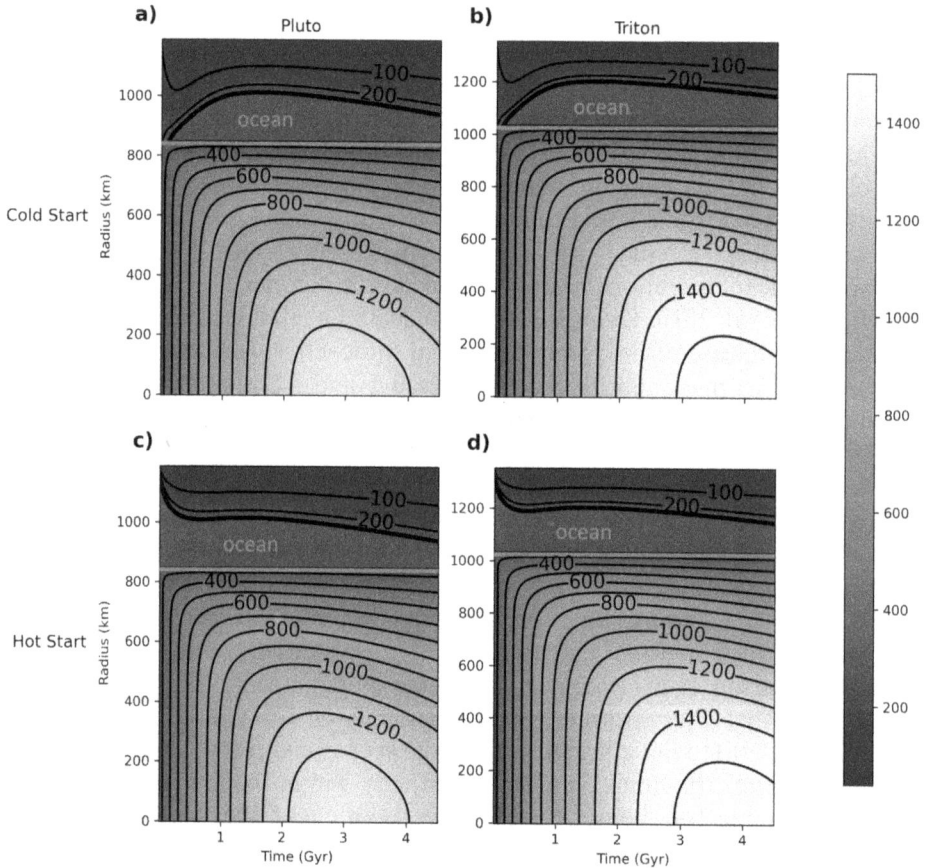

Figure 2.2. Conductive thermal evolution models for "cold start" (a and b) and"hot start" (c and d) Pluto and Triton. Model is described in Bierson et al. (2018). Colors and numbers denote temperature (in K). Solid black line denotes the base of the ice shell. Red line denotes the interface between rock and H$_2$O.

Because of the massive energy release that occurred during Triton's initial capture, its thermal evolution is expected to include an initial ocean and to most closely resemble the "hot start" case described for Pluto. Figures 2.2(b) and 2.2(d) show conductive thermal evolution in "hot start" and "cold start" cases for Triton, both of which yield similar present-day ice shell and ocean thicknesses to Pluto. While for Pluto the "hot start" is expected to begin with silicate temperatures near the melting temperature of water, for Triton the silicates may initially have been in excess of 1000 K (McKinnon et al. 1995; Nimmo & Spencer 2015). Furthermore, Triton's larger size results in higher peak core temperatures. As a consequence, hydrothermal alteration (Section 2.2.1.1) and core evolution in general (e.g., melting of metal and sulfides) is expected to have been more pervasive on Triton than Pluto.

For Triton as for Pluto the question of whether the ice shell is conductive or convective needs to be addressed. If Triton's ice shell is conductive, a subsurface ocean can be maintained over all of solar system history (Gaeman et al. 2012; Nimmo & Spencer 2015), especially if NH_3 is present (Hammond et al. 2018). At present, there are no Triton-specific thermal models that incorporate ice shell convection, but by analogy to Pluto we may expect that if the ice shell is convective the ocean will rapidly freeze out. Note, however, that the models shown here neglect tidal heating in the ocean, which for Triton may be an additional heat source comparable to or even greater than radiogenic heating (Section 2.3.1.3).

Another caveat with these models is that they assume relatively slow, conductive removal of heat (either radiogenic or primordial) from the silicate core. If deep hydrothermal circulation (i.e., water advecting heat from the interior) occurs, this process might remove heat more rapidly than shown in Figure 2.2, permitting a core heat flow closer to steady-state (at least in the past). Similarly, in carbon-rich cases, the presence of graphite could substantially increase the core thermal conductivity, again permitting more rapid core cooling. On the other hand, exothermic reactions in the core (e.g., serpentinization) would slow the core cooling rate. At elevated core temperatures *deserpentization*, although endothermic, would release hot, supercritical aqueous fluids, which could provide a geological heat pulse to the overlying ocean and ice shell above (McKinnon et al. 2023).

Summary Triton (almost certainly) and Pluto (probably) both started life with liquid water oceans. Radiogenic heating alone is likely enough to maintain such oceans to the present-day, as long as the ice shells are not convecting and core heat flows are somewhat greater than steady-state (as predicted by conductive cooling models). Little of the early tidal or accretional heating of either body remains at the present day, but in the case of Triton obliquity heating in any subsurface ocean could be a heat source comparable to radiogenic decay. Theoretical models of these bodies permit, but do not require, the presence of subsurface oceans.

2.3.2 Observations

There are various techniques for detecting subsurface oceans on icy moons (Nimmo & Pappalardo 2016), including measurements of induction, tidal response, librations, obliquity, and static gravity. Unfortunately, almost none of these are applicable, even in principle, to Pluto and Triton.

2.3.2.1 Pluto

Pluto's surface geology includes large numbers of normal faults, with the inferred extension being a minimum of 0.3%–0.4% on the imaged hemisphere (Bierson et al. 2020). One simple way of generating such extension is to appeal to refreezing of a water ocean (Nimmo 2004a). As water is converted to ice, the increase in volume means the surface radius must increase, leading to extension. In the absence of an ocean, Pluto's slow cooling with time would lead to contraction, which is not observed in surface tectonics. A "cold start" Pluto would likewise lead to initial ice shell thinning and surface compression, which is why Bierson et al. (2020) favour a "hot start" scenario.

Another corollary of ocean freezing is that the ocean beneath becomes pressurized, which can in principle propel water to the surface, powering cryovolcanism (Manga & Wang 2007). Individual cryovolcanic constructs have been tentatively identified on Pluto (Moore et al. 2016; Singer et al. 2022), and some of the surface fractures may be associated with relatively recent cryovolcanic flows or venting (Cruikshank et al. 2019, 2021).

If the putative oceans of Pluto or Triton froze entirely, ice II/III would be expected to form at the base (and certainly so for Triton, with its greater pressure scale). Because ice II/III has a higher density than liquid water, this would cause net compression (Hammond et al. 2016) which, at least on Pluto, is not observed. This absence argues for the presence of an ocean.

Another argument concerns the large Sputnik Planitia basin, which is located close to the anti-Charon point. Although this location could simply be a coincidence, it could also be explained if Sputnik Planitia were a mass excess, similar to the "mascon" basins on the Moon. Such a feature would reorient Pluto, moving itself closer to the anti-Charon point (Keane et al. 2016; Nimmo et al. 2016). Since Sputnik Planitia is a deep basin, the mass excess must be large and beneath the surface. If Sputnik Planitia is an impact basin (Johnson et al. 2016), the ice shell beneath the impact site will have been thinned, replacing ice with denser water and naturally forming a mass excess. Thus, the location of Sputnik Planitia can be readily explained if a subsurface ocean is present (Nimmo et al. 2016). We note for completeness that it has been suggested the inferred gravity anomaly at Sputnik Planitia is in fact due to the implanted silicate core of the basin-forming impactor (Ballantyne et al. 2022), and not the presence of an ocean.

Finally, Pluto does not exhibit a measurable tidal/rotational bulge (Section 2.2.1.1). Although the present-day bulge would not be detectable, some bodies (like Iapetus; Castillo-Rogez et al. 2007) display a "fossil bulge" which froze in at an early stage in the body's evolution when it was rotating more rapidly. On Pluto, models show that such a bulge does not survive in the presence of an ocean (Robuchon & Nimmo 2011). A fossil bulge would be strong evidence against an ocean (i.e., the interior would be cold and relatively rigid), but no such bulge is observed.

Thus, although the observations are consistent with the existence of a subsurface ocean on Pluto, none of the arguments presented above is bullet-proof. Given our current state of knowledge Pluto remains firmly in the "candidate ocean world" category.

2.3.2.2 Triton

At Triton the observations are different, but the conclusion is similar. Most of Triton's tectonic features are enigmatic, with few obviously extensional features recognized (Schenk et al. 2021). Triton's geysers could be sourced from a subsurface ocean, but they could equally well be shallow, insolation-driven phenomena (Hofgartner et al. 2022). Triton possesses very little relief compared to most other icy bodies, including Pluto, with a topographic amplitude of about ±1 km on the imaged hemisphere (Schenk et al. 2021). Other icy bodies with comparably subdued relief are Europa and Titan, both ocean worlds. The latter is also subject to surface erosion, so the former is more directly comparable. This low relief suggests that Triton has a thin lithosphere and experiences relatively high heat fluxes. Such characteristics could be indicative of a subsurface ocean, but they could alternatively be a consequence of solid-state convection in a mobile ice layer. Similar arguments can be made concerning Triton's young surface age, another characteristic it shares with Europa. We discuss these issues further below (Section 2.4.3.2).

Summary The observational evidence for subsurface oceans on Pluto and Triton is modest. At Pluto, extensional tectonics, cryovolcanism, the absence of a fossil bulge and the location of Sputnik Planitia are all consistent with, but do not require, the presence of an ocean. For Triton, the young surface age and low relief suggest high heat fluxes and a thin lithosphere, but these do not necessarily imply an ocean.

2.4 Lithospheric Structure and History

Between the surface (which we can observe) and the deep interior lies the lithosphere, the cold, mechanically strong portion of the ice shell. To use surface observations to make inferences about the deep interior, the role of the lithosphere must be understood. In particular, it is the lithosphere that controls how rapidly heat escapes from the interior to the surface, and the extent to which surface deformation may occur.

We begin with some definitions. The elastic thickness is the part of the lithosphere that responds elastically on geological timescales (or that part which is *effectively elastic*), and the brittle–ductile transition (BDT) is the depth at which the stresses required to cause ductile deformation drop below those required to cause motion along pre-existing faults. While obviously related, the lithosphere, elastic thickness and BDT depth are not necessarily the same (Melosh 2011; Turcotte & Schubert 2014; Watts & Burov 2003), and can be probed with different techniques. On Earth, the lithosphere can be up to 100–200 km thick, while on continents the BDT is at about 20–30 km and the effective elastic thickness is typically 5–15 km (e.g., Maggi et al. 2000).

The base of the elastic part of an icy lithosphere may be taken to be at a temperature of roughly 120–160 K (Conrad et al. 2019) or 40%–60% of the melting temperature (depending on timescale or strain rate). So if the thickness of the elastic layer can be measured and the thermal conductivity is known, the thermal gradient and heat flux can be inferred. Unfortunately, a measurement of the elastic thickness does not necessarily provide a constraint on the total shell thickness. If the shell is conductive, then the total shell will be roughly 2–3 times the thickness of the elastic layer. However, if the shell is convecting, then the total ice shell thickness could be much greater, with the lithosphere acting as the stagnant lid on top. Lithosphere thickness can only be used to place a lower bound on the total ice shell thickness.

A complication regarding effective elastic, or lithosphere, thickness and heat flux estimates is that they may be relevant to the time (and context) of deformation, rather than the present-day epoch. Thus, a rifting episode that occurred 1 Gyr ago may record a lower elastic thickness than the present day, while crater relaxation is most affected by the highest heat fluxes, even if these are of short duration (e.g., Dombard & McKinnon 2006). For Triton, this issue is not important because all geological features appear to be young (as young as ~10 Myr) and thus record present-day conditions.

Below, we consider potential lithospheric structures and how they may be reconciled with the available observational constraints.

2.4.1 Conduction

The simplest possibility is that the ice shells of Pluto and Triton are homogeneous and conductive throughout. In that case, neglecting heat sources in the ice shell, the conductive temperature profile is given by Equation (2.5). The ice shell thickness then depends simply on the rate at which heat is supplied to the base of the ice shell (see below).

There are at least three possible complications to this simple picture. The first is that the uppermost part of the ice shell may be porous, reducing the conductivity and increasing the temperature gradient. On Pluto the porous layer could be around 60 km thick (see Figure 2(b) in Nimmo & McKinnon 2021) and would result in a slightly thinner shell overall (e.g., Bierson et al. 2018). A porous layer would also be mechanically weaker than the intact ice below (Keller et al. 1999), with implications for, (e.g.), elastic thickness, convective yielding and fault mechanics. Such a porous layer is not expected on Triton because its extremely young surface age means impact-derived porosity has not had time to develop.

A second complication is that, if the ice shell is sufficiently thick and an ocean is absent, the base of the shell will undergo a phase change to a higher-pressure ice phase. The ice-Ih/ice-III/water and ice-Ih/ice-II/ice-III triple points are both at about 0.21 GPa (Choukroun & Grasset 2010). For our nominal models (Figure 2.1) the pressures at top of the silicate core are 0.23 and 0.27 GPa for Pluto and Triton, respectively, so a thin layer of a high-pressure ices could be present. Ice II and ice III have thermal conductivities roughly 70% and 40% that of Ice I (Ross & Kargel 1998), so these would act as an insulating basal layer. For less dense Pluto cores (e.g., McKinnon et al. 2017; Kimura & Kamata 2020), the pressure at the top of the core does not reach the triple-point pressures above. As noted in Section 2.3.2.1, formation of these higher-pressure ice phases would lead to surface compression, evidence for which (on Pluto) is not observed.

A more dramatic insulating layer will exist if clathrates (ice cage structures trapping other molecules like CH_4) are present. These have conductivities an order of magnitude smaller than ice I (Ross & Kargel 1998) and so can have a large effect on the lithospheric temperature structure, if present (Kamata et al. 2019).

2.4.2 Convection

As discussed in Section 2.3.1.2, a rigid lithosphere (or stagnant lid) could form the top of an ice shell which is experiencing convection beneath. The thickness of this lithosphere depends on the vigour of convection and goes as $Ra^{-1/3}$. A thin lithosphere implies a correspondingly higher heat flux. For instance, a convecting ice shell with a 50 km thick stagnant lid would imply a heat flux of roughly 20 mW m^{-2} (Equation (2.6)). A conductive ice shell with a total thickness of 50 km would yield almost the same flux.

For Pluto, a 20 mW m^{-2} heat flux far exceeds the likely heat sources (except at early, post-giant impact times) and should be regarded as highly implausible. Furthermore, observations (Section 2.4.3.1) suggest that convection is not occurring on Pluto. For Triton, 20 mW m^{-2} is still greater than the likely heat sources in recent epochs. In the case of Triton, it is also hard to reconcile the presence of a thick lithosphere with the young surface age, unless the resurfacing is being accomplished by cryovolcanism (see Section 2.4.3.2).

One interesting consequence of convection is that it might result in any clathrates that form being entrained into the ice shell (Carnahan et al. 2022). Because clathrates are stiffer than water ice, this entrainment would increase the bulk shell viscosity and thus could eventually shut convection down.

2.4.3 Observations

2.4.3.1 Pluto

One constraint on the elastic thickness of Pluto comes from the observation that extensional structures (fossae) there do not show any signs of rift-flank uplift. Since a rift imposes a negative (upwards) load on the lithosphere, one would expect to see the rift flanks bowing upwards. The fact that no such uplift is seen means the stress imposed by the rift topography is insufficient to cause the lithosphere to deform detectably. This observation allows a lower bound to be placed on the elastic thickness (Conrad et al. 2019). In the case of Pluto, this lower bound is 8 km at the time of deformation. The corresponding upper bound on heat flux is 66–85 mW m^{-2}, depending on the strain rate assumed.

A second constraint on Pluto's lithosphere comes from the inference that the nitrogen layer in the Sputnik Planitia basin is convecting (McKinnon et al. 2016). Individual convection cells can clearly be seen in Figure 2.3a. For convection to occur, the heat flux into the base of the nitrogen layer must exceed some critical value. Unfortunately, neither the layer thickness nor the rheology of solid nitrogen are well known; nonetheless, the anticipated present-day heat flux of a few mW m^{-2}

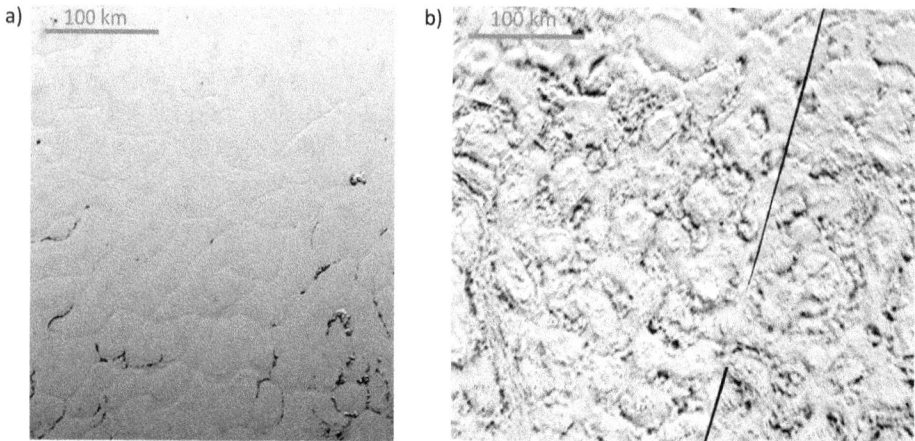

Figure 2.3. (a) Convecting nitrogen plains on Pluto (McKinnon et al. 2016). (b) Cantaloupe terrain on Triton (Schenk & Jackson 1993). Note the similarities between the two images—at least in plan.

is likely sufficient to drive the kind of convection observed. Further experiments on the rheology of solid nitrogen (Yamashita et al. 2010) would be highly desirable.

If Sputnik Planitia represents a large gravity anomaly (see Section 2.3.2.1) it must be mostly uncompensated, implying a present-day elastic thickness of at least 70 km (Nimmo et al. 2016) and a correspondingly lower heat flux (less than about 9 mW m^{-2}). This value is only a little larger than the expected maximum heat flux from the thermal evolution models above about 6 mW m^{-2}. Kihoulou et al. (2022) investigate the viscoelastic evolution of the Sputnik Planitia basin and conclude that either a relatively thin (~100 km) shell or a thicker shell with a cold base are possible.

The survival of an area of thinned ice at Sputnik Planitia (Nimmo et al. 2016; Johnson et al. 2016), if correct, places an important constraint on the ice shell viscosity structure. Shell thickness variations give rise to pressure variations which drive lateral flow of ice, eventually equalizing the thickness. The rate at which this happens depends on the viscosity structure of the ice shell (Nimmo 2004b). A convecting ice shell should exhibit little or no lateral thickness variations. A conductive shell with a basal temperature of 270 K would still relax very fast compared to the age of Sputnik Planitia, presumed to be $\geqslant \sim 4$ Gyr (Nimmo et al. 2016; Kamata et al. 2019; Kihoulou et al. 2022). A basal temperature of around 200 K would allow lateral shell thickness variations to persist (Nimmo et al. 2016); an insulating basal clathrate layer would have the same effect (Kamata et al. 2019). Thus, if Sputnik Planitia is an area of thinner ice, Pluto's ice shell must not only be conductive but cold.

A final constraint comes from two large (diameter of 115 and 140 km), apparently relaxed craters (McKinnon et al. 2023). Relaxation of initially deep craters happens at a rate that depends on the heat flux and the surface temperature. Unfortunately, the latter quantity is uncertain because there might be a near-surface, highly porous layer that increases the effective surface temperature. Even with an effective surface temperature of 100 K, relaxation requires heat fluxes in excess of 10 mW m^{-2}, somewhat higher than expected from the thermal models. Lower effective surface temperatures would require even higher heat fluxes to cause relaxation. It is hard to reconcile these values with the lower heat fluxes inferred above, but the explanation for the discrepancy is not obvious. Perhaps the heat flux is spatially variable and these craters happen to be located close to a hot spot. Alternately, early heat flows could have been higher due to rapid removal of core heat, e.g., via hydrothermal circulation (Section 2.3.1.3).

2.4.3.2 Triton

Resurfacing The most puzzling aspect of Triton's lithosphere is its youthful surface (Schenk & Zahnle 2007). Resurfacing has taken place on a ~10 Myr timescale, but how this occurred is unclear. The most likely candidates are cryovolcanism, tectonism, and crustal/lithospheric overturn. Here, we focus on the latter possibility but include brief discussions of the other two possibilities. The tectonics of Triton and Pluto are discussed briefly in Collins et al. (2009) and Moore et al. (2016), respectively, while cryovolcanism is discussed in Chapter 3.

The so-called cantaloupe terrain on Triton, shown in Figure 2.3b, bears some similarities to the convecting nitrogen plains on Pluto (Figure 2.3a). However, the interpretation here is that the features are the result of compositional convection (diapirism; Schenk & Jackson 1993), rather than thermal convection. Because the surface is young and the separation of these cells is known (roughly 50 km), a calculation of the Rayleigh–Taylor instability timescale (Turcotte & Schubert 2014) shows that the mean crustal viscosity must not exceed about 10^{22} Pa s, and that the unstable layer is about 20 km thick. Obviously this is a highly simplified analysis, and more sophisticated approaches using a vertically varying viscosity have been unable to generate active compositional diapirs (Ohara & Dombard 2019). An alternative possibility is that the cantaloupe terrain is a surface expression of subsurface thermal convection. This suggestion has the advantage that thermal overturn is continuous, whereas compositional overturn only happens once (and why should it have happened in the last 10 Myr?). But it would require near-surface deformation which is not expected in standard stagnant-lid convection.

If some kind of overturn is responsible, then the 10 Myr age can be converted into a heat flux constraint. Assuming that the resurfacing involves overturn of the ice shell to some depth h, then the implied heat flux is ~$h\Delta T \rho_i C_p / \Delta t$, where ΔT is the mean change in shell temperature during overturn, C_p is the specific heat capacity of the shell, and Δt is the overturn time. If the overturn involves replacement of cold ice with warm ice, ΔT is probably of order 100 K, and based on Schenk & Jackson (1993) the overturning layer thickness is roughly 20 km. The resulting heat flux is then ~12 mW m^{-2}, significantly exceeding the present-day radiogenic heat flux but comparable to the the expected obliquity tidal heating in a subsurface ocean (Section 2.3.1.3). If the age of the cantaloupe terrain is older (Mah & Brasser 2019), then the inferred heat flux would be lower. Note that these calculations implicitly assume continuous overturn, which for compositional diapirism would be incorrect (overturn happens only once).

Replacement of the entire lithosphere, rather than just the top 20 km, would increase the implied heat flux by an order of magnitude. Such high heat fluxes would cause a rapid freezing of the ocean and cessation of ocean tidal heating. Long-term maintenance of such heat fluxes in the absence of comparable heat sources is not energetically possible. A brief interval of overturn might be possible (see below), but would require us to be seeing Triton at a special time.

While no numerical studies of thermal convection on Triton have been carried out, parallel studies of convection on Pluto (Robuchon & Nimmo 2011) show it is possible for reasonable ice viscosities (see above). The principal problem with thermal convection is that it typically forms a thick stagnant lid which does not experience either deformation or overturn, and thus will not modify the surface (Solomatov 1995). Only if the ice has a low yield strength (e.g., due to pervasive fracturing) can complete overturn and resurfacing occur; this mobile-lid convection may happen episodically and only in localized regions (e.g., O'Neill & Nimmo 2010). Nimmo & Spencer (2015) used scaling laws to investigate yielding on Triton, but a numerical study of whether mobile-lid convection could plausibly produce the cantaloupe terrain, and more generally Triton's young surface age, would be useful.

The cantaloupe terrain is an example of where we are hampered by the lack of compositional information at Triton. While we know that weak, potentially convecting ice (e.g., CO_2) is present at Triton's surface (Cruikshank et al. 1993), we have no idea of its spatial distribution, and thus no idea whether it could explain the cantaloupe terrain characteristics. Such surficial convection, however, does have the advantage that it will not give rise to an energy crisis in the same way that lithospheric overturn does (see above). A near-surface convecting layer of (for example) CO_2 ice could explain the young surface without requiring the ice shell beneath to be convective.

However, although the morphology and spatial scale of the cantaloupe terrain does bear some resemblance to the convecting nitrogen on Pluto (Figure 2.3), the cantaloupe terrain appears more degraded and exhibits raised rims, opposite to the nitrogen plains. It seems very unlikely that active cantaloupe convection is taking place. If so, given the young surface age, we would then have to explain why convection stopped only recently, another question without an obvious answer.

An alternative, discussed in more detail in Chapter 3 is that cryovolcanism could be responsible for the resurfacing observed. In this context, cryovolcanism means effusive, low-viscosity cryolava outflows of the kind argued for by Croft et al. (1995) and Schenk et al. (2021). Because of the high density of water relative to ice, the mechanism by which eruption of water to the surface occurs is unclear. Possibilities include exsolution of dissolved gases (Crawford & Stevenson 1988), pressurization of the ocean due to ice shell refreezing (as noted earlier; Manga & Wang 2007), and density reduction with increased NH_3 concentration (Croft & Lunine 1988). As an order-of-magnitude estimate, assume that a 2 km thickness of flows is enough to erase existing craters. Depositing such a thickness in the last 10 Myr yields a heat flux of about 5 mW m^{-2}, comparable to the maximum expected heat flux from models without tidal heating (Section 2.3.1.3). So cryogenic resurfacing is energetically plausible, but the specific mechanism is enigmatic.

In the absence of obliquity tidal heating, Triton's ice shell should be thickening at a rate of about 30 km Gyr^{-1} (see Figure 2.2). Quantitative studies of the kind performed by Manga & Wang (2007) for Enceladus and Europa have not yet been carried out for Pluto, so it is not clear whether or how much water would be erupted to the surface as a consequence of freezing-induced pressurization. We also note that if obliquity tidal heating is operating, the freezing timescale would be significantly lengthened, in which case the young surface age would likely become harder to explain.

The final alternative, that of tectonic resurfacing, is hard to quantify—although Triton looks nothing like Ganymede, which does appear to be tectonically resurfaced across much of its surface (Pappalardo et al. 2004). Apart from thermal or compositional convection, the primary sources of stress are probably shell thickening, obliquity tides and perhaps non-synchronous rotation. Shell thickening can generate MPa-level stresses (e.g., Nimmo 2004a) but should yield a global network of extensional fractures, which (in contrast to Pluto) are not observed. Obliquity tides (Jara-Orué & Vermeersen 2011) yield smaller stresses, roughly 20 kPa, which is comparable to eccentricity tidal stresses on Europa. Whether such stresses can drive significant deformation is unclear. Ridges on Triton have been attributed to eccentricity tides during circularization (Prockter et al.2005), but this suggestion is hard (if not impossible) to reconcile with the apparently young surface age. Another potential source of stress is non-synchronous rotation, but this is not obviously consistent with the concentration of craters at the apex of motion (Zahnle et al. 2001).

Heat Flux. For Triton, the only available heat flux estimate appears to be from two degraded troughs interpreted to be graben (Croft et al. 1995). Based on the assumption that the graben width (10–18 km) is determined by the depth at which the two bounding faults intersect, and that this depth defines the BDT, a heat flux can be inferred (Martin-Herrero et al. 2018). For faults dipping at 60°, the inferred heat fluxes are 60–90 mW m^{-2} and 6–14 mW m^{-2} for water ice and $NH_3 \cdot 2H_2O$ ice, respectively. The large difference in values arises because of the large difference in thermal conductivity between the two end-members. Since there are currently no detections of NH_3 on Triton and, even if it were present, it should be volumetrically minor, the latter heat fluxes should be regarded as unlikely. This analysis makes several questionable assumptions, but taken at face value it suggests heat fluxes greatly exceeding the expected present-day values.

A present-day heat flux of $\sim$10 mW m^{-2} seems most likely for Triton, with some combination of crustal or lithospheric overturn and cryovolcanism as the likely transport mechanism. Either has the advantage of also potentially explaining Triton's young surface age. Quantitative investigations of these mechanisms are so far lacking, and would be of great interest.

2.4.3.3 Inferred Lithospheric Structures

Pluto

For Pluto, the likely existence of a gravity anomaly at Sputnik Planitia and the absence of rift-flank uplift suggest a large elastic thickness, perhaps >70 km at the

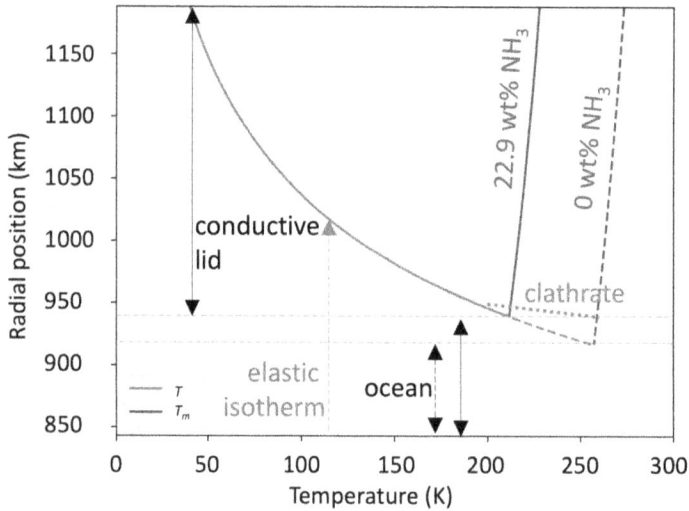

Figure 2.4. Hypothetical Pluto conductive lithospheric structures, with a heat flux of 3 mW m^{-2} and ice shell 269 km thick (0.16 GPa). Melting curves are from Leliwa-Kopystyński et al. (2002). A clathrate layer at the base of the shell results in the overlying ice shell being colder on average and less likely to convect. The same result occurs if the ocean is NH$_3$-rich. T is the conductive temperature profile and T_m the melting curve.

present day. Such a value is also consistent with the rugged topography observed, but not with two relaxed impact craters (which are, however, likely ancient in comparison). The ice shell must be conductive and cold to prevent lateral flow; possible explanations include an ammonia-rich ocean, or a clathrate layer at the base of the shell.

We concluded in Section 2.3 that subsurface oceans are likely a feature of both Pluto and (with much less certainty) Triton. Pluto is only heated by radiogenic decay at present, and covers its ocean with a thick, conductive ice shell.

Figure 2.4 shows a hypothetical temperature structure for Pluto's lithosphere. It shows a purely conductive situation, with the curvature arising primarily from the assumed temperature dependence of the thermal conductivity. The base of the ice shell is set when the ice shell temperature (red line) equals the melting curve (blue lines). The implied present-day elastic thickness (depth to the elastic isotherm) is roughly 171 km and the surface heat flux is 3 mW m^{-2}. Based on the lateral flow arguments above, a homogeneous ice shell over a pure water ocean (dashed red line) is unlikely. Two alternatives are that either the ocean is NH$_3$-rich (solid red line) or that there is an insulating basal layer of clathrates (dotted red line). The solid and dashed blue lines show the melting curves for an NH$_3$-rich and a pure water ocean, respectively. Porosity in the near-surface (not shown) would increase the shell temperature and decrease the equilibrium shell thickness (Figure 2 of Nimmo & McKinnon 2021).

Triton

The sparse observational constraints on Triton hint at higher heat fluxes and either crustal/lithospheric overturn or cryovolcanism as the cause of the young surface age.

The associated heat loss may be balanced mainly by obliquity tidal heating in an ocean. Complete lithospheric overturn is not favored because of the excessive implied heat fluxes, unless it is episodic and we happen to be seeing Triton at a special time. Crustal overturn of a high-mobility ice (like CO_2 or N_2) does not lead to the same energy budget problem but leaves open the question of why overturn (at least in the cantaloupe terrain) appears to have stopped recently. Cryovolcanism, perhaps driven by slow thickening of a conductive ice shell, appears a more likely candidate for explaining the young surface age.

2.5 Summary and Future Work

Pluto and Triton have similar radii and bulk densities, suggesting a roughly 2:1 rock/ice ratio in their interiors. Triton is slightly more rock-rich than Pluto, perhaps because of early volatile loss as a result of the intense tidal heating it experienced. Triton is almost certainly fully differentiated (ice/rock/metallic core) at the present day, and Pluto probably is as well, though completion of this differentiation likely post-dated the Charon-forming impact. Carbon compounds could make up a significant fraction of these bodies' mass, but direct evidence for such compounds is lacking.

Theory (for Triton) and observations (for Pluto) suggest that both started with subsurface oceans. Such oceans are expected to have survived until the present day unless the ice shells were convective. On Pluto, the dominance of extensional features, putative cryovolcanic features, absence of a fossil bulge, the location of Sputnik Planitia, and survival of presumed lateral shell thickness variations all point to a present-day ocean surmounted by a cold, rigid lithosphere. For Triton little direct evidence for a subsurface ocean exists, but obliquity tidal heating in such an ocean may help explain Triton's young surface age.

Given their bulk similarities, the fact that the surfaces of Pluto and Triton look so different is surprising. Pluto has ancient ($\sim$4 Gyr), rugged terrains, abundant extensional tectonics and putative cryovolcanoes; Triton's surface is very young, smooth, and possesses fewer identifiable geological features apart from possible diapirs and cryovolcanic flows. Pluto's appearance is consistent with slow cooling from an initial hot state. Triton's history is much more enigmatic; its young surface might be the result of crustal overturn of weak ices or cryovolcanic resurfacing, but little quantitative analysis has been done to date. Heat fluxes on present-day Pluto are probably a few mW m^{-2} while at Triton they could be higher, perhaps $\sim$10 mW m^{-2} if ocean obliquity tides are important.

The ultimate explanation for the differences in surface appearance remains elusive. Differences in initial conditions may or may not have been critical, depending on the evolutionary changes experienced by Triton due to extreme early heating. One key difference is that Triton today can access an extra heat source—obliquity tidal heating—that is not available at Pluto. Triton's relatively muted relief compared to Pluto is qualitatively consistent with a higher surface heat flux. This heating could conceivably drive convection of near-surface weak ices on Triton—but without better spectroscopic data it is hard to be certain. Alternatively, cryovolcanic resurfacing could be taking place—but it is not clear that the implied shell thickening can be reconciled with the obliquity tidal heating. Triton's young surface age remains a major puzzle.

2.6 Future Work

Near-term work on Pluto and Triton is likely to be driven by theory or laboratory experiments, rather than new observation. Quantitative modeling of either near-surface or mobile-lid convection/diapirism on Triton would be very valuable (Section 2.4.3.2), as would explorations of Triton's thermal evolution incorporating ocean tidal dissipation (Section 2.3.1.3). Laboratory experiments on the rheology of candidate convecting ices such as N_2 or CO_2 would be useful for both Pluto and Triton studies. JWST is beginning to provide very high-quality spectroscopic data that will reveal new aspects of the surfaces and interiors of Kuiper Belt Objects (e.g., Brown & Fraser 2023). However, because Pluto or Triton are only a few pixels wide, the information derived will necessarily be limited, in that sense. Further remote sensing of other large KBO's (e.g., Eris) will help put our more detailed knowledge of Pluto and Triton in context.

Ultimately, many of the unanswered questions identified here will only be answered by new spacecraft missions.[2] For Pluto, no such mission is likely in the forseeable future, but Triton was identified as a potential target for the *New Frontiers 7* call by the recent Decadal Survey (National Academies of Sciences, Engineering, and Medicine 2023). As this review has shown, Triton remains an extremely puzzling object, so even the distant prospect of a second spacecraft visit should provide encouragement to future researchers.

References

Agnor, C. B., & Hamilton, D. P. 2006, Natur, 441, 192

Altwegg, K., Balsiger, H., Hänni, N., et al. 2020, NatAs, 4, 533

Anderson, J., Jacobson, R., McElrath, T., Moore, W., Schubert, G., & Thomas, P. 2001, Icar, 153, 157

Arakawa, S., Hyodo, R., & Genda, H. 2019, NatAs, 3, 802

Bagheri, A., Khan, A., Deschamps, F., Samuel, H., Kruglyakov, M., & Giardini, D. 2022, Icar, 376, 871

Ballantyne, H., Asphaug, E., Denton, C. A., Emsenhuber, A., & Jutzi, M. 2022, Tech. rep. Copernicus Meetings

Bierson, C., & Nimmo, F. 2019, Icar, 326, 10

Bierson, C., Nimmo, F., & McKinnon, W. 2018, Icar, 309, 207

Bierson, C. J., Nimmo, F., & Stern, S. A. 2020, NatGe, 13, 468

Bills, B. G., & Nimmo, F. 2011, Icar, 214, 351

Brown, M. E., & Fraser, W. C. 2023, PSJ, 4, 130

Brown, R., Johnson, T., Goguen, J., Schubert, G., & Ross, M. N. 1991, Sci, 251, 1465

Canup, R. M. 2005, Sci, 307, 546

---

[2] Examples include one of the (unselected) finalists in the most recent *Discovery* competition (Mitchell et al. 2019) or a Pluto Orbiter (Howett et al. 2020) studied for the Decadal Survey.

Canup, R. M. 2011, AJ, 141, 35

Carnahan, E., Vance, S. D., Hesse, M. A., Journaux, B., & Sotin, C. 2022, GeoRL, 49, e97602

Carslaw, H. S., & Jaeger, J. C. 1986, Conduction of Heat in Solids (Oxford: Oxford Univ. Press)

Castillo-Rogez, J., Matson, D., Sotin, C., Johnson, T., Lunine, J., & Thomas, P. 2007, Icar, 190, 179

Chen, E., Nimmo, F., & Glatzmaier, G. 2014, Icar, 229, 11

Choukroun, M., & Grasset, O. 2010, JCP, 133, 144502

Cioria, C., & Mitri, G. 2022, Icar, 388, 234

Collins, G. C., McKinnon, W. B., Moore, J. M., et al. 2009, Planetary Tectonics (Cambridge: Cambridge Univ. Press) 264

Conrad, J., Nimmo, F., Schenk, P., et al. 2019, Icar, 328, 210

Crawford, G. D., & Stevenson, D. J. 1988, Icar, 73, 66

Croft, S., Lunine, J., & Kargel, J. 1988, Icar, 73, 279

Croft, S., Kargel, J., Kirk, R. L., et al. 1995, Neptune Triton (Tuscon, AZ: Univ. Arizona Press) 879

Cruikshank, D. P., Roush, T. L., Owen, T. C., et al. 1993, Sci, 261, 742

Cruikshank, D. P., Roush, T. L., Owen, T. C., Quirico, E., & Bergh, C. D. 1998, Solar System Ices: Based on Reviews Presented at the Int. Symp. (Berlin: Springer) 655

Cruikshank, D. P., Umurhan, O. M., Beyer, R. A., et al. 2019, Icar, 330, 155

Cruikshank, D. P., Dalle Ore, C. M., Scipioni, F., et al. 2021, Icar, 356, 786

Ćuk, M., & Gladman, B. J. 2005, ApJ, 626, L113

Denton, C. A., Johnson, B. C., Wakita, S., et al. 2021, GeoRL, 48, e2020GL091

Dombard, A. J., & McKinnon, W. B. 2006, JGRE, 111, E01001

Durham, W., & Stern, L. 2001, AREPS, 29, 295

Fray, N., Bardyn, A., Cottin, H., et al. 2017, MNRAS, 469, S506

Gaeman, J., Hier-Majumder, S., Roberts, J. H., et al. 2012, Icar, 220, 339

Gao, P., & Stevenson, D. J. 2013, Icar, 226, 1185

Glein, C. R., & Waite, J. H. Jr 2018, Icar, 313, 79

Goldreich, P., Murray, N., Longaretti, P., & Banfield, D. 1989, Sci, 245, 500

Goldsby, D., & Kohlstedt, D. L. 2001, JGRB, 106, 11017

Grimm, R. E., & McSween, H. Y. Jr 1989, Icar, 82, 244

Grundy, W., Buie, M., & Spencer, J. 2002, AJ, 124, 2273

Grundy, W., Binzel, R., Buratti, B., et al. 2016, Sci, 351, aad9189

Hammond, N. P., Barr, A. C., & Parmentier, E. M. 2016, GeoRL, 43, 6775

Hammond, N. P., Parmenteir, E., & Barr, A. C. 2018, JGRE, 123, 3105

Hemingway, D., Iess, L., Tajeddine, R., & Tobie, G. 2018, Enceladus and the Icy Moons of Saturn, P. M. Schenk et al (eds) (Tuscon, AZ: Univ. Arizona Press) 57

Hofgartner, J. D., Birch, S. P., Castillo, J., et al. 2022, Icar, 375, 835

Hogenboom, D., Kargel, J., Consolmagno, G., et al. 1997, Icar, 128, 171

Howett, C. J. A., Robbins, S., Fielhauer, K., & Apland, C. 2020, Outer Planets Assessment Group (Fall 2020) (Houston, TX: Lunar and Planetary Institute) Abstract #6032

Hussmann, H., Sohl, F., & Spohn, T. 2006, Icar, 185, 258

Hussmann, H., Choblet, G., Lainey, V., et al. 2010, SSRv, 153, 317

Jara-Orué, H. M., Vermeersen, B. L., et al. 2011, Icar, 215, 417

Johansen, A., Low, M.-M. M., Lacerda, P., & Bizzarro, M. 2015, SciA, 1, 1500109

Johnson, B. C., Bowling, T. J., Trowbridge, A. J., & Freed, A. M. 2016, GeoRL, 43, 10

Kamata, S., Nimmo, F., Sekine, Y., et al. 2019, NatGe, 12, 407

Kargel, J. S. 1992, Icar, 100, 556

Keane, J. T., Matsuyama, I., Kamata, S., & Steckloff, J. K. 2016, Natur, 540, 90

Keller, T., Motschmann, U., & Engelhard, L. 1999, GeopP, 47, 509

Kihoulou, M., Kalousová, K., & Souček, O. 2022, JGRE, 127, e2022JE007

Kimura, J., & Kamata, S. 2020, P&SS, 181, 828

Klinger, J. 1980, Sci, 209, 271

Le Roy, L., Altwegg, K., Balsiger, H., et al. 2015, A&A, 583, A1

Leliwa-Kopystyński, J., Maruyama, M., & Nakajima, T. 2002, Icar, 159, 518

Loveless, S., Prialnik, D., & Podolak, M. 2022, ApJ, 927, 178

Lunine, J. I., & Nolan, M. C. 1992, Icar, 100, 221

Maggi, A., Jackson, J., Mckenzie, D., & Priestley, K. 2000, Geo, 28, 495

Mah, J., & Brasser, R. 2019, MNRAS, 486, 836

Manga, M., & Wang, C.-Y. 2007, GeoRL, 34, L07202

Martin-Herrero, A., Romeo, I., & Ruiz, J. 2018, P&SS, 160, 19

McComas, D., Elliott, H., Weidner, S., et al. 2016, JGRA, 121, 4232

McKinnon, W., Lunine, J., & Banfield, D. 1995, Neptune and Triton (Tuscon, AZ: Univ. Arizona Press) 807

McKinnon, W. B. 1984, Natur, 311, 355

McKinnon, W. B., & Kirk, R. L. 2014, Triton, Encyclopedia of the Solar System (Amsterdam: Elsevier) 181

McKinnon, W. B., & Mueller, S. 1988, Natur, 335, 240

McKinnon, W. B., Simonelli, D. P., & Schubert, G. 1997, Pluto and Charon (Tuscon, AZ: Univ. Arizona Press) 295

McKinnon, W. B., Nimmo, F., Wong, T., et al. 2016, Natur, 534, 82

McKinnon, W. B., Stern, S., Weaver, H., et al. 2017, Icar, 287, 2

McKinnon, W. B., Glein, C. R., Bertrand, T., & Rhoden, A. R. 2021, The Pluto System After New Horizons, ed. S. A. Stern, J. M. Moore, W. M. Grundy, L. A. Young, & R. P. Binzel, (Tuscon AZ.: Univ. Arizona Press) 507

McKinnon, W. B., Bland, M. T., Singer, K. N., Schenk, P. M., & Robbins, S. J. 2023, JGRE, 128, e2023JE007

Melosh, H. J. 2011, Planetary Surface Processes (Cambridge: Cambridge Univ. Press) Vol. 13

Merk, R., & Prialnik, D. 2006, Icar, 183, 283

Mitchell, K. L., Prockter, L. M., Frazier, W. E., et al. 2019, 50th Lunar and Planetary Science Conf. (Houston, TX: Lunar and Planetary Institute) Abstract #3200

Moore, J. M., & McKinnon, W. B. 2021, AREPS, 49, 173

Moore, J. M., McKinnon, W. B., Spencer, J. R., et al. 2016, Sci, 351, 1284

Morbidelli, A., & Nesvorný, D. 2020, The Trans-Neptunian Solar System (Amsterdam: Elsevier) 25

Mumma, M. J., & Charnley, S. B. 2011, ARA&A, 49, 471

National Academies of Sciences, Engineering, and Medicine 2023, Origins, Worlds, and Life: A Decadal Strategy for Planetary Science and Astrobiology 2023–2032

Nesvorný, D. 2018, ARA&A, 56, 137

Nimmo, F. 2004a, JGRE, 109, E12001

Nimmo, F. 2004b, Icar, 168, 205

Nimmo, F., & McKinnon, W. B. 2021, The Pluto System After New Horizons (Tuscon, AZ: Univ. Arizona Press) 89

Nimmo, F., & Pappalardo, R. T. 2016, JGRE, 121, 1378

Nimmo, F., & Spencer, J. 2015, Icar, 246, 2

Nimmo, F., Hamilton, D., McKinnon, W., et al. 2016, Natur, 540, 94

Nimmo, F., Umurhan, O., Lisse, C. M., et al. 2017, Icar, 287, 12

Nogueira, E., Brasser, R., & Gomes, R. 2011, Icar, 214, 113

Ohara, S., & Dombard, A. 2019, Proc. 50th Lunar and Planetary Science Conf. (Houston, TX: LPI) 18

Ormel, C., Dullemond, C., & Spaans, M. 2010, ApL, 714, L103

O'Neill, C., & Nimmo, F. 2010, NatGe, 3, 88

Pappalardo, R. T., Collins, G. C., Head, J., et al. 2004, Jupiter: The Planet, Satellites and Magnetosphere (Cambridge: Cambridge Univ. Press) 363

Prockter, L. M., Nimmo, F., & Pappalardo, R. T. 2005, GeoRL, 32, L14202

Protopapa, S., Grundy, W., Reuter, D., et al. 2017, Icar, 287, 218

Robuchon, G., & Nimmo, F. 2011, Icar, 216, 426

Ross, M. N., & Schubert, G. 1990, GeoRL, 17, 1749

Ross, R. G., & Kargel, J. S. 1998, Solar System Ices: Based on Reviews Presented at the International Symposium "Solar System Ices" held in Toulouse, France, on March 27–30, 1995 (Berlin: Springer) 33

Rovira-Navarro, M., Matsuyama, I., & Hay, H. C. C. 2023, PSJ, 4, 23

Rufu, R., & Canup, R. M. 2017, AJ, 154, 208

Schenk, P., & Jackson, M. 1993, Geo, 21, 299

Schenk, P. M., & Zahnle, K. 2007, Icar, 192, 135

Schenk, P. M., Beyer, R. A., McKinnon, W. B., et al. 2018, Icar, 314, 400

Schenk, P. M., Beddingfield, C. B., Bertrand, T., et al. 2021, RemS, 13, 3476

Schultz, P. H., & Gault, D. E. 1975, Moon, 12, 159

Shock, E. L., & McKinnon, W. B. 1993, Icar, 106, 464

Singer, K., Greenstreet, S., Schenk, P., Robbins, S., & Bray, V. 2021, The Pluto System After New Horizons (Tuscon, AZ: Univ. Arizona Press) 121

Singer, K. N., White, O. L., Schmitt, B., et al. 2022, NatCo, 13, 1542

Solomatov, V. 1995, PhFl, 7, 266

Stern, S. A., & McKinnon, W. B. 2000, AJ, 119, 945

Stern, S. A., Grundy, W. M., McKinnon, W. B., Weaver, H. A., & Young, L. A. 2018, ARA&A, 56, 357

Thomas, P. 2000, Icar, 148, 587

Turcotte, D., & Schubert, G. 2014, Geodynamics, (Cambridge: Cambridge Univ. Press)

Tyler, R. H. 2008, Natur, 456, 770

Watts, A., & Burov, E. 2003, E&PSL, 213, 113

Wisdom, J. 2008, Icar, 193, 637

Wong, M. H., Lunine, J. I., Atreya, S. K., et al. 2008, RvMG, 68, 219

Yamashita, Y., Kato, M., & Arakawa, M. 2010, Icar, 207, 972

Yao, C., Deschamps, F., Lowman, J., Sanchez-Valle, C., & Tackley, P. 2014, JGRE, 119, 1895

Zahnle, K., Schenk, P., Sobieszczyk, S., Dones, L., & Levison, H. F. 2001, Icar, 153, 111

Chapter 3

Triton and Pluto: Cryovolcanism in the Outer Reaches of Our Solar System

Lynnae C Quick, Kelsi N Singer, Caitlin Ahrens, Candice J Hansen, Jason D Hofgartner, Karl L Mitchell, Louise M Prockter, Paul M Schenk and Orkan Umurhan

Throughout their geological histories, the surfaces of Triton and Pluto have been shaped by cryovolcanism. We present a review of cryovolcanic processes and features on these worlds and look ahead to measurements from future spacecraft that would elucidate the role that cryovolcanism has played in resurfacing these bodies over the age of the solar system. In the case of Triton, a dedicated flyby or orbital mission could shed light on whether active cryovolcanism is currently occurring and constrain the origin and distribution of plumes at the surface. An orbital mission to Pluto would allow improved constraints to be placed on the rates of occurrence of surface-subsurface exchange, the mobility of melt, and conditions under which it may extrude onto the surface. Further investigations of both worlds would provide additional constraints on expected geological activity rates on small, icy bodies for which tidal heating by solid body tides is negligible or non-existent. In doing so, these studies would enable us to obtain a more complete picture of the various manifestations of cryovolcanism throughout our solar system.

3.1 Cryovolcanism in the Solar System

Cryovolcanism is volcanism at low temperatures, much colder than silicate volcanism, that involves low melting point materials. The migration of cryogenic solutions, slurry-like suspensions and sub-solidus material between the interiors and surfaces of ice-rich planetary bodies has shaped many of the worlds in our solar system. Extrusive cryovolcanism describes processes associated with the eruption of solutions, slurry-like suspensions, and sub-solidus flows at the surfaces of icy worlds. It includes both effusive processes, such as the emplacement of cryolava flows and cryolava domes, and explosive processes such as geyser-like eruptions of fine particles in volatile-rich sprays. Intrusive cryovolcanism, or cryomagmatism, describes the generation and mobilization of cryogenic fluids in the interiors of these bodies and may include diking, diapirism and sill emplacement. Cryovolcanic fluids may include pure water, briny solutions, or partially frozen mixtures of water and freezing-point depressants such as salts, mineral acids, or compounds that are usually found in the gas phase on Earth such as ammonia or methanol. While this chapter will focus on water-based cryovolcanism, cryovolcanism is a key mechanism for the release of internal heat on volatile-rich worlds, regardless of its manifestation. We note that in the context of this chapter, where we review cryovolcanic processes on bodies that are not only rich in water ice but also in other frozen volatile species such as N_2, CO, and CH_4, that the word "ice" is used to refer to a variety of frozen compounds that are found at their surfaces.

Spacecraft data have revealed that cryovolcanism has shaped the surfaces of a variety of worlds. Cryovolcanic features on dwarf planet Ceres include the muddy cryovolcano Ahuna Mons and the wispy Vinalia Faculae (Ruesch et al. 2016, 2019; Quick et al. 2019). Galileo images revealed candidate surface flows on Jupiter's moon Europa (Fagents 2003; Miyamoto et al. 2005; Prockter & Schenk 2005; Prockter 2017; Lesage et al. 2021), and domes that may have formed from the eruption of viscous cryolava (Fagents 2003; Pappalardo & Barr 2004; Miyamoto et al. 2005; Quick et al. 2017, 2022) or by the intrusion of melt in the shallow subsurface (Figueredo et al. 2002; Pappalardo & Barr 2004; Manga & Michaut 2017). Observations from the Hubble Space Telescope and the Keck Observatory revealed possible detections of 50–200 km tall plumes on this enigmatic moon (Roth et al. 2014; Sparks et al. 2016, 2017; Paganini et al. 2019), and low-albedo deposits that lie along lineated features and surround a subset of lenticulae have been designated as potential deposits left behind by small (<30 km) plumes (Fagents et al. 2000; Quick et al. 2013; Quick & Hedman 2020). The Cassini spacecraft observed spectacular water plumes spewing from the south pole of Enceladus (Dougherty et al. 2006; Porco et al. 2006; Spencer et al. 2006, 2009), while radar imagery revealed that Doom Mons and Erebor Mons are two of the tallest peaks and the most distinguishable cryovolcanoes on Titan. Mohini Fluctus, an approximately 200-km long, northeast-trending cryolava flow, appears to originate from Doom Mons, which is flanked to its east by the cryovolcanic caldera Sotra Patera; lobate flows have also been imaged in association with Erebor Mons (Lopes et al. 2013). Voyager 2 imaged several potential indicators of past cryovolcanism on the surfaces of Ariel and Miranda, most notably viscous flows that lie adjacent to coronae and ridges and that cover deep graben floors (Jankowski & Squyres 1988; Schenk 1991; Cartwright et al. 2020; Beddingfield et al. 2021). Charon's southern hemisphere plains unit, Vulcan Planitia, may represent an extensive cryolava flow field (Moore et al. 2016; Beyer et al. 2019). Cryolava flows on Titan, Ariel, Miranda, and Charon may have an ammonia-rich composition, while those on Ceres and Europa may be best described as "briny", with cryovolcanic fluids on the former being composed of a mud-brine admixture. In this chapter we will review, in detail, the ways in which cryovolcanism has shaped the surfaces of Triton and Pluto.

3.2 Powering Cryovolcanism on Triton and Pluto: Internal Heating Mechanisms

If Triton had persisted as an uncaptured Kuiper Belt Object, its internal heat would have resulted primarily from radioisotope decay, similar to relict ocean world Ceres. However, Triton's capture into orbit around Neptune led to an initially highly eccentric orbit and tidal heating for at least a billion years, which diminished in magnitude as its orbit circularized and it became tidally locked (Ross & Schubert 1990; McKinnon et al. 1995; Agnor & Hamilton 2006). As a result of Triton's highly inclined, $-23°$ orbit, tidal heating is likely to have persisted to the present day due to obliquity tides; the obliquity tidal heating that Triton experiences is greater, by at least an order of magnitude, than the obliquity tidal heating experienced by other icy moons (Chen et al. 2014; Nimmo & Spencer 2015). If past orbital circularization permitted Triton's eccentricity to proceed slowly to its current value, the post-capture tidal dissipation that it experienced may have caused heat to concentrate in its interior, slowing and potentially reversing the growth of the ice shell, and allowing for the preservation of a subsurface ocean (Gaeman et al. 2012). Cosmochemical arguments, supported by the apparent rigidity and strength of putative cryovolcanic features (Kargel & Strom 1990; Kargel et al. 1991), indicate the likely presence of ammonia (or ammonium) and methanol in Triton's ice shell and ocean. Together with other "antifreeze" compounds, such as chlorides, ammonia would help maintain the ocean in a liquid state. Although some ammonia is likely to have been trapped within the top few kilometers of Triton's surface as the ice shell formed, most would have been excluded from the ice shell as it thickened. This would have resulted in increased concentrations of ammonia in the subsurface ocean, preventing it from freezing and permitting it to grow up to 50 km thick (Hammond et al. 2018).

Many unknowns remain concerning mechanisms for internal heating on Pluto. The surface is home to young terrains that appear to have been created from the emplacement of new material that migrated from the interior to the surface (as described below). However, the heat sources that would allow for the mobility of icy subsurface material in the geologically recent past have yet to be determined. We note that until the effects of obliquity tidal heating were considered, the internal heat sources that allowed for the youthful surfaces that Voyager 2 imaged on Triton were also unknown (Chen et al. 2014; Nimmo & Spencer 2015). Pluto's bulk interior is composed of approximately two-thirds rock (McKinnon et al. 2017), and radiogenic decay has likely produced heat flows of $\lesssim 5$ mW m^{-2} for most of its history (McKinnon et al. 1997; Robuchon & Nimmo 2011). Nevertheless, models still show that under certain conditions, radiogenic heating could allow for a deep subsurface ocean to persist until the present (Robuchon & Nimmo 2011; Bierson et al. 2018; Kamata et al. 2019; Bierson et al. 2020; Kimura & Kamata 2020; Nimmo & McKinnon 2021). Although Charon is thought to have formed by a giant, grazing impact (Canup 2011; Canup et al. 2021), models predict that tidal evolution of the Pluto-Charon system would have progressed rapidly, reaching the dually locked final tidal end state less than 100 Myr after the impact (Cheng et al. 2014).

Thus, tides would not play a major role in internal heating for most of Pluto's history. Elevated internal heating could have been produced by differentiation, but that heat is likely to have only been present early in Pluto's history. At ~2 km s^{-1}, typical impact velocities in the current Pluto system are slower than much of the inner solar system and much slower than most icy moons (Greenstreet et al. 2015). The low velocities and scarcity of large recent impacts mean that impact heating is also unlikely to drive much internal heating on Pluto. Given the lack of obvious, strong internal heat generating mechanisms, theories have emerged that call for the trapping of heat in Pluto's interior, allowing it to build up over time. For example, an insulating clathrate layer was first proposed as a way to allow for additional ocean uplift under the Sputnik Basin, forming a positive gravity anomaly and allowing the basin to migrate to its current position at Pluto's equator (Kamata et al. 2019). This is one example of how the specific internal structure of Pluto could allow for the additional accumulation of heat which could facilitate cryovolcanism.

3.3 Cryomagmatism: Melt Generation, Migration, and Composition

The primary mechanisms by which cryomagmas may form in the ice shells of worlds such as Triton and Pluto include localized melting in the ice shell, leading to the formation of discrete fluid pockets (i.e., fluid reservoirs), and expansive melting associated with ocean formation. As water-based fluids are negatively buoyant with respect to the ice shells through which they must ascend, it can be quite difficult for ocean-derived fluids to reach the surfaces of icy worlds, especially in the case of large bodies such as Pluto and Triton which have relatively thick crusts and relatively high gravities when compared to other icy worlds in our solar system (Manga & Wang 2007; Conrad et al. 2016). As such, cryomagmas on these worlds are likely to originate from discrete melt reservoirs that are perched within their ice shells. These reservoirs may be emplaced as liquid intrusions that originate from the subsurface ocean, or they may form *in situ* from local melting of the ice shell facilitated by brine mobilization, enhanced tidal dissipation in the case of Triton, or diapiric ascent of warm ice. These melt reservoirs are expected to exist at shallow crustal levels (e.g., see Fagents 2003; Schmidt et al. 2011; Chivers et al. 2021). If this is the case for Triton and Pluto, intra-crustal melt reservoirs would represent relatively shallow cryomagma source regions from which melt could ascend shorter distances to the surface. As on other icy bodies, it is expected that the compositions of crustal fluid reservoirs in Triton and Pluto would be representative of their modes of emplacement. For example, reservoirs that are emplaced as liquid intrusions of oceanic material will have initial compositions that are similar to the subsurface oceans, while fluid reservoirs that are emplaced as a result of *in situ* melting in the ice shell may have initial compositions that reflect both the surrounding country ice and the body which facilitated the localized melting, for example an ascending diapir of warm ice, or nearby brines (Fagents et al. 2022). In both cases, melt is likely to be composed of some non-H_2O materials, the concentration of which will increase as the reservoir cools over time and fractionation occurs. Larger reservoirs, which can hold on to more heat owing to their low surface area to volume ratios, will persist for longer than their smaller counterparts, especially if they contain a significant concentration of ammonia or other freezing-point depressants as part of their non-H_2O component.

3.3.1 Composition of Cryomagmatic Melts on Triton

The single flyby of Triton by Voyager 2 resulted in the collection of only limited compositional data of the surface and no direct subsurface measurements. Hence constraints on cryomagma composition and rheology have been derived from theory, ground-based observations, and from the limited imagery and surface compositional information provided by the spacecraft. We comment on those constraints here.

Although H_2O is likely to be the most abundant component in Triton's cryomagmas, these cryomagmas are unlikely to be composed of pure water. Multiple lines of evidence support the existence of ammonia (or ammonium) and other antifreeze materials in Triton's ocean, which would enable the formation of viscous cryomagmas whose rheologies may be more akin to terrestrial basalt than water (Kargel et al. 1991; Kargel 1992; Hogenboom et al. 1997; McKinnon & Kirk 2007). Triton's cryomagmas may also be contain volatiles such as CH_4, N_2, CO, or CO_2 (Croft et al. 1995), salts and/or carbonates (Castillo-Rogez et al. 2022, and references therein). Indeed, cosmochemical arguments and the proposed rigidity of cryovolcanic features at Triton's surface are indicative of ammonia–water cryolavas, while thick lobate flows imaged by Voyager 2 may have similar rheologies and compositions as flows on Ariel and Miranda (Kargel & Strom 1990). Taken together, this suggests that the most prevalent cryomagmas on Triton are NH_3–$2H_2O$ slurries and ammonia-water-ethanol mixtures.

The challenge of overcoming the negative buoyancy of water-based fluids could limit the formation of buoyant melts at the base of Triton's ice shell. Thus, the potential for cryomagmas to ascend and erupt has not been well-constrained. While ammonia-water is less dense than liquid water, it is more dense than even fully compacted water-ice, which could make diapiric rise and magma chamber formation difficult (Head & Wilson 1992). Nevertheless, ice shell convection may allow heat to be transported to the near-surface, which could lead to phase-changed induced activity. Due to its resemblance to exposed terrestrial salt diapirs, Triton's Cantaloupe Terrain (Figure 3.1) has been interpreted as the possible result of such processes. These organized cellular depressions, which range from 30–50 K in diameter, extend across much of the trailing hemisphere. While formation processes such as sublimation and scarp regression have been suggested for cantaloupe terrain, Schenk & Jackson (1993) proposed that the cells resulted from compositionally driven convection in a layered ice shell ~20 km thick, that has a maximum viscosity of 10^{22} Pa s, and is partially composed of ice phases other than water. If terrestrial experience is any indicator, direct transport from ocean-to-surface via fractures, even through the ductile lower ice shell, is possible, because strain rates associated with dike propagation would be much greater than tectonic strain rates (Rivalta et al. 2015, and references therein). However, as with activity on Earth, sustained eruptive activity may not be guaranteed unless sufficient volatiles are exsolved to

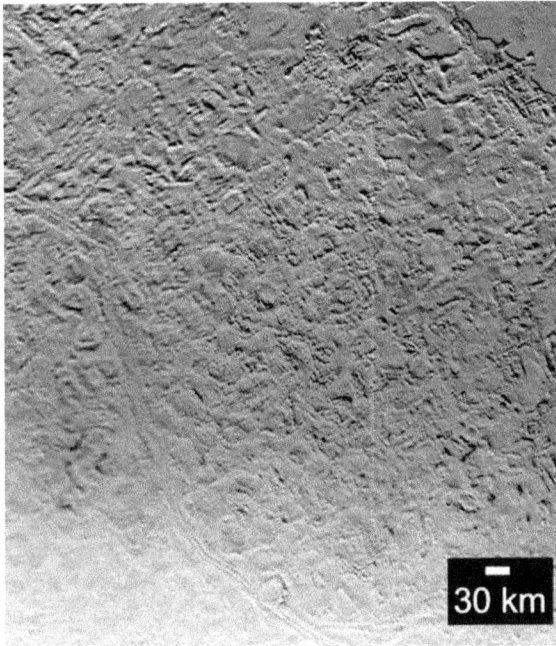

Figure 3.1. Triton's Cantaloupe Terrain consists of cellular depressions that formed as a result of convection driven by compositional differences in the ice shell.

lead to expansion of the rising melt, or unless the melt source, whether the ocean or a discrete melt pocket, is significantly pressurized due to freezing (Fagents 2003; Manga & Wang 2007). Once flow is established, pressurized pathways could be maintained if the reservoir is large enough, and if fluid ascends to the surface quickly enough, to prevent freezing en route (Quick & Marsh 2016).

3.3.2 Composition of Cryomagmatic Melts on Pluto

Similar to the case for Triton, the overall subsurface plumbing systems, source regions, and driving forces for cryovolcanism on Pluto are not well constrained. As previously mentioned, traditional models for internal heating and long-term evolution of Pluto allow for liquid water in the interior but mobile melt could also take the form of a slushy water-ice mixture or a solid-state flow. As is the case of Voyager 2 at Triton, since New Horizons made no direct subsurface measurements at Pluto, constraints on cryomagma viscosity and composition can only be inferred from data taken near putative cryovolcanic features on the surface.

Spectral signatures associated with ammonia were detected in areas of Pluto near large extensional faults and provided some of the impetus to investigate whether other features were cryovolcanically emplaced (e.g., Virgil Fossae; Dalle Ore et al. 2019). However, ammonia is not found in the vicinity of other large features with morphologies suggestive of cryovolcanic emplacement (e.g., the structures southwest of Sputnik Planitia which are described in more detail below; Singer et al. 2022). The presence of ammonia would reduce the melting temperature of water by up to ~100 degrees at low pressures (Johnson & Nicol 1987); further reductions in melt temperature could occur if multiple antifreeze components are present (Kargel 1998). Indeed, despite the significant temperature difference between Pluto's 44 K surface and the 176 K ammonia-water eutectic temperature, features indicative of surface flow are still present on Pluto's surface (see Section 3.6.1).

Episodic pressure gradients would be required for material to have been driven to Pluto's surface on multiple occasions and in the geologically recent past. If the melt source region is a relatively deep subsurface ocean or even a melt pocket at tens of kilometers deep, reaching the surface would require considerable transport distances. Assuming a mostly water-ice lithosphere, the overburden pressure at 150 km depth on Pluto, accounting for a surface gravity of 0.62 m s^{-2}, is approximately 100 MPa. On Earth magma ascent is driven by buoyancy because liquid "magma" has a lower density than the surrounding bedrock. Under certain conditions, such as would be the case if a positively buoyant diapir of warm ice or melt containing a significant fraction of ammonia were to ascend through the cold, icy crust, density-driven cryomagma ascent could also occur on Pluto. Notwithstanding, given Pluto's lithospheric structure and the likely composition of cryomagmas, density-driven ascent is unlikely to be a strong driver of cryovolcanic activity on the dwarf planet. While pressurization due to the gradual freezing of a subsurface ocean has been explored as one mechanism that could facilitate cryovolcanism on Pluto, the relatively high overburden pressures that would be encountered suggest that ocean pressurization alone is inadequate to permit fracturing and ascent of material directly from a subsurface ocean (Conrad et al. 2016). However topographic variations, such as those that are due to the presence of Sputnik basin, cause stress gradients on the dwarf planet (McGovern & White 2019; Martin & Binzel 2021). These stress gradients, which may cause pressure differences that lead to volatile exsolution of near-surface cryomagmas (Neveu et al. 2015), could facilitate cryomagma ascent. Tectonic fracturing of Pluto's crust in certain locations has also been invoked as a possible conduit for cryomagmatic ascent.

3.4 Cryovolcanism on Triton and Pluto

The surfaces of Triton and Pluto host a variety of features whose morphological characteristics have been interpreted as analogous to volcanic landforms on Earth and elsewhere in the solar system. As both worlds were observed during rapid flybys at distances >10,000 km by Voyager 2 and New Horizons, respectively, mapping data are limited and highly variable across their surfaces (Schenk et al. 2018; Schenk et al. 2021). Corresponding image pixel scales >325 and >90 m, respectively, provided limited morphologic and topographic constraints. Furthermore, Voyager 2 lacked a mapping spectrometer at Triton, and while New Horizons gathered spectral composition data, freeze-out and deposition of volatile ices from Pluto's atmosphere may have served to mask bulk crustal composition.

3.5 Triton: A Cryovolcanic World?

Triton's surface contains a number of unique, putative cryovolcanic features (Figure 3.2). Four large (100–400 km), quasi-circular, walled landforms are found at low latitudes, distributed geographically in two pairs (Figures 3.3 and 3.4). These contain relatively flat, smooth plains bounded by inward-facing, crenulated scarps several hundred meters high, with floors containing clusters of pits or small dome-like structures. The walled plains landforms have been interpreted as collapse calderas with floors infilled by cryovolcanic deposits, while pit clusters on the floors are interpreted to have resulted from late-stage outgassing or drain-back of partly congealed cryolavas; pit clusters have also been interpreted as source vents (Croft et al. 1995). The apparent flatness and floor pits of Triton's smooth plains deposits has been suggested to be the result of low-viscosity cryolavas (Kargel and Strom 1990; Croft et al. 1995), although alternative formation models have been proposed, such as glacial processes or wide-scale volatile sublimation (e.g., Croft 1990; Sulcanese et al. 2023).

Perhaps the most extensive and convincing putative cryovolcanic materials on Triton are centered on the quasi-circular 90-km-wide Leviathan Patera, and the

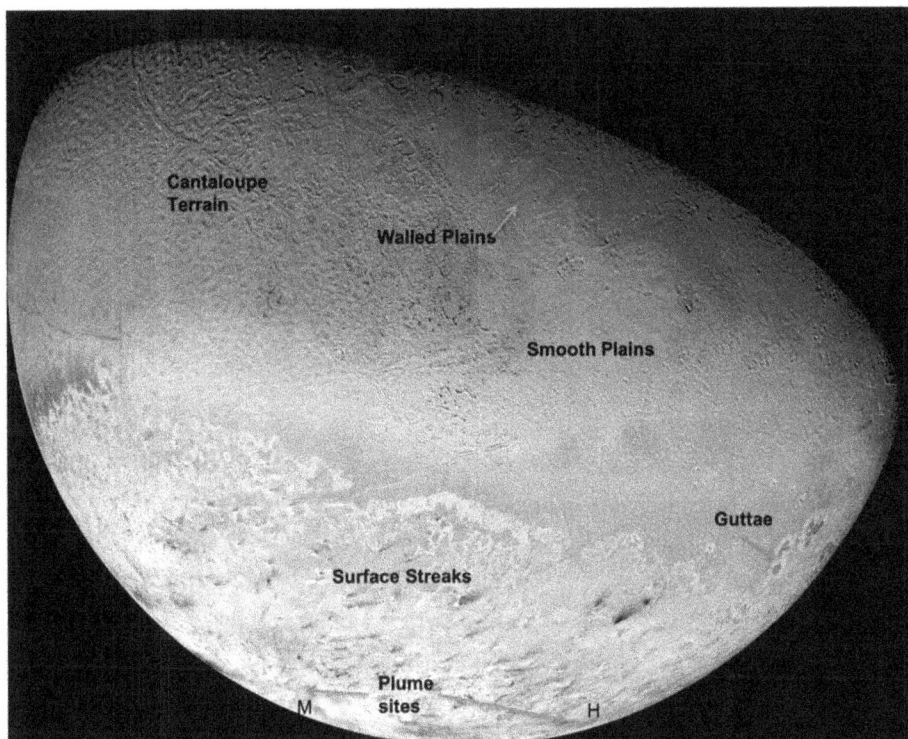

Figure 3.2. This mosaic of Triton shows the various terrain types, and the putative cryovolcanic features located within them, that were imaged on the sub-Neptunian hemisphere by Voyager 2. Triton's Fans (labeled "Surface Streaks") are widely distributed across the bright Southern Hemisphere Terrain (SHT). The sites of the two confirmed plumes are marked, with "M" for Mahilani and "H" for Hili.

Figure 3.3. Triton's surface exhibits an abundance of candidate cryovolcanic features such as large walled plains (e.g., Ruach Planitia), candidate calderas surrounded by annular smooth plains (e.g., Leviathan Patera), and pit chains (e.g., Set and Kraken Catenae). See also Figure 3.4.

elongate pit chains Set and Kraken Catenae (Figure 3.3). This >1000 km region is dominated by smooth (at >500 m scale) plains and the non-impact walled depressions of the patera and catenae. Although it is not formed on any topographic edifice, Leviathan Patera has internal relief of ~1 km and features several low domes on its floor that are reminiscent of terrestrial resurgent caldera domes (Schenk et al. 2021). Additional smooth plains materials are distributed in elongated lanes or quasi-circular patches centered on individual pits or groups of pits typically 20–50 km wide (Figure 3.3). Several of the larger walled pits in Kraken Catena have dark aureole deposits. These generally have feathered or indistinct margins and appear to stand higher than the surrounding terrain (Croft et al. 1995; Schenk et al. 2021). Triton's chains of pits strongly resemble the pit chains formed over terrestrial subsurface dike intrusions, such as Kilauea, further implying a cryovolcanic origin through explosive eruption of low-viscosity cryolavas onto the surface. On the basis of their ring-like shapes, morphologies, gradational boundaries, and apparent softening of topography near their margins, Croft et al. (1995) proposed that the 100–200-km widths of the rings are comparable to the distances cryoclastic material could be ejected from H_2O-rich or NH_3-H_2O cryomagmas containing dissolved nitrogen or methane. To the degree allowed by the Voyager data, there is no

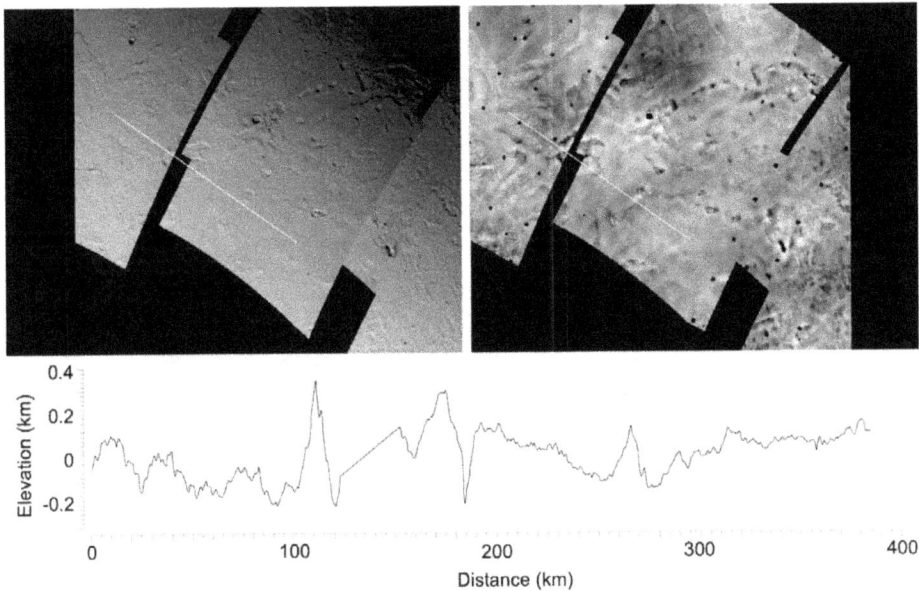

Figure 3.4. Topography of the candidate cryovolcanic region of Triton's surface shown in Figure 3.3.

evidence of any large-scale doming or depression across this vast region (Schenk et al. 2021; Figure 3.4).

Possibly the most enigmatic of Triton's geological features are the "guttae". Found close to the edge of the bright South Polar Terrain, these strange, lobate features have dark centers surrounded by bright aureoles. The aureoles tend to be a fairly constant 20–30 km in width, and the margins of both the centers and the aureoles are relatively distinct at observed resolutions. The smooth surfaces and lobate edges of the guttae, along with their inferred height of tens of meters, suggest that they formed by viscous flows (Croft et al. 1995), or alternatively from surface collapse over "hotspots" of liquid water (Croft 1990). The association of the guttae with the edge of the South Polar Terrain may also imply that the guttae are the remnants of layered frozen volatiles that have been eroded over time.

A sparse number of double ridges (fractures bounded by two ridges that are uniform in height and width) have been identified on Triton and have been suggested to have formed by viscous flooding of graben (Smith et al. 1989), or as fissure eruptions above dikes (Croft 1990). If analogous to Europa's similar but smaller and far more ubiquitous double ridges, they are more likely to result from frictional melting along cracks due to tidal deformation (Prockter et al. 2005); however, the origins of Europa's ridges are themselves still far from certain.

3.5.1 Triton's Plumes: A First Glimpse of Extraterrestrial Geysers

Voyager 2 discovered two confirmed plumes erupting from Triton's surface in 1989. Initially dubbed "east plume" and "west plume" they were later named "Hili" and

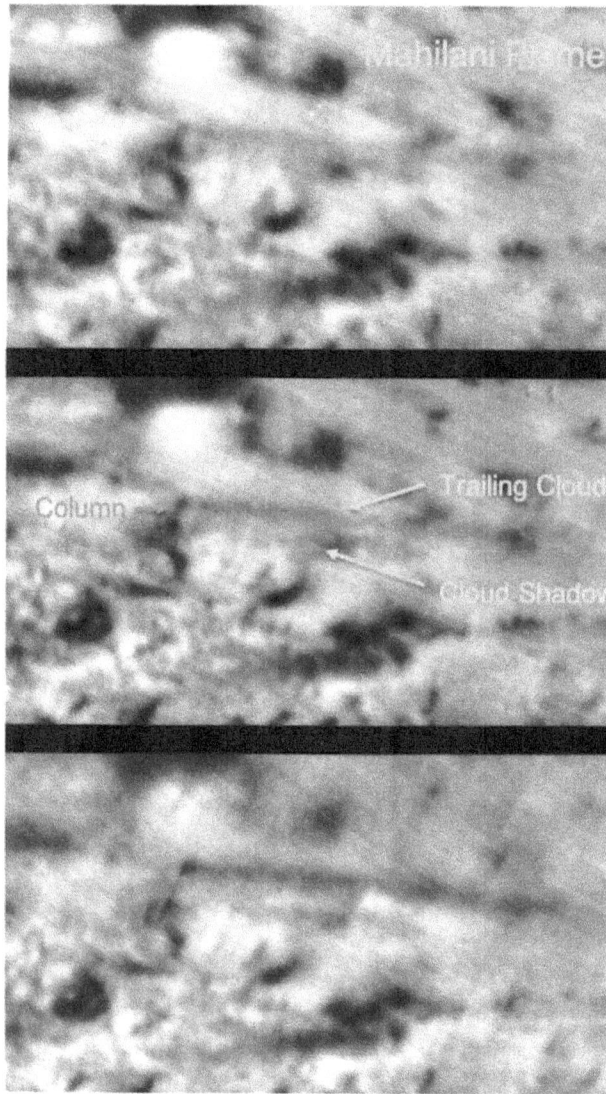

Figure 3.5. Three images of the Mahilani plume, shown here, were used to derive the column height of 8 km from parallax (Smith et al. 1989). Once the column reaches approximately 8 km, the winds blow the plume constituents to the west, stretching over 150 km. The optical density of the plume particles is sufficient to cast a shadow. A data compression artifact causes the plume to look narrow in the third, highest resolution, panel.

"Mahilani", respectively (Figures 3.2 & 3.5 and Table 3.1). These were the first active plumes to be discovered on an icy world (Smith et al. 1989; Soderblom et al. 1990). Two additional features that are presumed to be plumes are located several tens of kilometers south of Hili. The columns connecting the observed clouds from these features to Triton's surface are not completely visible down to the surface because they intersect the limb. The apparent diameter of plume columns on Triton

Table 3.1. Triton's Plumes

Plume name	Original designation	Latitude	Longitude	Height	Trailing cloud length	Status
Hili	East plume	57° S	38° E	8 km	>100 km	Confirmed
Mahilani	West plume	49° S	2° E	8 km	90 to >150 km	Confirmed
_____	East plume companion #1	59° S	43° E	8 km	>100 km	Unconfirmed
_____	East plume companion #2	60° S	44° E	8 km	>100 km	Unconfirmed
_____	Limb haze with overshoot	62° S	67° E	6.5 km	unknown	Unconfirmed
_____	Fixed Terminator cloud	37° S	47° W	>13 km	>400 km	Unconfirmed
_____	Crescent streak #1	40° S	144° E	1—3 km	77—254 km	Unconfirmed
_____	Crescent streak #2	41° S	144° E	1—3 km	77—254 km	Unconfirmed
_____	Crescent streak #3	43° S	133° E	1—3 km	77—254 km	Unconfirmed
_____	Crescent streak #4	47° S	159° E	1—3 km	77—254 km	Unconfirmed
_____	Crescent streak #5	49° S	154° E	1—3 km	77—254 km	Unconfirmed
_____	Crescent streak #6	51° S	156° E	1—3 km	77—254 km	Unconfirmed

Parameters extracted from Hansen et al. (1990) and Table 3.1 of Kirk et al. (1995).

ranges from <2 to 3 km but could be smaller if the rising columns are sheathed by descending material.

Triton's plumes were imaged multiple times with spatial resolutions ranging from 0.8 to 4 km pixel^{-1}. Parallax was used to determine that the columns of both Hili and Mahilani reach elevations up to 8 ± 1 km, before bending quite sharply into long, diffuse, dark clouds (Soderblom et al. 1990). The highest resolution images reveal an approximately 1 km overshoot at the tops of the plumes before winds carry plume particles off to the west at the 8 km altitude. The trailing cloud of Hili stretches horizontally over 100 km from the top of the plume, while the Mahilani trailing cloud extent was observed to vary in length from approximately 90 km to over 150 km within a 45-minute timeframe (Table 3.1), implying a wind velocity of approximately 15 m s^{-1} at 8 km altitude (Soderblom et al. 1990). Plume cloud particles were observed to remain suspended over the entire length of the trailing cloud and were consistent with particle sizes of ~5 μm (Sagan & Chyba 1990; Soderblom et al. 1990).

Portions of Mahilani's trailing cloud were optically dense enough to cast a shadow, providing independent confirmation of its approximately 8 km height

(Figure 3.5). The cloud trail and cloud shadow were only a few percent darker than Triton's surface. In order to be compatible with the cloud shadow measurements, Mahilani's particles could be very dark dust with an optical depth of $\sim$0.05, a bright frost with an optical depth of 0.10, or a mixture of the two, depending on particle size (Soderblom et al. 1990). Considering the width of the trailing cloud, the maximum altitude the particles could have settled over its entire length, and a wind velocity of 15 m s^{-1}, Soderblom et al. (1990) calculated a mass flux of $<$10 kg s^{-1} for dark dust and 20 kg s^{-1} for ice crystals. Depending on the temperature of the source reservoir and the ratio of gas to ice, the flux of vapor could be as high as 400 kg s^{-1} (see Soderblom et al. 1990 for the assumed values and calculations).

Other potential plume detections remain ambiguous. While other clouds and hazes were detected along the terminator, it is unknown whether they have connections to plumes erupting at Triton's surface. Six long bright streaks were observed in high phase angle images at 1–3 km altitude (Table 3.1). These streaks were located between 40° S to 51° S, were 77–254 km long, $\sim$10 km wide, and their direction of motion appeared eastward (Hansen et al. 1990). The optical densities of the particles within these streaks may have been high enough for them to possibly cast shadows on Triton's surface, and their altitude and direction are consistent with the atmospheric model described in Ingersoll (1990). If all ambiguous interpretations are attributed to eruptions, then Triton could have had at least 12 active plumes, in the area imaged, at the time of the Voyager flyby (Table 3.1).

Three established hypotheses exist for the eruptive generation of Triton's plumes. These hypotheses are presented in thorough reviews by Kirk et al. (1995) and Hofgartner et al. (2022), and we summarize them here (Figure 3.6). The solar-driven hypothesis was the initially preferred formation hypothesis for Triton's plumes and is thus the most studied model for plume generation on the icy moon (Figure 3.6, left; Soderblom et al. 1990; Hansen et al. 1990; Kirk et al. 1990; Brown et al. 1990). In this model, nitrogen-ice at Triton's surface, which is more transparent to incident solar radiation than emitted thermal radiation, acts as a greenhouse. The greenhouse effect increases the temperature beneath the surface, which, due to nitrogen's

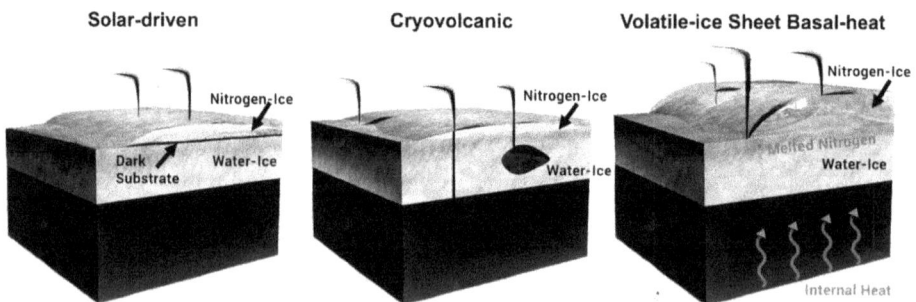

Figure 3.6. Artistic depiction of the three eruption hypotheses for Triton's plumes. Features are not shown to scale. Image reprinted from Hofgartner et al. (2022), Copyright (2022), with permission from Elsevier; image created by Lizbeth de la Torre and Lisa Poje at the Jet Propulsion Laboratory in consultation with the authors and used with permission.

volatility, significantly increases the nitrogen vapor pressure. This results in the pressurized nitrogen vapor explosively erupting as a plume. If verified as such, solar-driven plume eruptions would not qualify as cryovolcanism based on most definitions (e.g., see Lopes et al. 2010).

A second hypothesis for Triton's plumes involves explosive cryovolcanism, as defined in the introduction to this chapter (Figure 3.6, center; Kirk et al. 1995). As described above and in Croft et al. (1995), numerous features on Triton may be attributed to cryovolcanism. At the time that Voyager 2 observed Triton, explosive cryovolcanism in the form of geyser-like plumes had not yet been definitively identified on any icy world. Since then, Cassini has directly observed the active south pole of Enceladus (Porco et al. 2006), and recent work has shown that: (1) Triton is a likely ocean world (Hendrix et al. 2019; Hansen et al. 2021) (2) with an estimated vapor mass flux >400 kg s^{-1}, the vapor mass flux of Triton's plumes is similar to the 200 kg s^{-1} vapor mass flux of Enceladus' cryovolcanic plume (Hansen et al. 2021), but greatly exceeds the approximately 0.2 kg s^{-1} vapor mass flux that would be associated with solar-driven eruptions on Mars, which serve as an analog for solar-driven eruptions on Triton (Table 3.2) (3) explosive cryovolcanism on Triton could be powered by internal heating from obliquity tides (Nimmo & Spencer 2015). These facts warrant a reevaluation of the possibility that Triton's plumes are cryovolcanic.

A third hypothesis for the generation of Triton's plumes suggests that the base of a thick (on the order of a kilometer) layer of N_2-ice is heated by the moon's internal energy and that plume eruptions occur when nitrogen that is warmer than the

Table 3.2. A Comparison of Properties for Mars', Triton's, and Enceladus' Plumes and the Surface Environments they Erupt into

	Mars	Triton	Enceladus
Erupting volatile	CO_2	N_2 or H_2O	H_2O
Surface gravity (m s^{-2})	3.72	0.779	0.113
Plume height (km)	0.08	8	1500
Source vent diameter (m)	<1	<3000	~9
Exit velocity (m s^{-1})	20–300	20–40	450
Vapor mass flux	0.15 kg s^{-1}	Up to 400 kg s^{-1}	200 kg s^{-1}
Particle mass flux	0.03–0.15 kg s^{-1}	<10 kg s^{-1}	~50 kg s^{-1}
Surface temperature (K)	140	38–42	76–170
Eruption duration	<2 h	~1–3 years	Continuous
Atmospheric column mass[a]	188 kg m^{-2}	1.8 kg m^{-2}	Trace
Atmospheric mass density[b]	1.9×10^{-2} kg m^{-3}	1×10^{-4} kg m^{-3}	Trace

Adapted from Table 1 of Hansen et al. (2021) using values from Ingersoll (1990), Soderblom et al. (1990), Barlow (2008), and Hansen et al. (2021) and references therein. Published by the American Astronomical Society. CC BY 4.0.
[a] Obtained by dividing average surface pressure by gravity. Average surface pressures for Mars and Triton are approximately 700 and 1.4 Pa, respectively (Yelle et al. 1995; Barlow 2008).
[b] Obtained for Mars by dividing Mars' atmospheric column mass by an ~10 km atmospheric scale height (Barlow 2008); value for Triton is from Soderblom et al. (1990) and is consistent with dividing Triton's atmospheric column mass by the 14.8 km atmospheric scale height given in Ingersoll (1990).

surrounding atmosphere is transported to Triton's surface (Figure 3.6, right). In one version of this proposed model the base of the nitrogen-ice layer melts and the melt is exposed to the atmosphere after buoyant vertical rise through the nitrogen-ice and/ or flow at the base to the margin (Brown & Kirk 1994; Kirk et al. 1995), resulting in both the nitrogen and heat being transported in the liquid phase. Another version of this model calls for the nitrogen and heat to be transported in the solid phase via solid-state convection (Duxbury & Brown 1997).

While the solar-driven hypothesis for Triton's plumes was preferred by Kirk et al. (1995), Hofgartner et al. (2022) argued that an eruptive hypothesis is favored to non-eruptive hypotheses (Ingersoll & Tryka 1990; Kirk et al. 1995). Building upon the extensive review of Kirk et al. (1995) and subsequent analyses by Schenk et al. (2021), they concluded that all three of the above hypotheses warrant further consideration, without preference. Note that of the hypotheses presented here, and according to the definition of cryovolcanism as defined in this chapter, only the second hypothesis invokes cryovolcanism for the generation of Triton's plumes. Thus, at this time, it is not definitively known whether Triton's plumes have a cryovolcanic origin.

3.5.2 Triton's Fans

In addition to plumes, Voyager 2 observed features that are interpreted to be deposits from past eruptions on Triton. These features are wedge-shaped with an apex and diffuse tail that appears to be wind-blown and have been termed "fans" (Figure 3.2). The fans vary significantly in length from less than 5 km to greater than 100 km; they also vary in aspect ratio as some are fairly narrow while others are almost as wide as they are long (e.g., Hansen et al. 1990; Kirk et al. 1995). Triton's fans appear to be much more abundant than the plumes. Indeed, approximately 120 fans were mapped on the Voyager 2 encounter hemisphere (approximate sub-Neptune hemisphere). Most fans are oriented with an apex-to-tail direction that trends approximately northeast. By contrast, the two confirmed plumes Hili and Mahilani have trailing clouds that trend in an approximately east-to-west direction (Hansen et al. 1990). Previous work has suggested that the difference in orientation between fans and plumes is the result of variations in wind direction with altitude; wind direction is also expected to vary seasonally on Triton and could affect fan and plume orientation (Ingersoll 1990; Kirk et al. 1995).

While the fans are dark relative to Triton's high-albedo surface, they are intrinsically bright in comparison to most planetary surfaces (Soderblom et al. 1990) and are thus similar in brightness contrast to the clouds associated with the Hili and Mahilani plumes. Like the plumes, the fans occur exclusively on Triton's southern hemisphere terrains (Figure 3.2; Hofgartner et al. 2022). However, they are distinct from the confirmed plumes in that they are widely distributed across this terrain, notably on the sub-Neptune hemisphere, which was observed at high spatial resolution.

While the age of Triton's fans has not been well-constrained, their wind-blown morphologies and the seasonal migration of volatile ices (i.e., N_2, CO, CH_4, etc.)

(e.g., see Smith et al. 1989; Bertrand et al. 2023) suggest geologically short lifetimes. Based on Voyager 2's observations of plumes and fans on Triton, the distinct properties of the fans that have been described here are consistent with the interpretation that they were emplaced as distinct deposits from former plumes. A related hypothesis is that a subset of fans are active plumes with trailing clouds that reach <1 km in altitude, such that parallax cannot be observed with Voyager 2 images (Hansen et al. 1990; Kirk et al. 1995). Although the association between Triton's fans and plumes remains unconfirmed, we regard such an association as likely. If either hypothesis is true, then fans offer additional constraints on the nature of Triton's plumes (e.g., Hofgartner et al. 2022).

Fans were located equatorward (i.e., northward) of the plumes (Figure 3.2; Hansen et al. 1990). We note that while the geographical distribution of Triton's fans is consistent with widespread fans on the Voyager 2 encounter hemisphere Southern Hemisphere Terrain, worsening Voyager 2 image quality toward Triton's south pole limited observations, and the identification of fans, to north of the plumes, resulting in an image bias. If Triton's plume eruptions are solar-driven, the identification of additional fans may also have been limited due to migration of the plumes and fans with changing seasons (Hofgartner et al. 2022).

Interesting comparisons can be drawn between Triton's plumes and fans and the plume of Enceladus and its resulting deposits. Triton likely has both active plumes and former plumes and these features are geographically widely distributed. By contrast, Enceladus' plume is more localized (i.e., it is commonly referred to a single plume comprised of multiple jets), showcased continuous eruptions over the >10-year epoch of Cassini observations (Table 3.2), and may have a much longer continuous eruption history (e.g., see Kite & Rubin 2016; Choblet et al. 2022 and references therein).

3.6 Pluto: Evidence for Cryovolcanism at Both Large and Small Scales

Pluto exhibits several unique terrains that aren't found elsewhere in the solar system (Stern et al. 2015; Moore et al. 2016; White et al. 2021) (Figure 3.7). One of these is a large region to the southwest of Sputnik Planitia with large topographic relief in the form of giant broad rises and rough texture at several scales (Figures 3.8 & 3.9). This area has few impact craters and has been hypothesized to have been resurfaced and built up by cryovolcanism (e.g., Moore et al. 2016; Ahrens 2020; Singer et al. 2022). At the largest scale, the broad rises are 50–100 km across and several-to-8 km high. A few of these broad rises are somewhat domical in shape, but most have more varied shapes and are often interconnected with each other. Intermediate-scale undulations (6–12 km in width and a few hundred meters to 1 km high) cover much of this region, including flatter areas, and on the slopes and tops of the broad rises. While the undulations do vary in size, their size and wavelengths seem to be generally consistent across the region. Superimposed on these features is yet a smaller scale texture that may be boulders or ridges approximately 1–2 km across (these features are nearing the image resolution limits, thus are difficult to positively identify).

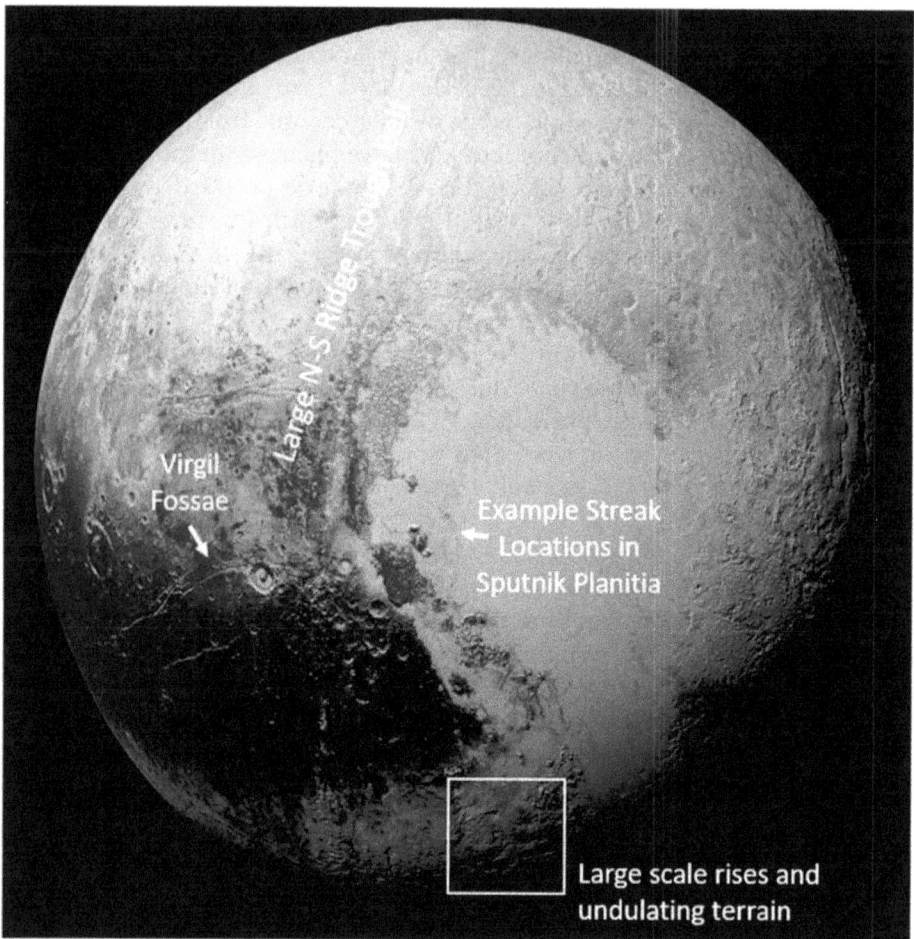

Figure 3.7. Pluto in enhanced color indicating the locations of several regions and geologic features that may have been cryovolcanically emplaced (image is Request ID P_Color taken by the New Horizons Multispectral Visible Imaging Camera [MVIC] camera).

Information on the bulk composition of these large structures is mostly indirect, although some spectral data from New Horizons does exist (Protopapa et al. 2017; Schmitt et al. 2017; Singer et al. 2022). Because nitrogen, methane, and carbon monoxide participate in Pluto's seasonal cycles of sublimation and redeposition as ice on the surface, these volatile ices significantly coat the surface and thus dominate the spectral signature, making it difficult to place constraints on what lies beneath. With Pluto's extreme obliquity, the cryovolcanoes Wright and Piccard Mons are within the "tropical arctic" zone (Moore et al. 2016; Binzel et al. 2017), in which oscillations in the multi-million-year cycles can influence the distribution of volatiles. However, the spectral data gathered by New Horizons does show some water ice in this area, in darker/warmer areas where the methane does not deposit.

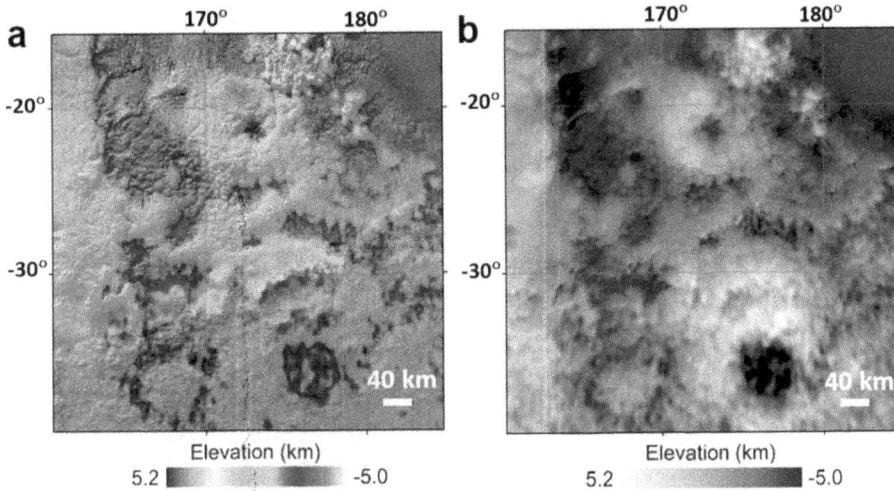

Figure 3.8. Putative cryovolcanic terrains southwest of Sputnik Planitia, Pluto. (a) color topography (Schenk et al. 2018), overlain on panchromatic basemap mosaic (image resolutions ranging from ~235–480 m px^{-1}); (b) greyscale topography alone. We show topography in both panels because this region was either very near the terminator or only seen in haze light during New Horizons' closest approach. Thus, the lighting geometry and deep shadows in some areas bias or obscure morphology. Simple cylindrical projection, shown with north up, and lighting from the upper left in panel (a). Topography shown with a linear stretch. Figure adapted from Singer et al. (2022). CC BY 4.0.

Nitrogen ice cannot maintain topography on the surface of Pluto. The rheological properties of methane are not well known, so while methane could be present (Howard et al. 2023), the characteristic sublimation textures observed in substantial methane deposits seen elsewhere on Pluto (e.g., the bladed terrain) are absent. Hence these structures may primarily consist of water ice, with other antifreeze ices or volatile ices as smaller constituents.

The few, if any, superposing craters suggest that Pluto's putative cryovolcanic region is relatively young, thus implying that conditions for material flow may have been present in the recent geologic past. These features do not appear to be erosional remnants, or to have been formed purely by collapse as there are no obvious signs of either of these processes occurring (e.g., no collapse terraces). Additionally, variations in color, texture, albedo, and topography across the region suggest that the underlying terrain was constructed via multiple emplacement episodes (see Singer et al. 2022 and references therein for details).

It is thermally challenging to mobilize water ice on or near the surface of Pluto. The current surface temperatures range from ~30–65 K (e.g., Young et al. 2021), potential heat sources are relatively minor (described above), and traditional models would place any extant internal ocean at the bottom of a thick ice shell >100 km beneath the surface. With its low average surface temperatures and low atmospheric pressure of ~10 microbars (Gladstone et al. 2016), pure liquid without a non-ice component cannot remain exposed on Pluto's surface for any appreciable amount of time. However locally (in dark areas) or temporally (during long summers) higher

Figure 3.9. Close up of intermediate-scale undulations and smaller-scale textures in the region southwest of Sputnik Planitia, Pluto. Image from the New Horizons observation request ID PEMV_P_MVIC_LORRI_CA (315 m px^{-1}; see Table S1 in Singer et al. 2022 for more details).

temperatures and surface pressures are possible (Earle et al. 2017; Stern et al. 2017). The large structures southwest of Sputnik Planitia are rough and not smooth at any observable scale in the New Horizons images, implying higher viscosity cryolavas or potential flow under a cooled, insulating carapace.

As previously described, one possible mechanism for enhanced heating near Pluto's surface is the presence of an insulating layer of clathrates that traps heat. If such an insulating layer exists, it could facilitate episodic cryovolcanic activity. If heat builds up under an insulating layer that occasionally faults or fractures, this could allow ascent of mobile material (Singer et al. 2022). If liquid solutions of ammonia or methane exist in Pluto's subsurface, diagenetic alteration of solid ice to include clathrates could occur (Fortes et al. 2007; Kamata et al. 2019).

While prospects for mud volcanism, and possible compositions of associated cryolavas, were first considered for Titan (Bradak & Kereszturi 2003; Fortes & Grindrod 2006; Solomonidou et al. 2013), this process has also been considered as a potential form of cryovolcanism on Pluto (Ahrens 2020). Mud volcanism can occur when thick sequences of accumulated sediments containing overpressurized mud layers are overlain by denser material (Fortes & Grindrod 2006; Skinner & Tanaka 2007). On Pluto, slushy cryolavas consisting of "muds" such as NH_3 and/or CH_4 may be produced as a result of interactions between NH_3- or CH_4-rich cryomagmas and ice (Cruikshank et al. 2019; Ahrens 2020). Models for mud volcanism suggest that cryolava composition will influence the dimensions of any cryovolcanoes that are formed. For example, ammonia-dominated source fluids may produce smaller cryovolcanoes and wider flow profiles than eruptions with methane-rich source fluids (Ahrens 2020).

Overall, the subsurface plumbing and driving stresses that may allow cryovolcanism on Pluto are not well understood. However, the existence of these unique surface features, and others that appear to be constructional (see below), imply that material is able to ascend to the surface to create new surface features and resurfaced terrain. Below we describe additional features on Pluto that may have a cryovolcanic origin.

3.6.1 Virgil Fossae

Pluto's Virgil Fossae region exhibits signs of low-viscosity flows across the surface, and crater statistics indicate that the region is relatively young compared to the age of the planet. (Figure 3.10; Cruikshank et al. 2019). The flows occur along the bottom of the fossae troughs, which display exposed water ice with a veneer of red, tholin-like, organic materials. In addition to direct evidence of water ice absorption bands, analysis of data acquired by New Horizons' imaging spectrometer (LEISA) also indicates the presence of NH_3 in some form (absorption bands at 1.65 and 2.2 μm), which may be associated with near-surface ammoniated salts, an H_2O-NH_3 hydrate complex, or some other ammoniated species (e.g., Dalle Ore et al. 2019;

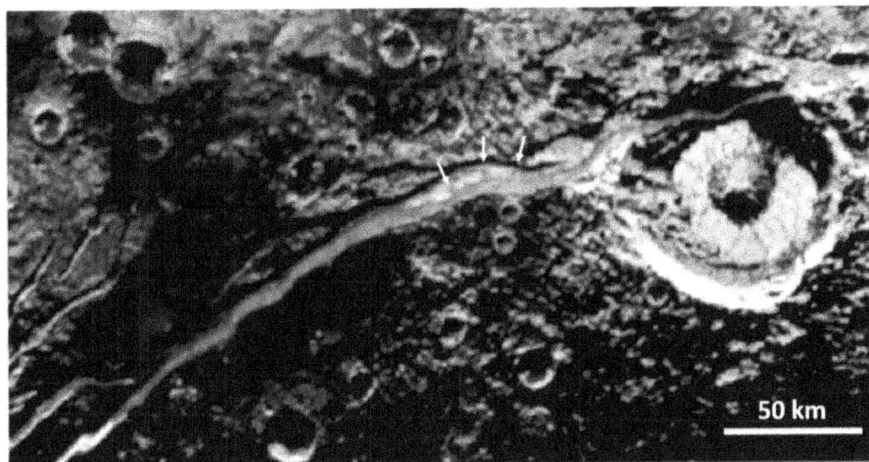

Figure 3.10. Color image from New Horizons of the Virgil Fossae region of Pluto. The red deposits in and around the fractures, and the white arrows indicate potential flow features (image is Request ID P_Color taken by the Multispectral Visible Imaging Camera [MVIC] camera). Adapted from Cruikshank et al. (2019), Copyright (2019), with permission from Elsevier.

Cruikshank et al. 2021; Protopapa et al. 2021; Cook et al. 2023). Due to limitations in LEISA's spectral range, the composition of the observed red-colored tholins in this area is unknown.

Cruikshank et al. (2019) hypothesized that in the past, low viscosity, slushy, aqueous solutions flowed within the fossae and their immediate surroundings. Here "slush" is defined as a fluid that contains both liquid and solid crystals, whose composite behavior exhibits a viscous, slurry-like rheology (e.g., see the experiments of Kargel et al. 1991). This flow was suggested to have been emplaced as a result of overpressurization of a subsurface ocean containing substantial amounts of ammonia and organic tholins. Cruikshank et al. (2019) showed that once a flow a few meters thick reached the surface, it could travel several tens of kilometers before freezing out.

A key constraint on any proposed model for fluid flow at Virgil Fossae is provided by the actual survivability of NH_3 on Pluto's surface, for which much uncertainty remains. Model calculations of NH_3 destruction driven by galactic cosmic-ray bombardment suggest an upper bound of 10^9 years before total removal of NH_3 from the top meter of Pluto's crust. On the other hand, lifetime estimates based on Lyman-alpha and solar wind charged particles suggest a far shorter range of lifetimes—from as little as 100 kyr to up to 100 Myr, depending upon the transparency of Pluto's atmosphere to UV photons, which only penetrate the surface down to an order of a few microns. If the original low-viscosity, NH_3-bearing flow was indeed several meters thick before freezing out, it may have been emplaced, at most, 10^9 years ago. Notwithstanding, Cruikshank et al. (2019) allow for the possibility that the Virgil Fossae region could be much younger.

3.6.2 Change Detection

The relatively short time span of the New Horizons flyby of Pluto decreases the probability of change detection. Nevertheless, a search for temporal changes was motivated by: (1) Pluto's extremely young terrains and geologic evidence for ongoing or recent resurfacing (e.g., Moore et al. 2016), (2) observed changes on Triton over similar timescales at similar spatial resolution (e.g., Soderblom et al. 1990), (3) reports of temporal changes in Earth-based observations of Pluto (e.g., Buie et al. 2010; Buratti et al. 2015), (4) the significant scientific value of observed changes. Thus, a thorough search for changes of Pluto's surface or atmosphere was conducted using New Horizons images. No differences that are strongly indicative of a temporal change were found (Hofgartner et al. 2018). Limits on temporal changes as a function of time interval, spatial resolution, and location were published. While numerous changes of appearance were identified, they were attributed to imaging artifacts and imaging parameters (e.g., imaging geometry as it relates to incidence, emission, and solar phase angles). Among the changes observed were features that changed in contrast relative to their surroundings from darker, to barely apparent to brighter, and features that brightened significantly at high solar phase angles (Hofgartner et al. 2018).

3.6.3 Absence of Plumes and Limited Analogous Fans

As discussed above, Triton had both active plumes during the Voyager 2 encounter and abundant fans which were interpreted to be deposits from prior plume eruptions. While plumes were not observed on Pluto during the New Horizons encounter, the presence of fans cannot be ruled out as dark streaks with varying degrees of resemblance to the fans on Triton were identified (Figure 3.11). Some streaks are adjacent to topographic barriers, such as hills, and may have been emplaced as a result of interactions between these barriers and winds on Pluto (e.g., Stern et al. 2015). Other streaks may have a primarily tectonic rather than wind-blown origin (e.g., Hofgartner et al. 2018). A few dark streaks on Sputnik Planitia may be most akin to Triton's fans (Hofgartner et al. 2018). These streaks are fan-shaped with wind-blown morphologies, have no apparent topographic or tectonic associations, occur on very high-albedo terrain, and have approximately similar orientations (Figure 3.11). Notwithstanding, these streaks are much less abundant than Triton's fans, with only a handful being identified on Pluto's Sputnik Planitia,

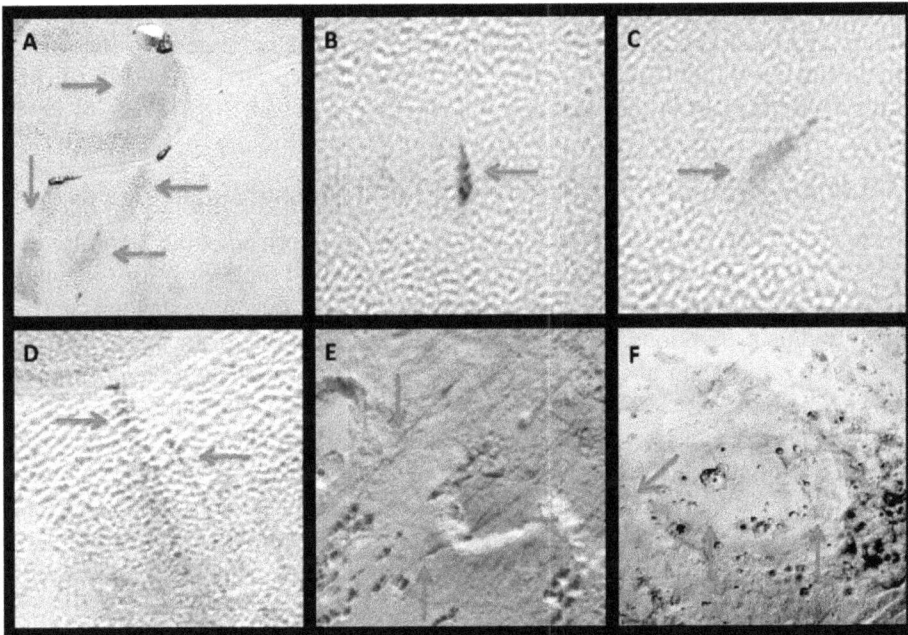

Figure 3.11. Dark surface streaks on Pluto. Blue arrows in panel **A** indicate streaks in Sputnik Planitia that are adjacent to hills and may be the result of interactions between the hills and winds (Stern et al. 2015). Red arrows in panels **A–D** indicate streaks in Sputnik Planitia that may be smaller versions of the streaks indicated by blue arrows in panel **A** or that may be the result of previous plumes. Green arrows in panel **E** point to dark streaks to the east of Sputnik Planitia that may be tectonically controlled or that may be a result of previous plumes. Green arrows in panel **F** indicate a few of the many dark streaks in and near Burney crater. Panel **A** is ∼70 km by 70 km, panels **B** and **C** are ∼16 km by 16 km, panel **D** is ∼18 km by 18 km, panel **E** is ∼150 km by 150 km, and panel **F** is ∼450 km wide by 350 km high. Figure reprinted from Hofgartner et al. (2018), Copyright (2018), with permission from Elsevier.

versus more than 100 fans being identified on Triton's Southern Hemisphere Terrain. Pluto's streaks are also generally smaller than Triton's fans. Taken together, the absence of active plumes on Pluto and the limited resemblance of Pluto's dark streaks to Triton's fans suggests that the differences between Triton and Pluto may not be simply a matter of luck of timing and/or of the locations observed. Rather, the processes that generate plumes and fans on Triton may operate quite differently on Pluto or may not occur at all. Limited datasets for both bodies preclude us from determining which is the case.

It is interesting to note that the seasonal timing of the encounters at Triton and Pluto were similar, with Voyager 2 encountering Triton at a subsolar latitude of 45°S, moving south toward solstice, and New Horizons encountering Pluto at 52° N moving northward. Notwithstanding, while both Triton's Southern Hemisphere Terrain and Pluto's Sputnik Planitia have both been predicted to be ice sheets produced from seasonal volatile transport (e.g., Bertrand et al. 2016, 2018, 2023) they have very different latitudinal extents. While Triton's Southern Hemisphere Terrain may extend to its south pole, Pluto's Sputnik Planitia has a maximum latitude of approximately 50° N. In the context of the solar-driven hypothesis for Triton's plumes, active plumes would not necessarily be expected on Pluto during the New Horizons encounter since Sputnik Planitia does not extend to the subsolar latitude; on the other hand, streaks might then be expected to be abundant on Pluto's Sputnik Planitia, but they are not (Hofgartner et al. 2018).

3.7 Looking Ahead: Future Mission and Measurement Needs

The existence and study of Triton's plumes and fans provides the opportunity to constrain whether cryovolcanism is a historic process on Triton, or if it is ongoing on the icy moon (Hofgartner et al. 2022). If the plumes and fans have a significant H_2O content and/or if eruptions occur at temperatures that are close to the H_2O melting temperature rather than at temperatures that are close to the melting temperatures of nitrogen, methane, etc., then that would support the cryovolcanic hypothesis for Triton's plumes. Conversely, a seasonal migration of the plumes and fans, or a strong association with volatiles, such as nitrogen, would be strong evidence against the cryovolcanic hypothesis. Thus, obtaining active plume and fan composition and surface distribution (i.e., location) during a season that differs from that of the Voyager 2 encounter would represent significant, constraining measurements by which the origin of Triton's plumes and fans could be ascertained. This could be achieved with remote sensing instruments on future flyby or orbiter missions (Hofgartner et al. 2022).

Additional investigations that could assist in determining the formation mechanisms for Triton's plumes and fans include ascertaining the presence and distribution of anomalous surface temperatures, which could lend support to the solar driven or basal heating hypotheses for the plumes, and the composition and thickness of the Southern Hemisphere Terrain, which is expected to vary based on the plume model in operation. These investigations and the implications of their outcomes were thoroughly reviewed in Hofgartner et al. (2022). Notably, as the solar-driven and basal heating models for Triton's plumes both require an abundance of nitrogen ice, plume eruptions in areas of the surface where there is a dearth of nitrogen ice would support a cryovolcanic origin. In addition, as the solar-driven model for Triton's plumes predicts that plumes would migrate according to seasonal changes in solar insolation, sustained eruptions at the locations of the Mahilani and Hili plumes, and eruptions that are not confined to latitudes that are consistent with seasonal insolation, would also lend support to the explosive cryovolcanism model for Triton's plumes.

The 2030s and 2040s represent an ideal time to explore Triton. During this epoch the subsolar latitude is over the Southern Hemisphere Terrain; afterwards, the subsolar latitude will migrate away from the Southern Hemisphere Terrain for approximately a century. Hence the last occurrence of solar-driven plumes for 100 years would occur during this epoch, providing the opportunity to test whether Triton's plumes are solar-driven. Triton also experiences solar illumination that would be favorable for global observations during this time (Hofgartner et al. 2022). High resolution observations of Triton's putative cryovolcanic features (Figures 3.2 & 3.3), including obtaining repeat observations for change detection, and additional compositional information for areas of the surface where cryovolcanic features have been observed, would allow improved constraints to be placed on their ages and formation mechanisms.

There are several measurements that a future orbiter in the Pluto-system could make that would greatly increase our knowledge of the potential for currently active processes on Pluto including the potential for cryovolcanism (Howett et al. 2021). A Pluto-system orbiter could conduct many measurements that a flyby mission is not suited for. Although some constraints have been placed on the bulk composition and structure of Pluto's surface and interior, acquiring more high-resolution spectral coverage and gravity measurements or ice-penetrating radar would greatly increase our knowledge of the surface and interior. Obtaining additional information about potential cryovolcanic and tectonic structures in both the shallow and deep subsurface would allow hypotheses regarding the mobility of subsurface material and the processes by which it may extrude onto Pluto's surface to be tested. As New Horizons observed only approximately 40% of Pluto's surface at moderate-to-high resolution, an orbiter would ensure that more of Pluto could be observed and that repeat observations of the same regions of the surface could be made at different lighting geometries. Future orbital missions could therefore substantially increase our knowledge of the icy world.

3.8 Conclusions

Both Triton and Pluto display varied, numerous, and extensive terrains suggestive of cryovolcanism. While geological features with analogous morphologies are present on both worlds (see Chapter 4), whether there is a single confirmed cryovolcanic feature that is common to both Triton and Pluto remains unknown. This is surprising, given the numerous similarities between these worlds that are discussed throughout this book. Obtaining a complete understanding of the cryovolcanic processes at work in our solar system hinges on gaining an understanding of the ways in which cryovolcanism manifests on Triton and Pluto. This will undoubtedly require additional spacecraft exploration of both bodies and would also benefit from exploration of worlds such as Eris, Sedna, and Makemake. Given the critical importance of subsurface reservoirs and oceans in distributing melt and enabling cryomagmatism within the interiors of these bodies, establishing the threshold at which icy worlds are too small to maintain internal oceans is also essential. We therefore look to the outer planets and small bodies communities to assist us in placing these key constraints on Triton, Pluto, and similarly sized worlds. We also look to studies of other potentially cryovolcanically active bodies such as Ceres, Enceladus, and Europa to shed light on the ways in which varied heating mechanisms (i.e., radiogenic versus tidal), and the presence or absence of insulating layers, may affect the distribution and longevity of cryovolcanism at the surfaces of our solar system's ocean worlds. Indeed, given the comparable sizes and equilibrium ice shell thicknesses of Triton and Europa, studies regarding surface-subsurface exchange on the latter, particularly as it relates to the state of the ocean, the gradual cooling of discrete crustal reservoirs, and the dynamics and heat transfer associated with the movement of melt in the interior, can inform our understanding of the ways in which these processes may manifest to produce eruptive events at Triton's surface. Likewise, continued studies of the ways in which melt migration has manifested within Ceres can shed light on the degree and frequency with which surface-subsurface exchange, leading to cryovolcanism, may manifest on Pluto and other small, icy worlds with limited internal heating. We therefore look forward to the continued identification, exploration, and discovery of the ocean worlds in our solar system, as it will lead to a holistic understanding of the ways in which volcanism may manifest on icy worlds throughout the galaxy.

Acknowledgements

We would like to thank Will Grundy for a very helpful and constructive review of this chapter. Karl Mitchell's contribution to this work was conducted at the Jet Propulsion Laboratory, California Institute of Technology, under a contract with the National Aeronautics and Space Administration (80NM0018D0004)

References

Agnor, C., & Hamilton, D. 2006, Natur, 441, 192

Ahrens, C. J. 2020, JVGR, 406, 107070

Barlow, N. A. 2008, Mars: An Introduction to its Interior, Surface and Atmosphere (New York: Cambridge Univ. Press)

Bertrand, T., & Forget, F. 2016, Natur, 540, 86

Bertrand, T., Forget, F., Sicardy, B., & Lllouch, E. 2018, EPSCA, 12, 601

Bertrand, T., Forget, F., & Lellouch, E. 2023, EGU General Assembly 2023 (*Vienna, Austria, 24–28 Apr 2023*) EGU23-12587 (Munich: EGU)

Beddingfield, C. B., & Cartwright, R. J. 2021, Icar, 367, 114583

Beyer, R. A., Spencer, J. R., McKinnon, W. B., et al. 2019, Icar, 323, 16

Bierson, C. J., Nimmo, F., & McKinnon, W. B. 2018, Icar, 309, 207

Bierson, C. J., Nimmo, F., & Stern, S. A. 2020, NatGe, 13, 468

Binzel, R. P., Earle, A. M., Buie, M. W., et al. 2017, Icar, 287, 30

Bradak, B., & Kereszturi, A. 2003, LPSC, 34,

Brown, R. H., Kirk, R. L., Johnson, T. V., & Soderblom, L. A. 1990, Sci, 250, 431

Brown, R. H., & Kirk, R. L. 1994, JGR, 99, 1965

Buie, M. W., Grundy, W. M., Young, E. F., et al. 2010, AJ, 139, 1128

Buratti, B. J., Hicks, M. D., Dalba, P. A., et al. 2015, ApJL, 804, L6

Canup, R. M. 2011, AJ., 141, 35

Canup, R. M., Kratter, K., & Neveu, M. M. 2021, The Pluto System, ed. S. A. Stern, L. A. Young, J. M. Moore, W. M. Grundy, & R. P. Binzel (Tucson, AZ: Univ. Arizona Press) 475

Cartwright, R. J., Beddingfield, C. B., Nordheim, T. A., et al. 2020, ApJL, 898, L22

Castillo-Rogez, J. C., Daswani, M. M., Glein, C. R., Vance, S. D., & Cochrane, C. J. 2022, GeoRL, 49, 16

Cheng, W. H., Lee, M. H., & Peale, S. J. 2014, Icar, 233, 242

Chen, E. M. A., Nimmo, F., & Glatzmaier, G. A. 2014, Icar, 229, 11

Chivers, C. J., Buffo, J. J., & Schmidt, B. E. 2021, JGRE., 126, e2020JE006692

Choblet, G., Tobie, G., Buch, A., et al. 2022, ExA, 54, 809

Conrad, J. W., Nimmo, F., & Singer, K. N. 2016, Modelling Cryovolcanism Due to Subsurface Ocean Freezing on Pluto and Charon, AGU Fall Meeting Abstracts. P44A-05

Cook, J. C., et al. 2023, Icar, 389, 115242

Croft, S. 1990, LPSC, 21, 246

Croft, S., et al. 1995, Neptune and Triton, (Tuscon, AZ: Univ. Arizona Press) 879

Cruikshank, D. P., et al. 2019, Icar, 330, 155

Cruikshank, D. P., Grundy, W. M., Protopapa, S., Schmitt, B., & Linscott, I. R. 2021, The Pluto System After New Horizons ed. S. A. Stern, et al. (Tucson, AZ: Univ. Arizona Press) 165

Dalle Ore, C. M., et al. 2019, SciA, 5, eaav5731

Dougherty, M. K., et al. 2006, Sci, 311, 576

Duxbury, N. S., & Brown, R. H. 1997, Icar, 125, 83

Earle, A. M., et al. 2017, Icar, 287, 37

Fagents, S. A., Greeley, R., Sullivan, R. J., Pappalardo, R. T., Prockter, L. M., & The Galileo SSI Team, 2000, Icar, 144, 54

Fagents, S. A. 2003, JGR, 108, 5139

Fagents, S. A., Lopes, R. M., Quick, L. C., & Gregg, T. K. 2022, Planetary Volcanism Across the Solar System (Amsterdam: Elsevier) 161

Figueredo, P. H., Chuang, F. C., Rathbun, J., Kirk, R. L., & Greeley, R. 2002, JGR, 107, 5026

Fortes, A., & Grindrod, P. 2006, Icar, 182, 550

Fortes, A., Grindrod, P., Trickett, S., & Vocadlo, L. 2007, Icar, 188, 139

Gaeman, J., Hier-Majumder, S., & Roberts, J. H. 2012, Icar, 220, 339

Gladstone, G. R., et al. 2016, Sci, 351, aad8866

Greenstreet, S., Gladman, B., & McKinnon, W. B. 2015, Icar, 258, 267

Hammond, N. P., Parmenteir, E. M., & Barr, A. C. 2018, JGRE, 123, 3105

Hansen, C. J., McEwen, A. S., Ingersoll, A. P., & Terrile, R. J. 1990, Sci, 250, 421

Hansen, C. J., Castillo-Rogez, J., Grundy, W., et al. 2021, PSJ, 2, 137

Head, J. W., & Wilson, L. 1992, JGRE, 97, 3877

Hendrix, A. R., Hurford, T. A., Barge, L. M., et al. 2019, AsBio, 19, 1

Hofgartner, J. D., Birch, S. P. D., Castillo, J., et al. 2022, Icar, 375, 114835

Hofgartner, J. D., Buratti, B. J., Devins, S. L., et al. The New Horizons Science Team 2018, Icar, 302, 273 2018

Hogenboom, D. L., Kargel, J. S., Consolmagno, G. J., et al. 1997, Icar, 128, 17

Howard, A. D., et al. 2023, Icar, 405, 115719

Howett, C. J. A., et al. 2021, PSJ, 2, 75

Ingersoll, A. P. 1990, Natur, 344, 315

Ingersoll, A. P., & Tryka, K. A. 1990, Sci, 250, 435

Jankowski, D. G., & Squyres, S. W. 1988, Sci, 241, 1322

Johnson, M. L., & Nicol, M. 1987, JGR, 92, 6339

Kamata, S., et al. 2019, NatGe, 12, 407

Kargel, J. S., & Strom, R. G. 1990, 21st Lunar and Planetary Science Conf. 599

Kargel, J. S., Croft., S. K., Lunine, J. I., & Lewis, J. S. 1991, Icar, 89, 93

Kargel, J. S. 1992, Icar, 100, 556

Kargel, J. S. 1998, Solar System Ices. Astrophysics and Space Science Library (Dordrecht: Springer) 3

Kimura, J., & Kamata, S. 2020, P&SS, 181, 104828

Kirk, R. L., Brown, R. H., & Soderblom, L. A. 1990, Sci, 250, 424

Kirk, R. L., Soderblom, L. A., Brown, R. H., Kieffer, S. W., & Kargel, J. S. 1995, Neptune and Triton, ed. D. P. Cruikshank (Tucson, AZ: Univ. Arizona Press) 949

Kite, E. S., & Rubin, A. M. 2016, PNAS, 113, 3972

Lesage, E., Schmidt, F., Andrieu, F., & Massol, H. 2021, Icar, 361, 114373

Lopes, R. M. C., Mitchell, K. L., Williams, D., & Mitri, G. 2010, What Is a Volcano? eds. E. Cañón-Tapia, & A. Szakács (Boulder, CO: Geological Society of America) 470

Lopes, R. M. C., Kirk, R. L., Mitchell, K. L., et al. 2013, JGRE, 118, 416

Manga, M., & Wang, C.-Y. 2007, GeoRL, 34, L07202

Manga, M., & Michaut, C. 2017, Icar, 286, 261

Martin, C. R., & Binzel, R. P. 2021, Icar, 356, 113763

McGovern, P. J., & White, O. L. 2019, 50th Lunar and Planetary Science Conference (Houston, TX: LPI)

McKinnon, W. B., Lunine, J., & Banfield, D. 1995, Neptune and Triton, ed. D. P. Cruikshank (Tucson, AZ: Univ. Arizona Press) 807

McKinnon, W. B., et al. 1997, Pluto and Charon, ed. S. A. Stern (Tuscon, AZ: Univ. Arizona Press) 295

McKinnon, W. B., & Kirk, R. L. 2007, Encyclopedia of the Solar System (2nd ed.; Amsterdam: Elsevier) 483

McKinnon, W. B., et al. 2017, Icar, 287, 2

Miyamoto, H., Mitri, G., Showman, A. P., & Dohm, J. M. 2005, Icar, 177, 413

Moore, J. M., McKinnon, W. B., Spencer, J. R., et al. 2016, Sci, 351, 1284

Neveu, M., et al. 2015, Icar, 246, 48

Nimmo, F., & Spencer, J. R. 2015, Icar, 246, 2

Nimmo, F., & McKinnon, W. B. 2021, ed. S. A. Stern, L. A. Young, J. M. Moore, W. M. Grundy, & R. P. Binzel The Pluto System After New Horizons (Tucson, AZ: Univ. Arizona Press) 89

Paganini, L., Villanueva, G. L., Roth, L., et al. 2019, NatAs, 4, 266

Pappalardo, R. T., & Barr, A. C. 2004, GeoRL, 31, L0 1701

Porco, C. C., Helfenstein, P., Thomas, P. C., et al. 2006, Sci, 311, 1393

Prockter, L., & Schenk, P. 2005, Icar, 177, 305

Prockter, L. M., Nimmo, F., & Pappalardo, R. T. 2005, GeoRL, 32, L14202

Prockter, L. M., Shirley, J. H., Dalton, J. B., & Kamp, L. 2017, Icar, 285, 27

Protopapa, S., et al. 2017, Icar, 287, 218

Protopapa, S., Cook, J. C., Grundy, W. M., et al. 2021, The Pluto System After New Horizons eds. S. A. Stern, et al. (Tucson, AZ: Univ. Arizona Press) 433

Quick, L. C., Barnouin, O. S., Prockter, L. M., & Patterson, G. W. 2013, P&SS, 86, 1.

Quick, L. C., & Marsh, B. D. 2016, JVGR, 319, 66

Quick, L. C., Glaze, L. S., & Baloga, S. M. 2017, Icar, 284, 477

Quick, L. C., Buczkowski, D. L., Scully, J. E. C., et al. 2019, Icar, 320, 119

Quick, L. C., & Hedman, M. M. 2020, Icar, 343, 113667

Quick, L. C., Fagents, S. A., Núñez, K. A., et al. 2022, Icar 387, 115185

Rivalta, E., Taisne, B., Bunger, A. P., & Katz, R. F. 2015, Tectp, 638, 1

Robuchon, G., & Nimmo, F. 2011, Icar, 216, 426

Ross, M. N., & Schubert, G. 1990, GeoRL, 17, 1749

Roth, L., Saur, J., Retherford, K. D., et al. 2014, Sci, 343, 171

Ruesch, O., Platz, T., Schenk, P., et al. 2016, Sci, 353, 6303

Ruesch, O., Genova, A., Neumann, W., et al. 2019, NatGe, 12, 505

Sagan, C., & Chyba, C. 1990, Natur, 346, 546

Schenk, P. M. 1991, JJGR, 96, 1887

Schenk, P., & Jackson, M. 1993, Geo, 21, 299

Schenk, P. M., Beyer, R. A., McKinnon, W. B., et al. 2018, Icar, 314, 400

Schenk, P. M. 2021, Remote Sens., 13, 3476

Schmidt, B. E., Blakenship, D. D., Patterson, G. W., & Schenk, P. M. 2011, Natur, 479, 502

Schmitt, B., et al. 2017, Icar, 287, 229

Singer, K. N., et al. 2022, NatCo, 13, 1542

Skinner, J., & Tanaka, K. 2007, Icar, 186, 41

Smith, B. A., Soderblom, L., Banfield, D., et al. 1989, Sci, 246, 1422

Soderblom, L. A., Kieffer, S. W., Becker, T. L., et al. 1990, Sci, 250, 410

Solomonidou, A., Coustenis, A., Hirtzig, M., et al. 2013, European Planetary Science Congress 2013 (Paris: ESA) EPSC2013–854–1

Sparks, W. B., Hand, K. P., McGrath, M. A., et al. 2016, ApJ, 829, 121

Sparks, W. B., Schmidt, B. E., McGrath, M. A., et al. 2017, ApJL, 839, L18

Spencer, J. R., Pearl, J. C., et al. 2006, Sci, 311, 1401

Spencer, J. R., Barr, A. C., Esposito, L. W., et al. 2009, in Enceladus: An Active Cryovolcanic Satellite. Saturn from Cassini-Huygens ed. M. K. Dougherty, L. W. Esposito, & S. M. Krimigis (New York: Springer) 681

Stern, S. A., et al. 2015, Sci, 350, aad1815

Stern, S. A., et al. 2017, Icar, 287, 47

Sulcanese, D., Cioria, C., Kokin, O., et al. 2023, Icar, 392, 115368

White, O. L., Moore, J. M., Howard, A. D., et al. 2021, The Pluto System After New Horizons eds. S. A. Stern, et al. (Tucson, AZ: Univ. Arizona Press) 55

Yelle, R. V., Lunine, J. I., Pollack, J. B., & Brown, R. H. 1995, Neptune and Triton ed. D. P. Cruikshank (Tucson, AZ: Univ. Arizona Press) 1031

Young, L. A., Bertrand, T., Trafton, L. M., et al. 2021, The Pluto System After New Horizons eds. S. A. Stern, et al. (Tucson, AZ: Univ. Arizona Press) 321

Chapter 4

Divergent: Triton and Pluto Morphology and Geology

Paul M Schenk

4.1 Introduction

Triton and Pluto form a natural pair in size and density but not in internal evolution and geologic history as expressed in morphology. Both are large enough and rich enough in low-melting point volatile ices to retain very thin atmospheres that have affected surface morphology in different ways. These same ices have been involved in internal heat transfer expressed as tectonics, volcanism, and overturn but in different ways due to very different thermal histories. This chapter will explore these differences, and similarities, and the constraints that limit our understanding of these morphologies.

Neptune's large satellite Triton (Figure 4.1) was revealed by Voyager 2 in August 1989 (Smith et al. 1989; Croft et al. 1995) to be a diverse, youthful, and active body. As a likely ocean world, it is now a major focus of planetary exploration (National Academy of Sciences 2022) and has been designated as the top priority target for future exploration of icy ocean worlds (e.g., Hendrix et al. 2019). Triton's ubiquitously very young surface (perhaps as young at $\sim$10 myr; Schenk & Zahnle 2007), outer ice-rich shell, geologic complexity (Croft et al. 1995; Schenk et al. 2021), and active atmospheric plumes (Soderblom et al. 1990) related to either solar or geothermal heating (Hofgartner 2022) all suggest that Triton, most likely a captured Kuiper belt dwarf planet, has been subject to high levels of internal heat, may be currently active, and likely has an internal ocean today (e.g., Gaeman et al. 2012; Hussmann et al. 2006). The known surface ices on Triton include water, methane, N_2, CO and CO_2 ices (Grundy et al. 2010). Unlike Cassini or New Horizons, Voyager did not carry a mapping infrared spectrometer and did not map out the distribution or geologic correlations of these materials. These factors also hint that Triton may have implications for the nature, diversity, and likelihood of habitable environments across the Galaxy.

doi:10.1088/2514-3433/ad5278ch4

(a)

(b)

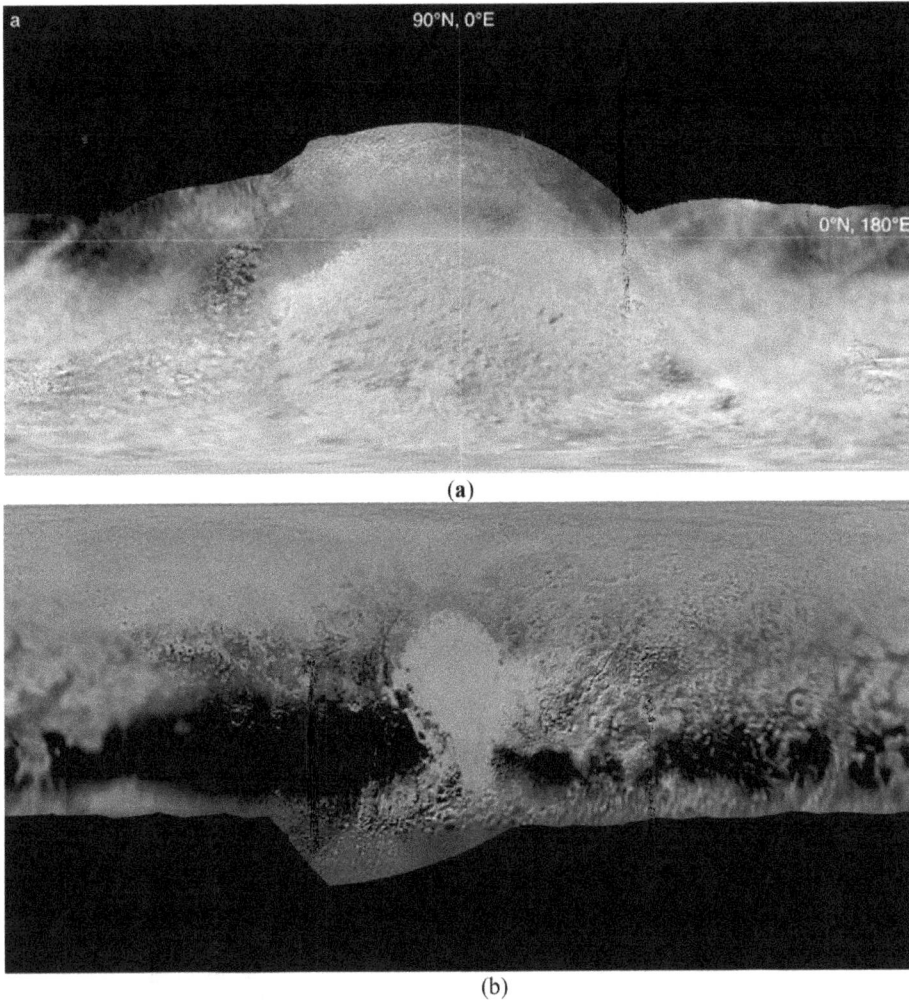

Figure 4.1. (a) Global color map mosaic of Triton. Color-composite uses orange, green, and UV filters for red, green, and blue colors. Simple cylindrical projection at 0.35 km pixel scales from −180° to 180° E longitude (same map projection used in all similar maps unless noted). Reproduced from Schenk et al. (2021). CC BY 4.0. (b) same for Pluto using near-IR, red, and blue channels.

Pluto was revealed by New Horizons in July 2015 (Moore et al. 2016), revealing an equally complex body (Figure 4.2) with surface ages of >4000 myr to negligible (Singer et al. 2019) indicating multiple preserved geologic episodes. These include older cratered plains and much younger glaciated terrains, a large but highly degraded ridge-trough system, (cryo)volcanic terrains that are very different in character than those on Triton, and of course the extremely young and likely still convecting nitrogen (+CO, CH$_4$) ice sheet, all superficially very different from what occurs on Triton or unique to Pluto. With a broadly similar volatile-rich surface composition

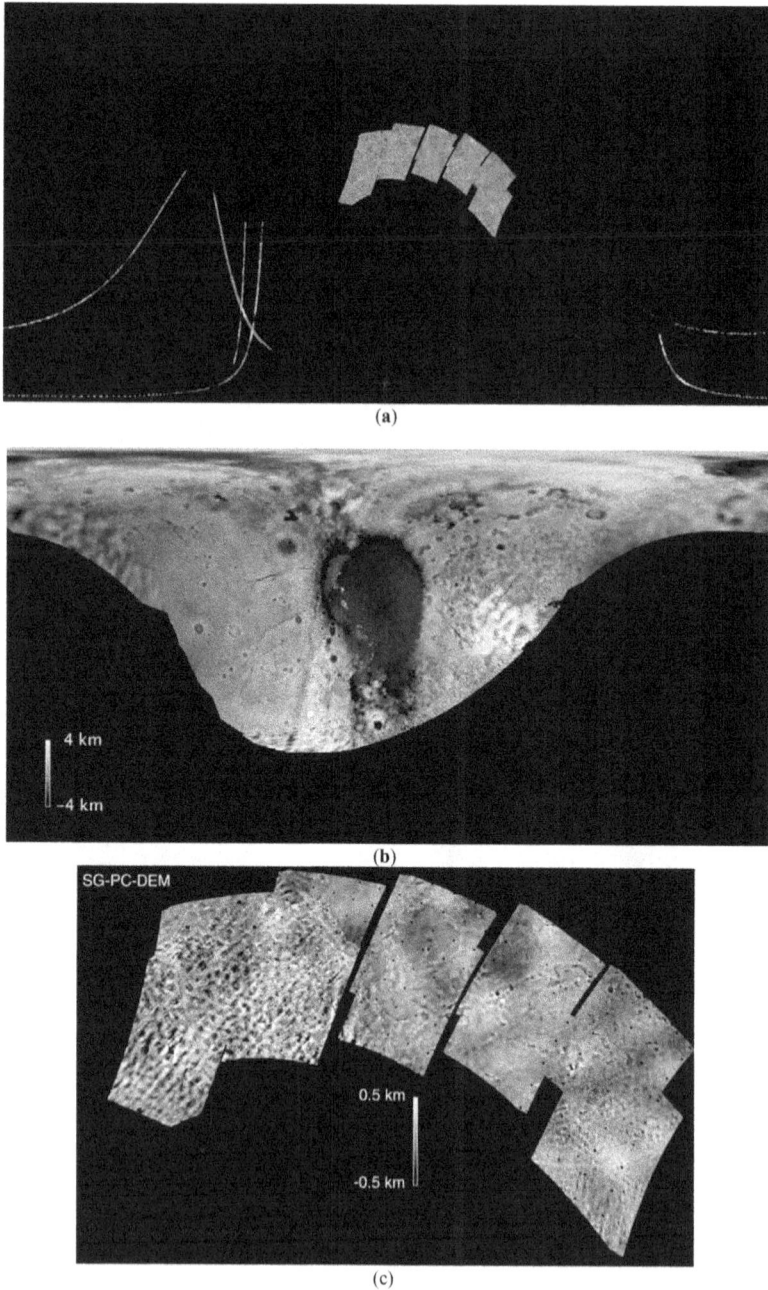

Figure 4.2. (a) Global map of stereogrammetric DEMs for Triton and curvilinear limb profile tracks. Reproduced from Schenk et al. (2021). CC BY 4.0. Only those stereo DEMs where geologic features are nominally resolved are shown. Map limits from −180° to 180° E longitude. (b) same for Pluto except limb profiles are not included. Adapted from Schenk et al (2018a), Copyright (2018), with permission from Elsevier. (c) enlargement of integrated highest resolution DEM of Triton highlighting cantaloupe terrains (left) and volcanic units (center). Reproduced from Schenk et al. (2021). CC BY 4.0.

(Grundy et al. 2016), surface–atmosphere interactions result in a strong latitudinal color and surface modification pattern (Schenk et al. 2018a; Protopapa et al. 2017) that may be related but is different than on Triton. One fundamental factor driving many of these differences in the observed records is the existence of a very large impact feature: the >1000 km Sputnik Planitia basin (Schenk et al. 2018a) the depressed topography of which drives much of Pluto's volatile redistribution (Bertrand et al. 2020). In contrast, no direct or indirect evidence for impact features larger than ~25 km or for terrains older than circa 10–100 myr is evident on Triton (Croft et al. 1995; Schenk et al. 2021).

The observational record and constraints in terms of global color mosaics and topography have most recently and comprehensively been reviewed for Triton by Schenk et al. (2021), and for Pluto by Schenk et al. (2018a). We do not repeat these details but briefly emphasize key aspects relative to our geologic mapping needs. The only mapping data we have for both are from rapid flybys from Voyager and New Horizons resulting in best imaging on one hemisphere (except where in shadow) and decreasing pixel scale around to the opposite hemisphere. Because New Horizons was closer and carried a camera with better IFOV and low-light sensitivity, basic mapping pixel scales are roughly a factor of 2 better on Pluto. Best hemispheric mapping pixel scales are ~1600–600 m for Triton and ~600–325 m for Pluto, with very limited high-resolution imaging of ~325 m for Triton and ~120–70 m for Pluto. Mapping of Pluto benefitted from additional multicolor and multispectral mapping cameras, while the Voyager camera used multiple filters but no multi-spectral capabilities. The primary differences are that the color mapping range is ~0.6–0.325 for Triton but 0.870–0.475 for Pluto, with only two points of overlap at orange-red and blue, limiting the ability to compare color units on the two bodies. These factors severely limit our ability to correlate ice composition with geology on Triton compared to Pluto.

The greater distance of Voyager from and inherently lower topography of Triton both conspire to severely limit geologically useful stereo topography of Triton to a small area <10% of the surface (Schenk et al. 2021) compared to circa 35%–40% of Pluto encompassing most of the encounter hemisphere (Schenk et al. 2018a). Known relief in the resolved mapping area of Triton is limited to +/−1 km (to 95%) compared to −3 to +6 km for Pluto. Triton stereo topography is supplemented by shape-from-shading mapping along Voyager's equator, highlighting geologic features but providing little constraint on long-distance topographic variations. At least some portions of the known major terrain types on Triton were mapped in stereo, except the enigmatic southern terrains although photogrammetry suggests low relief there as well (Schenk et al. 2021).

4.2 Triton–Pluto Comparisons

One of the major surprises of the New Horizons encounter with Pluto was that there appeared to be little obvious commonality with Triton morphology (e.g., Schenk & Zahnle 2007; Schenk et al. 2021). The similarities in size, density, and surface composition between Triton and Pluto (Grundy et al. 2010, 2016) and the possibility both may have been or currently are ocean worlds (Gaeman et al. 2012; Hussmann et al. 2006) allowed for the possibility that some geologic processes might have been shared by the two Kuiper Belt region dwarf planets (McKinnon, 1984), despite their likely very different thermal histories (Triton being captured, e.g., Moore et al. 2015) and Pluto forming in a giant collision (e.g., Canup 2011). Although no compositional mapping comparable to that at Pluto (Grundy et al. 2016) was acquired by Voyager, the current and broadly similar global mapping (Schenk et al. 2021, 2018a) allows us to reexamine potential Triton and Pluto comparisons.

4.2.1 (Cryo)Volcanism

Both Pluto (Moore et al. 2016; Schenk et al. 2018a; Singer et al. 2022) and Triton (Smith et al. 1989; Croft et al. 1995; Schenk et al. 2021) exhibit terrains and features that are considered to form by icy volcanic resurfacing, but with very different morphologies. The extensive plains of Cipango Planum (Figure 4.3) on Triton east of cantaloupe terrain (described below) are easily the most recognizable volcanic landforms beyond Mars in that they bear a strong resemblance in morphology and spatial relationships to basaltic complexes on Earth (Kuntz et al. 2021). The plains feature the large circular depression Leviathan Patera (Figure 4.4), as well as numerous widely distributed circular and oval pits up to ~25 km wide. These pits are both aligned in chains radiating from Leviathan Patera and as single structures in various locations in Cipango Planum (Figure 4.3). Cipango Planum covers an area

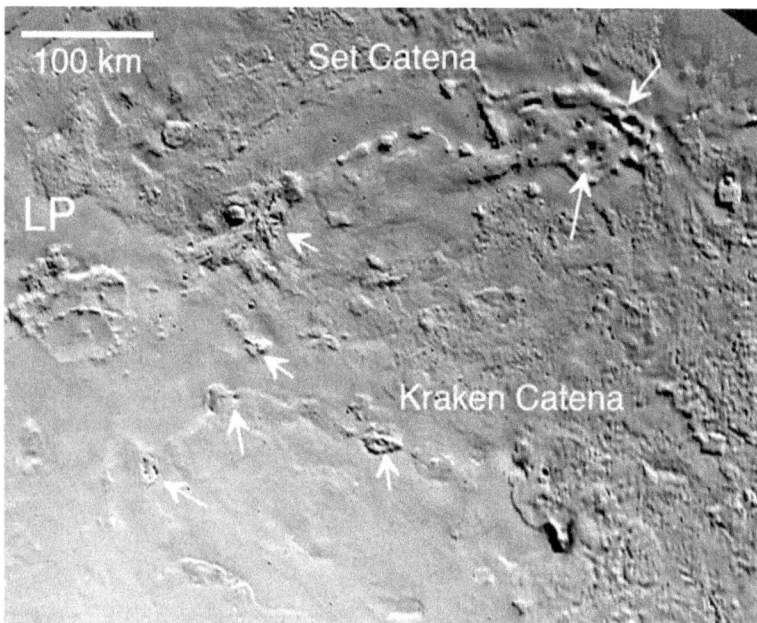

Figure 4.3. Volcanic plains of Cipango Planum showing spatial association of Leviathan Patera (LP) and Set and Kraken Catena pit chains. Arrows highlight knobs that may be partially buried. Scene width ~540 km. Reproduced from Schenk et al. (2021). CC BY 4.0.

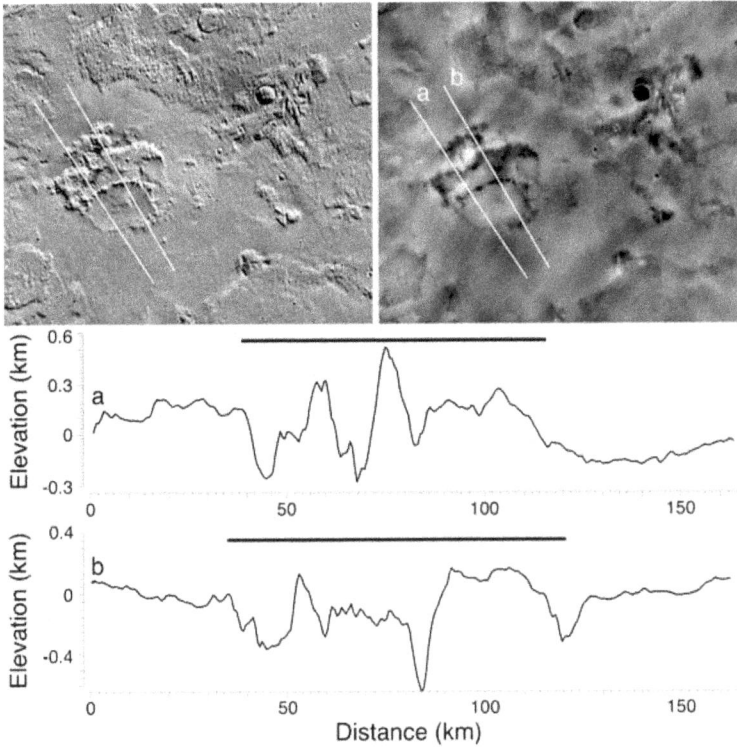

Figure 4.4. Topographic profiles across Leviathan Patera, highlighting scarp-bounded margin, broad domes in parts of the floor and high promontory on northwest patera floor. (Top) Image mosaic and (right) DEM scene widths are ~300 km, from SG-PC-DEM in Schenk et al. (2021). CC BY 4.0.

of at least 700 × 700 km, although the northern boundary was in darkness during Voyager. The eastern boundary is gradational, transitioning into terrains described as 'etched plains' (see below). Numerous authors have concluded that all these features are volcanic in origin and are probably part of one major event (Croft et al. 1995; Schenk et al. 2021), but it is stressed here again that the mechanism of resurfacing in unconstrained (see Chapter 3 for discussions on (cryo)volcanic processes on both bodies). While the pits and patera are more likely explosive or collapse related, the smooth plains involve volcanic effusion, explosive ash deposition, or both, as suggested by dark diffuse haloes around some of the larger rimmed pits (e.g., Croft et al. 1995).

The ~80-km-wide Leviathan Patera appears to form the structural center of Cipango Planum (Figure 4.3). The northern half of the floor of this circular structure is mostly a low plain ~450 m deep and includes several rim reentrants and an irregular mesa or plateau several hundred meters high. A promontory that forms part of the northwestern section of Leviathan Patera is ~1 km high (Figure 4.4) and may be the highest known individual geologic feature in the topographic mapping area. The southeastern floor of Leviathan Patera is occupied by a broad rise 35 × 53 km across and ∼400 m high (approximately level with surrounding plains), separated from the rest of the floor by a chain of linear depressions ∼100–300 m deep.

The curvilinear alignment of pits radiating from the central Leviathan Patera complex (Figure 3), with its low floor and interior domes, recalls such volcanic terrains as the Kilauea caldera and its East Rift and the Craters of the Moon rift in Idaho (Kuntz et al. 2021), albeit in its own distinctive Tritonian style. One assumes from the analogy to basaltic complexes that these pits combine explosive or airfall eruption styles with collapse of vent enlargement processes. The morphologic analogy to basaltic volcanic rift features on Earth would seem to imply similarly relatively low viscosities on Triton. With compositional mapping lacking for Triton, these materials could be ammonia- or methanol-rich (e.g., Hammond et al. 2018), or other combinations of mobilized ice phases as fluids or fluid-solid mixtures, each of which would likely have different viscosities. As resolutions are no better than ∼340 m, we cannot yet identify flow fronts, burial, or other structures that would indicate mode of emplacement.

The broadly circular shape, low relief, and internal domes of Leviathan Patera are reminiscent of resurgent silicic calderas on Earth (Hildreth et al. 1984) and could indicate that volcanism in Cipango Planum occurred in several phases: low viscosity or high effusion rate resurfacing of the plains (or pyroclastic deposition), later stage explosive eruptions or pit collapse along the pit chains and rimmed pits, and later stage extrusions of relatively higher viscosity icy materials, or lower extrusion rates, as low eruption rates can also lead to thicker volcanic flow morphologies (Macdonald 1953; Fink & Griffiths 1998).

In addition to Leviathan Patera, there are several oval closed depressions ≤100 km across (Figure 4.5) located predominantly along the Voyager terminator. Paterae take a variety of morphologic styles from relatively well defined and sharp-edged to more irregularly shaped walled depressions (Figure 4.5) but are normally distinguished by their non-circular shapes. Paterae edges are defined by scarps or ridges <200 to 500 m in height, which can vary around the circumference of the feature. The floors are either level or are domed as high as ∼100 m. These patera structures lack the crenulated or scalloped margins of the walled planitia (discussed below) and are distinctly non-impact related. This and their distinctive oval or kidney shapes suggest structural control by an endogenic caldera-forming process such as magmatism or collapse, perhaps analogous to uneroded silicic caldera on Earth.

The Plutonian massifs Wright and Piccard Montes and hummocky textures of Hyecho Planum (on the southwestern flank of Sputnik Planitia) in which they sit have been interpreted as volcanic features (Moore et al. 2016; Schenk et al. 2018a; Singer et al. 2022). Despite the similarities in bulk and surficial compositions of the two bodies, these volcanic features and units could hardly be more different in style and morphology from the plains, pit chains and caldera-like structures in Cipango Planum on Triton (Figure 4.3). Hummocky textures with hundreds of broadly similar-sized rounded mounds on rolling plains together with arcuate scarps and the large circular 5 km high massifs of Wright and Piccard Mons dominate Hyecho Planum (Figure 4.6). Like the features of Cipango Planum, the terrains of Hyecho

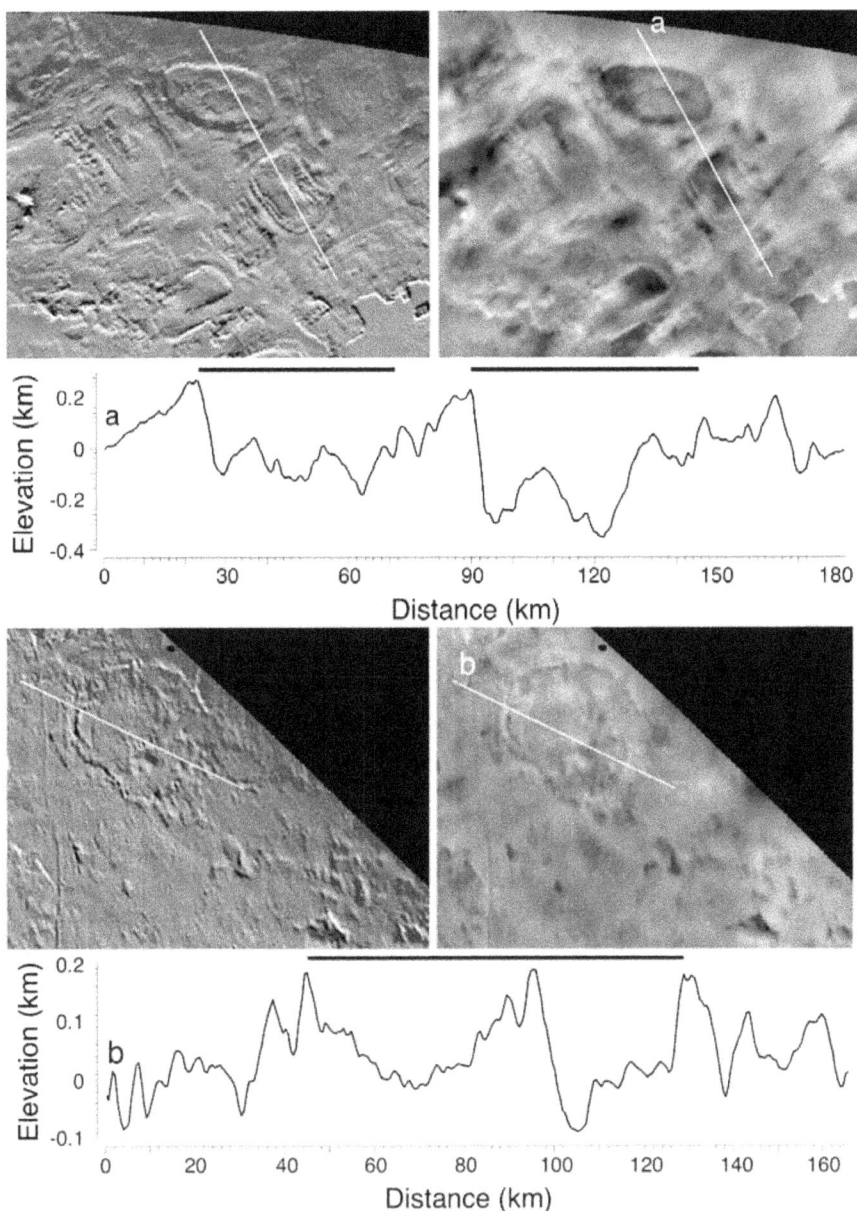

Figure 4.5. Topographic profiles of three paterae on Triton. (Top) Image mosaic and (right) DEM scene width (a) is ~220 km wide, northwest of Tuonela Planitia and centered at 40° N, 5° E. (top) Image mosaic and (right) DEM scene width (b) is ~275 km across and is centered northeast of Cipango Planum at 14° N, 60° E, from DEM in Schenk et al. (2021). CC BY 4.0.

Planum are very young (Singer et al. 2019; Singer et al. 2022) but the high relief and distinctive hummocky textures of Hyecho Planum (Figure 4.6) are very suggestive of a different style of eruption than on Triton or on the Saturnian or Uranian moons

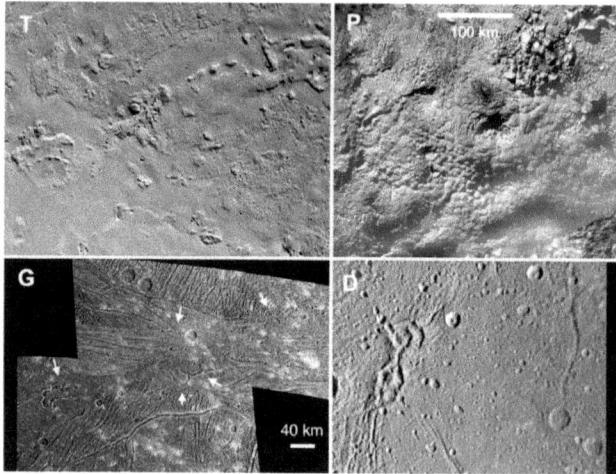

Figure 4.6. Clockwise from upper left: Volcanic terrains on (T) Triton, (P) Pluto, (D) Dione, and (G) Ganymede. Images shown to same scale (0.350 km/pixel); scene width ∼400 km. Triton view shows pits and plains of Cipango Planum including caldera-like Leviathan Patera (far center left); Pluto view shows Wright Mons (center) and hummocky plains of Hyecho Planitia; Dione view shows oval caldera-like Metiscus and Murranus Paterae at center left located in the center of smooth plains; Ganymede view shows several crenulate open caldera-like features (arrows) associated with a band pf smooth bright terrain (note different scale in this mosaic).

(e.g., Schenk et al. 2018b). The broad low relief of the structures within Cipango Planum at Triton and resemblance to basaltic eruption complexes on Earth suggest extrusion or eruption of lower viscosity materials and perhaps high eruption rates (except in mixed ice phases such as ammonium hydrate, rather than actual basalt). The locally high relief at Pluto is perhaps more analogous morphologically to dacitic or other higher viscosity materials on Earth, but in a style peculiar to Pluto and its ices; the origin of the densely spaced mounds as localized extrusive deposits is still debated (Singer et al. 2022; Howard et al. 2023) but they are not suggestive of low viscosity extrusion.

It therefore seems likely that both Triton and Pluto experienced extensive volcanic activity driven by internal heat, but the contrasting morphologies strongly indicate that emplacement processes and compositions of materials involved were very different. The lack of similar morphologies on other icy bodies, even on similar or larger-sized bodies like Ganymede, Titan, or Europa, suggests that the more complex ice chemistries present on these two outer bodies are partly responsible for their more complex morphologies. Indeed, if any heat sources are present in other Kuiper Belt bodies, then we might expect equally complex volcanic landforms there. Irregularly shaped depressions typically 20–60 km wide are found in association with smoother lanes of bright terrains on Ganymede (Howard et al. 2023) and even a pair of such features in the smooth plains of Dione (Schenk & Moore 1995; Figure 4.6) may be indirectly related to Leviathan Patera, though neither type are associated with pits or pit chains. Collectively they suggest that caldera-like collapse or explosive processes are a likely explanation for such irregular rimmed depressions on all icy bodies.

Although these comparisons are handicapped by our poor understanding of volcanic petrology of mixed and eutectic ice phases on icy worlds (Chapter 3), the formation of low and high relief volcanic materials on these icy bodies suggests that broadly similar ranges of low- to high-viscosity materials can be generated and erupt out there as well. Although CO, CO_2, and CH_4 have been identified spectroscopically on Triton and Pluto, direct compositional information on the emplaced volcanic materials (as opposed to surficial or seasonal frosts) is lacking here and on all icy bodies. Further, the viscosities of emplaced materials on these bodies can only be guessed at as none have been observed in the act and are dependent on factors such as supercooling and crystallinity. A pure ammonium hydrate fluid may have viscosities as low or lower than terrestrial basalts, or as high as andesite if those other factors are significant (e.g., Kargel & Croft 1989). Other or more complex phase mixtures are even more poorly understood.

An alternative measure of flow properties is yield strength, which is a measure of resistance to flow that incorporates both the flow and its frozen outer carapace (as discussed for icy bodies by Schenk 1990). This measure is perhaps easier to estimate in that it is not dependent on unknowable flow rates or durations on these bodies. Thicker flows tend to have higher yield strengths indicating a more resistant flow composition, though here again different eruption rates, temperatures, and crystallinities can result in thicker flows.

4.2.2 Walled Planitia

Walled plains (planitia) occur on both worlds. The two large walled plains Tuonela and Ruach Planitia (Figure 4.7) on Triton are distinct features characterized by a sinuous inward-facing scarp surrounding a low plain (Schenk et al. 2021). These scarps cut into both cantaloupe terrain and the western portions of Cipango Planum (described above). The sinuous scarps suggest erosional retreat into those terrains. The DEM indicates variable scarp heights relative to surrounding terrains of ∼50 m up to ∼300 m, with an average of circa 150 m (Figure 4.7).

Textures across the planitia include marginally resolved small-scale knobs, depressions, and occasional mesa-like outliers (Figure 4.7) localized near the center or northern margins of the Planitia. These could also be evidence of erosional scarp retreat due to sublimation (Moore & Spencer 1990), or they could be related to emplacement of endogenic materials involving pitting by collapse or venting. The largest pit in the center of Ruach Planitia is almost 10 km wide and ∼350 m deep and is surrounded by smaller pits, linear troughs, and scarps <100 m deep. The pits in the northern corner of Tuonela Planitia are more complex, with a low dome centered within an oval depression. The highly concentrated distribution of these pits within one specific region on each Planitia, suggests either eruptive vent centers or sublimation of the floor materials initiated near their centers, rather than impact. We have no direct observational constraints on the composition of Triton's planitia or any volatile that might be causing scarp retreat if that is the origin of the scarps.

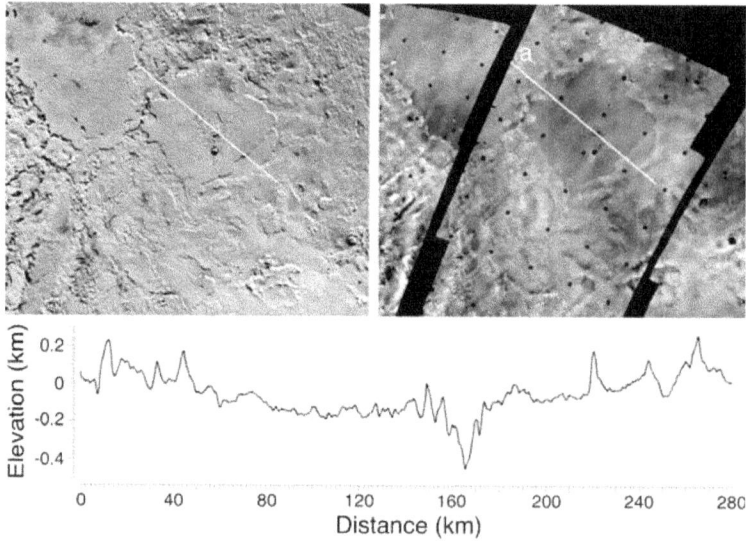

Figure 4.7. Topographic profile across the margin and central pits of the walled Ruach Planitia. Subtle undulations on the floor of the walled plains are within the stereoscopic uncertainty limits of the DEM and might not be real. (top) Image mosaic and (right) DEM scene width is ~480 km. Arrows indicate crenulated margins. Reproduced from Schenk et al. (2021). CC BY 4.0.

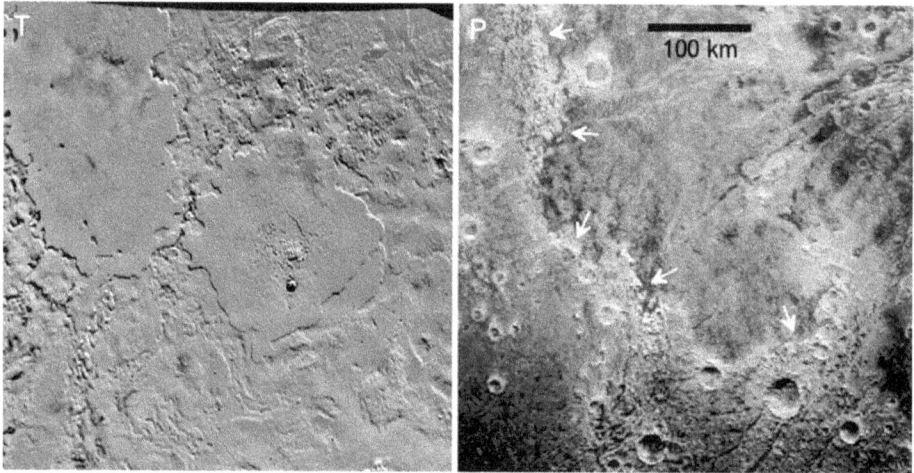

Figure 4.8. Walled plains on (T) Triton and (P) Pluto. Images shown to same scale (0.35 km/pixel) and same dimensions; scene width ~400 km. Reproduced from Schenk et al. (2021). CC BY 4.0.

On Pluto, the closest analog in scale and outline to the two walled plains of Triton appears to be Piri Planitia (Figure 4.8). Piri Planitia is partially enclosed by a scalloped and/or crenulate scarp, which is several hundred meters high and characterized by alcoves and stranded mesa-like outliers suggestive of sublimational erosive retreat of a resistant scarp unit (Moore et al. 2017). The scarps enclosing

Ruach and Tuonela Planitia on Triton are more curvilinear and generally lacks erosional-style alcoves, although some stranded, barely resolved mesa-like outliers in Tuonela Planitia have been noted. Whether these differences are related to compositional differences and sublimational behavior or to Triton's youth is unknown, given the lack of compositional constraints and limited resolutions on the putative eruptive materials.

4.2.3 Nitrogen Ice Sheets

The presence of nitrogen ice on the surfaces of both Triton and Pluto leads to inevitable comparisons but the surface distribution of the phase is very different. On Pluto most surface nitrogen is in the bright ice sheet confined within the 1000-by-1300 km 3 km deep Sputnik Planitia impact basin (Moore et al. 2016; Schenk et al. 2018a). The thickness of the ice sheet is inferred to be approximately 10 km thick from convective modelling (McKinnon et al. 2016) and impact basin shape models. Additional surficial nitrogen ice covers Pluto in latitudinal bands along the $\sim$35° N zone and within the Tombaugh Regio highland along the southeast margin of Sputnik Planitia, leading to active glaciation and return flow into the ice sheet (see next section). The ice sheet likely formed due to the topographic depression and preferential condensation of this ice at the lower elevation and higher atmospheric pressure at those depths.

On Triton, we do not have resolved maps of nitrogen distribution but infer from the absence of large circular or oval bright units in the global map of Triton (Figure 4.1) that similar >500-km-wide deep N_2 ice sheets do not exist there (lack of north polar coverage notwithstanding). This is not surprising given the very young crater retention ages on Triton and thus lack of preserved large ancient basins. It is also consistent with the extremely low topography on Triton which would discourage the accumulation of large N_2 deposits there given the large seasonal variations. In contrast to Pluto, the nitrogen ice on Triton (e.g., Grundy et al. 2010) must be lateral and probably discretely dispersed.

The entire southern hemisphere or southern terrains (ST; Figure 4.9) has been commonly referred to as a 'polar cap' of nitrogen of significant thickness but this framework is neither likely (Howard et al. 2023) nor supported by the observations (Schenk et al. 2021). First, there is no evidence that these terrains are elevated above (or below) the global mean (granting that topographic data are very limited) but rather have limited relief of <1 km, though surface slopes may be locally significant.

Because most of the ST was observed at lower resolutions of >1 km pixel scales and in non-shadow and marginal-stereo conditions, morphologies and relief are not well expressed or easily interpretable. Secondly, the STs are heterogenous at all scales and are not monolithic but consist of at least four distinct terrains (Figure 4.9; Schenk et al. 2021): an extension of cantaloupe terrain (*ct*) west of $\sim$340° E and north of $\sim$20° S; macular terrains (*mt*) of amoeboid-shaped spots (referred to as "maculae" or "guttae") separated by brighter materials of variable albedo, generally north of $\sim$40° S; lineated terrains (*lt*) south of the *mt* consisting of bright and dark curvilinear and irregular surface markings; and bright lobate terrains (*blt*) with

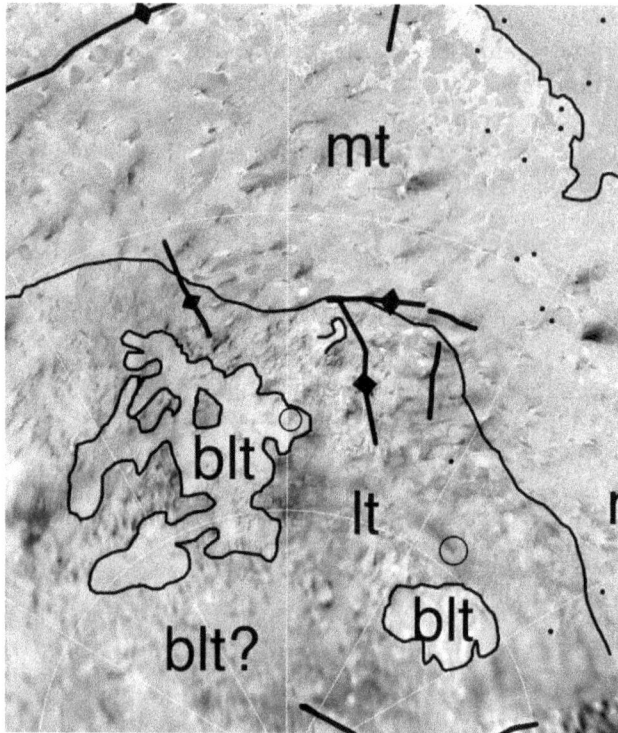

Figure 4.9. Polar stereographic map of part of Triton southern hemisphere; (*mt*) macular terrain, (*lt*) lineated terrain, and (*blt*) bright lobate terrains that appear to embay lineated terrains in two large units. Note diffuse dark fan-shaped deposits in (*mt*). Open circles indicate known plume sources, heavy lines are linear features, heavy dashed lines are approximate locations of limb haze, and black dots are (not-to-scale) impact craters. Remapped from Kargel & Croft (1989) and Hofgartner (2022). Some terrain boundaries are indistinct in the images and all boundaries are approximate, intended as a guide for the reader. Central longitude 0° E is at the top. Enlargement of central section of figure from Schenk et al. (2021), showing southernmost regions. Reproduced from Schenk et al. (2021). CC BY 4.0.

occasional faint surface markings, embedded within the lineated terrains. Due to the lower resolutions and the lack of low-Sun shading variations, the three terrains particular to southern latitudes (*mt, lt, blt*) have no obvious planetary analogs elsewhere in the solar system, including Pluto. Those terrains closest to the south pole itself and south of $\sim$60° S are especially enigmatic due to lower resolution, lack of shading and oblique viewing, and are not classified.

The macular terrains (Figures 4.9 and 4.10) consist of amoeboid-shaped and mostly featureless darker units within a heterogenous but generally brighter unit. Most of the diffuse dark fans (e.g., Hofgartner 2022) occur in this terrain. The dark fans (Figure 4.9) are deposited on both lower albedo maculae and the intervening bright materials but are usually darker on the maculae than the bright materials. The fans are likely airborne deposits (Hofgartner 2022) given their diffuse margins and indiscriminate crossing of maculae margins, which also suggests that the maculae are not high enough to form topographic barriers to deposition.

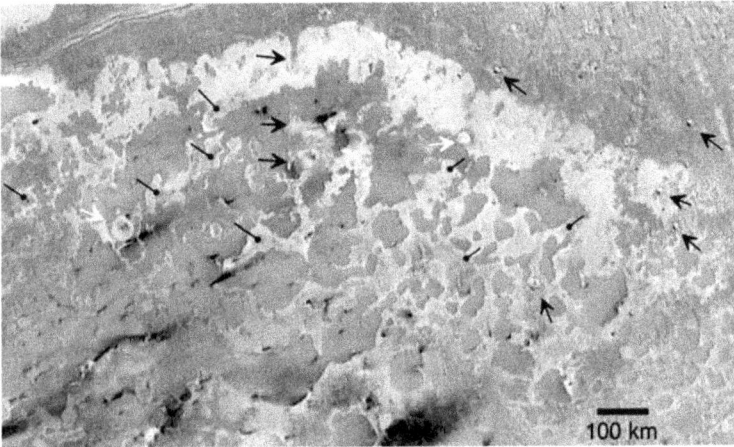

Figure 4.10. Enlargement of global cylindrical map showing macular southern terrains near the equator of Triton. The terrain consists of dark maculae units of various sizes up to ~100 km across. Between the maculae are bright materials, which are often concentrated around the edges of the patches and form diffuse contacts with a darker substrate exposed between the maculae (smaller black lines). A few preexisting features are exposed within this dark substrate (a narrow fracture (dark arrows) and two apparently degraded craters at the left and near center (white arrows)) that may have been exhumed when the patches formed by scarp retreat. North is up.

Most of the impact craters within the macular terrains (Figure 4.10) are fresh looking but several are more degraded, embayed, or infilled with the inter-macular bright materials. A few linear structures can also be traced well into the macular terrains. These craters and linea form in the inter-macular bright materials, suggesting either re-exposure of an older surface due to retreat of the maculae or incomplete burial by later-formed maculae. The brighter inter-macular materials are not always uniformly bright; in some areas, these units are brightest near the edges of the maculae, suggesting scarp retreat, control of deposition by local winds at possibly elevated maculae margins, or variable elevation on the underlying bright material substrate (Figure 4.10). A sequence in which a top-lying dark unit is being eroded by scarp retreat, exposing an underlying dark unit, with fringes of residual volatile brighter ice has been proposed (Schenk et al. 2021).

The lineated terrains (*lt*; Figure 4.11) are even more enigmatic than macular terrains, but we infer from the weak stereo images (Figure 4.11) and the correlation of dark albedo with elevated septa in the cantaloupe terrain (Schenk et al. 2021) that at least some of the irregular dark patches and linear dark markings could be the crests of higher features in this terrain if topographic control of deposition holds true here as well. Topographic control of volatile ice distribution on Pluto is well documented (Schenk et al. 2018a; Protopapa et al. 2017; Bertrand et al. 2020) and is likely on Triton as well. On Triton, the much narrower topographic range could result in more complex albedo patterns but the poorer topographic coverage precludes detailed mapping of this association.

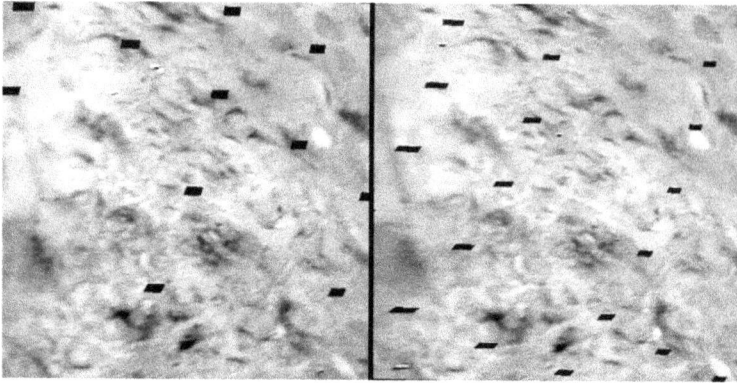

Figure 4.11. Best Voyager stereo pair of contact between lineated ST (left) and macular ST (upper right), with Hili plume and bright lobate materials at far left center. The more densely marked lineated terrains at center may be more rugged than the macular terrains at upper right with some of the darker deposits appearing to be higher standing, and narrow lobate brighter materials at lower left suggest a downslope flow. Stereo pairs such as these are difficult to interpret, however, due to the resolution, high solar elevation and low parallax. Scene width is ∼400 km, centered at 44° S, 0° E; north is to the right. Black squares are reseaux marks in images where data are nulled. Reproduced from Schenk et al. (2021). CC BY 4.0.

The complex albedo patterns of the lineated terrains (Figure 4.11) also suggest a complex interplay of surface ices of different composition. The narrow dark linear features may be elevated ridges covered by darker frosts, as on areas of cantaloup terrains (Figures 4.9, 4.11) and extensively on Pluto (Schenk et al. 2018a; Bertrand et al. 2020). Some of the brighter features within the lineated terrains have lobate finger-like patterns suggestive of downslope flow of bright materials, some of which form within larger dark patches that may be elevated (based on the possible correlation of some dark material with elevation (Figure 4.11)). In other cases, the dark linear markings appear to be shading of scarps due to illumination, all of which makes interpretation of these terrains difficult in available imaging.

A variety of candidate mobile or volatile ice deposits have been identified in Triton's ST (Schenk et al. 2021). Inter-macular bright materials fringing the dark maculae are variable in brightness (Figure 4.10) and may be residual deposits from scarp retreat. The localized crater filling deposits on Triton (Figure 4.10) resemble on a smaller scale the altitude-controlled deposition process for volatile ices that has been described on Pluto (Bertrand et al. 2020) and suggests it may also work on Triton despite the generally much lower topographic relief. Narrow sinuous bright features a few 10s of kilometers in length in the lineated southern terrains may be downslope ice flows (Figure 4.11). The brightest units within the STs appear to be confined within several large bright lobate deposits 250–500 km across the shapes of which appear to be controlled by local topography (Figure 4.9). Whether any of these deposits include nitrogen, methane or another volatile ice is currently unknown, but a possible correlation with detached haze layers may indicate volatile-atmospheric exchange (Schenk et al. 2021).

Some nitrogen and other volatile ices are present within the ST (Grundy et al. 2010), but without resolved IR-spectroscopy, it is not possible to map explicitly where they are distributed with respect to geologic units and albedo features. The bright lobate terrains (*blt*; Figures 4.9 and 4.10) may be the best candidates for large, extended, volatile ice deposits within the ST of thicknesses great enough to induce local flow. These bright lobate materials are characterized by relatively uniform albedos and lobate margins with the lineated terrains, suggesting embayment of locally low topography though no topographic data is available due to the poorly resolved surface textures. The faint surface markings in the *blt* consist of several clusters of darker spots and lineations with unresolved relief.

The best resolved example of bright lobate terrain Is ≥400 km across (Figure 4.9) and centered at 51° S, 0° E. Another large bright lobate unit centered at 65° S, 48° E, ~250 km across and featuring several curvilinear bright lines, but it is resolved only in the 4 km whole disk images. Possible additional examples of *blt* are centered near 73° S, 343° E and 60° S, 320° E in the 4 km pixel^{-1} approach imaging. If a correlation of plume location and *blt* margins (Schenk et al. 2021) is valid, it could be consistent with a hypothesis of basal melting beneath a volatile ice unit (Schenk et al. 2021; Hofgartner 2022) but any conclusions are frustrated by the very oblique low-resolution viewing of this area, which prevents examination and mapping of the plume site vent and local geology. Observations of other plumes with future mission observations, including topographic constraints, should allow a test of this and other hypotheses (Hofgartner 2022).

The non-volatile "bedrock" of the STs could be of comparable age as terrains to the north, as they have comparable superposed crater spatial densities associated with the planetocentric population that may have dominated Triton (Schenk & Zahnle 2007). Preservation of this cratering alone suggests that at least the macular terrains are longer-lived non-seasonal deposits (Moore & Spencer 1990). These factors as well as the complexity and sharp delineation of features within the ST lead to the conclusion that the ST is a permanent geologic terrain, as proposed by (Moore & Spencer 1990), although interspersed with local and extended seasonal or slowly modifying volatile ice deposits. Hence, we do not refer to a hemispheric scale "south polar cap" but to "southern terrains" pending more robust data and assessments of these latitudes.

4.2.4 Surface Modification by Lateral Flow: "Glaciation"

The presence of broad sinuous channels on both Pluto and Triton (Schenk et al. 2021; Moore & Spencer 1990) and on comparable scales (Figures 12 and 13) suggests another analog process. North of Ruach Planitia lies a complex knobby terrain cut by several sinuous and intersecting channels <10 km wide and ~200 m deep (Figure 4.12), one of which merges with northern Ruach Planitia. These are the only such examples recognized on Triton and suggest that northern areas in darkness in 1989 are geologically complex. These are broadly similar in morphology to large sinuous troughs that appear to flow southward into the basin and partially cut into the northwestern rim of Sputnik Planitia basin (Figure 4.13) and suggest

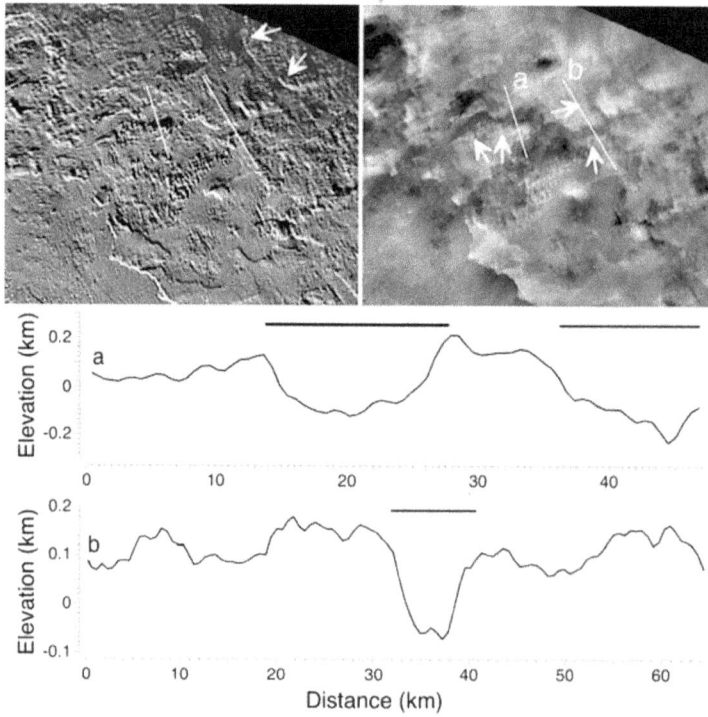

Figure 4.12. Topographic profiles across sinuous channel-like depressions north of Ruach Planitia. Note also shallow walled plains south of indicated channels. (top) Image mosaic and (right) DEM scene widths are ~150 km from SG-PC-DEM in Figure 4.8. Note different horizontal scales in profiles. Horizontal bars indicate extent of relevant features. Reproduced from Schenk et al. (2021). CC BY 4.0.

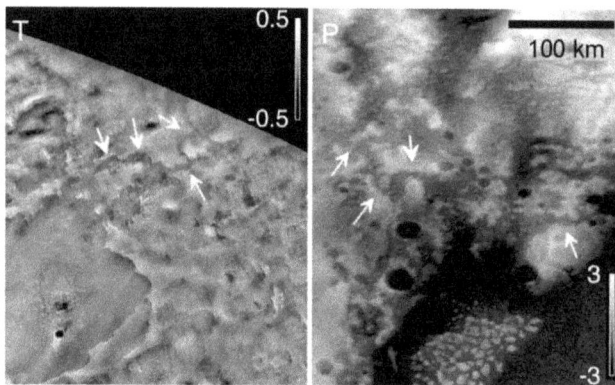

Figure 4.13. DEMs of (T) Triton and (P) Pluto showing sinuous channel-like features (arrows). Images shown to same scale (0.350 km pixel[-1]) and scenes are ~300 km wide. Topography used in these views to highlight channel-like morphologies. Reproduced from Schenk et al. (2021). CC BY 4.0.

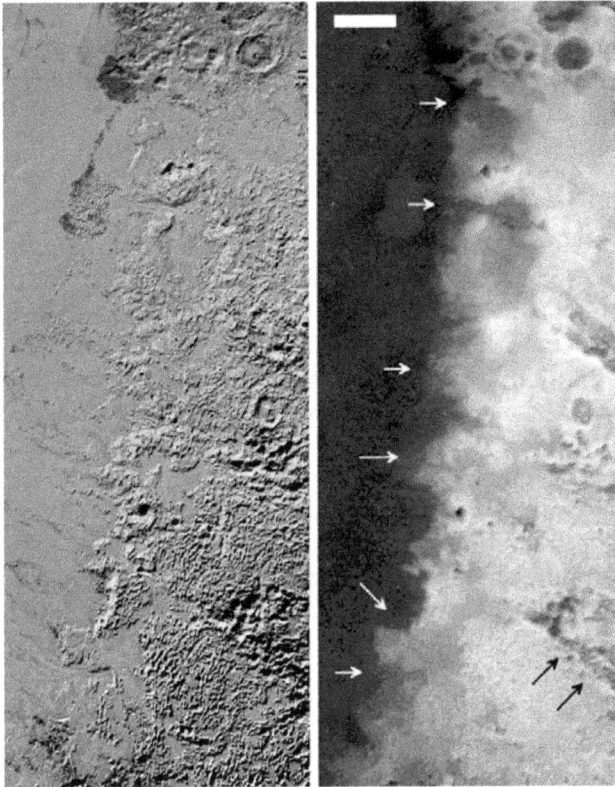

Figure 4.14. Glacial flow deposits and features on the southeastern boundary of Sputnik Planitia with Tombaugh Regio, Pluto. Left is image mosaic, right is DEM showing range from −3 to +3 km. White arrows highlight areas where glaciers discharge into Sputnik Planitia, and dark arrows highlight some deep pits not partially filled with smooth glacial material.

significant trough cutting on both bodies (although the true extent is unclear on both worlds).

A more explicitly glacial landform occurs on Pluto in the form of sinuous partially filled channels in western Tombaugh Regio (Howard et al. 2017), in which glacier-like valley fill actively discharges into and appears to be recharging parts of the Sputnik Planitia ice sheet (Figure 4.14). Likely composed of nitrogen eroding a water-ice rich bedrock, these are the only known active extraterrestrial glaciers and suggest that the unfilled valleys at northern Sputnik Planitia were once N_2-glacier filled. This terrain is also noteworthy for numerous closed quasi-circular depressions a few to ∼10 km across, some of which appear to be filled with a low-lying bright material that may be topographically trapped glacial deposits, while other depressions are unfilled (Howard et al. 2017).

There is no definitive evidence for glacial-like flow elsewhere on Triton other than the example in Figure 4.12. Sinuous features are abundant but, in each case, they lack topographic or morphologic recharge or discharge associations. Further, the relief is so low on Triton (most surfaces having <500 m amplitude) that glacial flow really has no real downslope 'throw' or catchment with which to erode and develop landforms. It is more likely that such enigmatic features are related to other forces, such as incipient folding, faulting, or diapirism, or even scarp retreat such as in cantaloupe and other terrains.

4.2.5 Convective Overturn in the Interior

Both Triton and Pluto exhibit features indicative of convective overturn in the solid state in their outer layers and at surprisingly similar scales, but in very different materials. The surface expression of convective upwelling in which solid ices rise upward through an outer layer as a lens or spheroid is very uncommon generally on icy satellites. The most obvious expression of this elsewhere is on Europa where oval lenticulae and chaos units are interpreted as the direct or indirect result of icy material from the lower crust rising and penetrating the upper crust (Singer et al. 2022). The Europa features are not well organized and tend to be widely scattered at different spacing intervals. On both Triton and Pluto this process is much more organized and coherent in expression.

Cantaloupe terrains on Triton (Figure 4.15) consists of topographically closed, oval to kidney-shaped cellular depressions (or cavi/cavus) 20–40 km wide over a contiguous area of at least 1400 by 2000 km (Croft et al. 1995; Schenk et al. 2021), though the full extent westward and northward is unknown. The cavi are deeper than the intervening ridges, or septa, except to the north where embaying smooth plains, interpreted as (cryo)volcanic, overlap with and partially fill the cells. The unmodified cavi we can measure reliably are 200–800 m deep relative to the septa, with mean depths of ≈400 m (Schenk et al. 2021). Septa could be contractional folds associated with diapiric overturn, but they could also be erosional remnants of the original surface layers. These nearly crater-free terrains have been attributed to a compositionally driven (or possibly thermal-driven) diapir-like crustal overturn

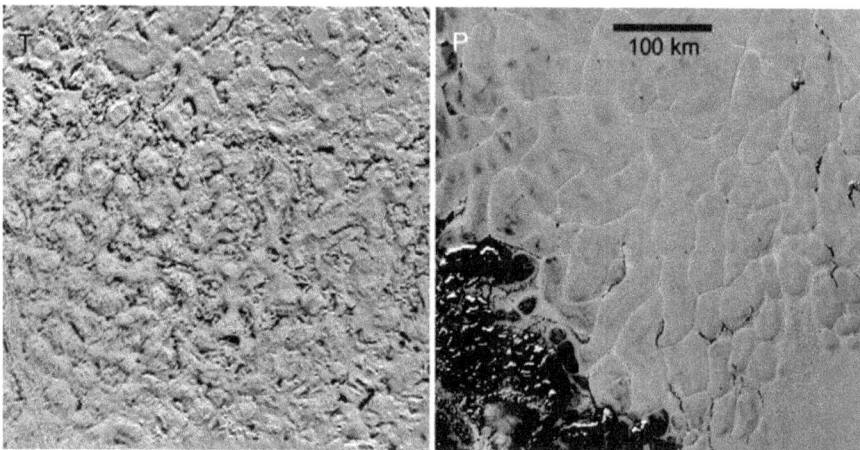

Figure 4.15. Cellular plains on (T) Triton and (P) Pluto. Images shown at same effective scale (0.35 km/pixel); scene width is ~450 km.

process (Schenk & Jackson 1993). The lack of resolved compositional constraints, however, hinders modeling and the driver behind this overturn remains poorly constrained, however. The maintenance of several hundred meters of relief across cantaloupe cells (Schenk et al. 2021) indicates that more refractory ices such as CH_4, CO_2, ammonia, or water ices are involved. The opposite is true of Pluto where relief across cell boundaries is lower.

The close similarity in planform shape and dimensions of the cavi depressions in Triton's cantaloupe terrain to the cell pattern in the N_2-CH_4 ice sheet of Pluto's Sputnik Planitia (Figure 4.15) is curious. Both are characterized by adjoining and tightly packed oval-to-kidney-shaped cells roughly 20–50 km across (with some larger examples up to 75 km long on Pluto), patterns that are associated with convective solid-state overturn in extended layers with large width-to-thickness ratios (McKinnon et al. 2016; Trowbridge et al. 2016). The key difference is that the cells in Sputnik Planitia are within the spectroscopically identified nitrogen ice sheet and are probably limited to only that layer. Here however the cell widths are more variable by a factor of 2, possibly reflecting variations in ice sheet thickness and hence convective dimensions, or perhaps variations in heat flow from beneath. On Triton, convective overturn likely involves a substantial fraction of the upper crustal and ice layers of differing composition or rheology (Schenk & Jackson 1993). On Triton, the involvement of higher melting point ices requires high heat flows to trigger overturn whereas the nitrogen ice sheet would require much lower temperatures to trigger. It is also interesting to note that cantaloupe terrain was not found elsewhere on Pluto, indicating that either Pluto's icy outer shell was too cold or thick to convect, that compositional layering requirements were not met, or such terrains were not in the New Horizons mapping areas. The complex floor morphologies of cantaloupe terrain cavi (Howard et al. 2017) also suggest that a combination of extrusion and/or scarp retreat may be involved which, except for small 10s–100s of meters-scale pitting, is not observed in the Plutonian case.

4.2.6 Erosive Processes: Etched Plains

Terrains east of Cipango Planum on Triton are complex in morphology, characterized by low plains marked by irregular knobs and sinuous scarps. In numerous locations, the sinuous scarps in these terrains form closed or nearly closed sigmoidal depressions (Figures 4.16 and 4.17) that are typically 5–15 km wide and 200–300 m deep. Broadly similar enclosed sigmoidal depressions are evident on Pluto, particularly northeast of Sputnik Planitia in Viking Planitia (Schenk et al. 2018a; Howard et al. 2017).

These closed and semi-closed depressions are usually considered as erosional in origin, formed by sublimation and scarp retreat (Howard et al. 2017). Erosive processes occur at a variety of scales on both worlds although resolution limits restrict our awareness of these to larger scales. The partially enclosed sigmoidal depressions of the eroded or "etched" terrains east of Cipango Planum on Triton (Figure 4.17) are significantly shallower and smaller in scale than those on Pluto, however. The depressions on Triton are <500 m deep and typically 5–10 km wide,

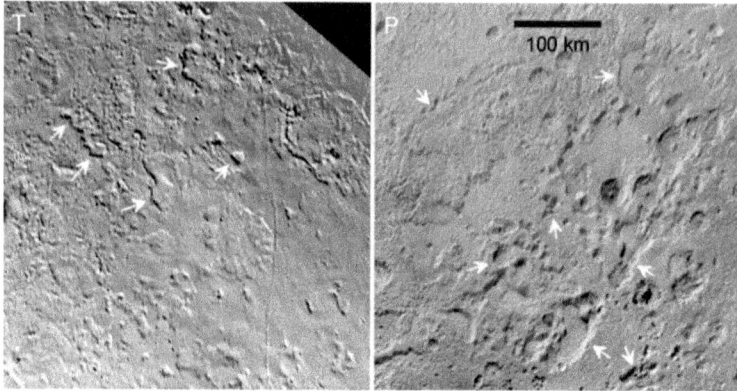

Figure 4.16. Etched plains on (T) Triton and (P) Pluto. Images shown to same scale (0.35 km pixel⁻¹) and same dimensions; scene width is 375 km. Reproduced from Schenk et al. (2021). CC BY 4.0.

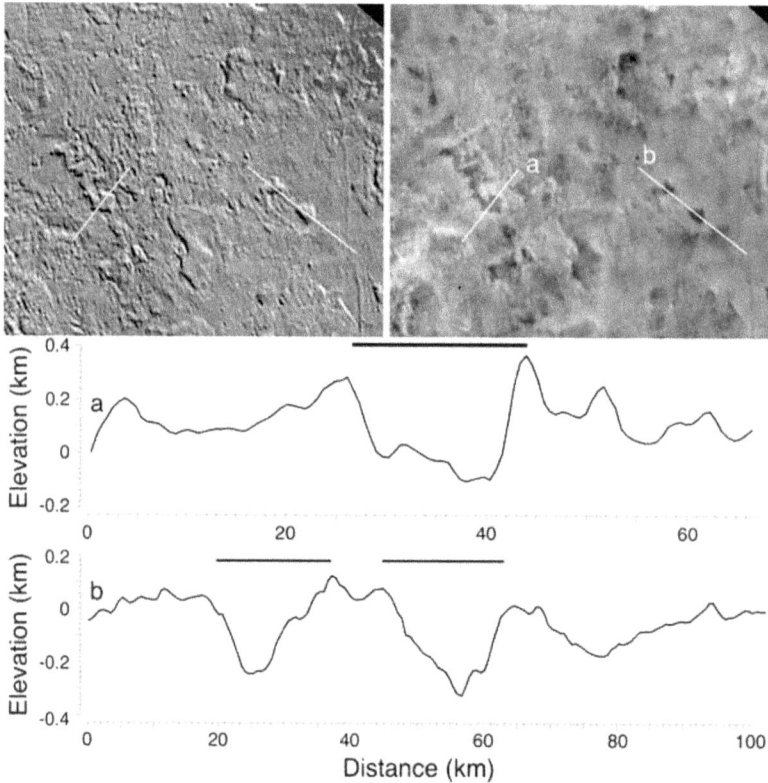

Figure 4.17. Topographic profiles across closed oval and sigmoidal troughs in the etched plains of Triton. (Top) Image mosaic and (right) DEM scene width is ~300 km and is centered at 15° N, 51° E. Reproduced from Schenk et al. (2021). CC BY 4.0.

whereas those on Pluto are 3–5 km deep and roughly 5–70 km in size (Figure 4.16). The shallower depths and smaller scales of the enclosed depressions on Triton could be due to slower erosion rates on Triton, the much younger ages of these terrains on Triton (resulting in less time for the process to proceed), or to differences in the thicknesses of the layers being sublimated.

4.3 Discussion and Conclusions

A detailed survey of the comparable though globally incomplete mapping of Triton and Pluto (e.g., Schenk et al. 2021, 2018a) demonstrates that these two very similar bodies with similar compositions have very different thermal and geologic histories. Triton topography rarely exceeds 1000 m in elevation, at least in the three major terrain units sampled by Voyager north of the equator. This is much lower than the $+/-5$ km amplitudes commonly found on Pluto. Triton may have among the lowest topographic amplitude of any icy world measured to-date (Schenk et al. 2021), although our sampling is rather sparse. Despite this some morphological similarities have been identified that likely relate to the similar bulk compositions which are dominated by water and more volatile ice phases (e.g., Grundy et al. 2010, 2016).

While all observed terrains on Triton are very young, consistent with ongoing high heat flow, Pluto displays a much broader range of surface ages and hence a much larger amplitude in topography. The large 3 km deep Sputnik Planitia impact basin is also a large sink for N_2 ice. Although Triton differs from Pluto in many respects, several features may be similar or at least analogous. Closed sigmoidal depressions associated with scarp retreat and sublimation occur on both bodies but are significantly shallower and smaller on Triton (Figures 4.16 and 4.17), suggesting differences in layer depths or process duration. Even N_2 glacial processes may be common on both, although the evidence for such on Triton is very limited and the very low topographic range also suggests that the process would be more limited there. The solid-state crustal convection inferred for Triton's crustal cantaloupe terrain finds no analog on Pluto, except in the much thinner nitrogen ice sheet of Sputnik Planitia. The horizontal scale of this overturn is very similar, despite the inferred compositional differences, probably a result of the lower viscosity and thinner layers involved in the Pluto case (e.g., McKinnon et al. 2016). The lack of such convection elsewhere on Pluto could imply that its ice shell has always been too thick and/or cold to convect. Volcanic features occur on both bodies but with fundamental differences (Figures 4.3–4.6). The much lower topographic relief and different morphological style (Figure 4.16) of Cipango Planum and Leviathan Patera on Triton, in contrast to the mounded textures and large and elevated volcanic edifices such as Wright or Piccard Montes on Pluto, is an indication that the viscosities or yield strengths associated with flows or deposits emplaced at Cipango Planum are lower (Kargel & Croft 1989; Schenk 1990; and/or effusion rates higher) than inferred for the Wright and Piccard massifs and Hyecho Planum mounds, thus implying a compositional difference for extruded materials (e.g., Schenk et al. 2021; Singer et al. 2022; Howard et al. 2023). Of course, a Tritonian Wright Mons could have formed in the more poorly or non-imaged areas of Triton, but it seems likely that the internal thermal and compositional requirements for these structures, whatever they may be, did not occur within Triton.

The geological and topographic differences between Pluto and Triton are equally compelling, such as the apparent absence of methane-rich Plutonian bladed terrains, polar domes, large ancient ridge-trough systems (e.g., Moore et al. 2016), and giant (or even just large) impact scars on Triton. The latter two examples are related to the much-longer recorded history of Pluto. The absence of Plutonian bladed terrain plateaus (which are 3-5 km high; Schenk et al. 2018a) may also be related to Triton's extreme youth, precluding time to either accumulate or erode them but also perhaps to different surface–atmosphere processes or different abundances of methane. Even when features are more similar (such as enclosed depressions), the smaller and shallower magnitude of such features on Triton further emphasizes that the geologic and thermal histories of the two bodies are fundamentally different but still share common materials.

The general finding of very low topographic amplitudes on Triton compared to Pluto (Schenk et al. 2021, 2018a) is consistent with the evolution of a larger icy world with a prolonged history of elevated heat flow, creating thermal conditions in which the icy shell cannot support and/or quickly erases large topographic loads. Future topographic mapping of Triton will require use of all available techniques, including match point radii, limb tracks (at higher resolution), photoclinometry and stereo-grammetry (and a laser altimeter if practical) to more fully elucidate the topographic properties and internal state of Triton and its ice shell and (probable) ocean.

Acknowledgments

Lunar and Planetary Institute contribution. The Lunar and Planetary Institute is operated by USRA under a cooperative agreement with the Science Mission Directorate of the National Aeronautics and Space Administration.

References

Bertrand, T., Foget, F., White, O., & Schmidt, B. 2020, JGR, 125, e2019jE006120

Canup, R. M. 2011, AJ, 141, 35

Croft, S. K., et al. 1995, Neptune and Triton (Tucson, AZ: Univ. Arizona Press) 879

Fink, J., & Griffiths, R. 1998, JGR., 103, 527

Gaeman, J., Hier-Majumder, S., & Roberts, J. H. 2012, Icar, 220, 339

Grundy, W. M., Binzel, R. P., Buratti, B. J., et al. 2016, Sci, 351, aad9189

Grundy, W. M., Young, L. A., Stansberry, J. A., et al. 2010, Icar, 205, 594

Hammond, N. P., Parmenteir, E. M., & Barr, A. C. 2018, JGRE, 123, 3105.

Hendrix, A. R., Hurford, T. A., Barge, L. M., et al. 2019, AsBio, 19, 1

Hildreth, W., Fierstein, J., & Calvert, A. 2017, JVGR, 335, 1

    Marsh, B. D. 1984, JGR, 89, 8245

Hofgartner, J. 2022, Icar, 375, 114835

Howard, A., et al. 2017, Icar, 287, 287

Howard, A., Moore, J., Umurhan, O., et al. 2023, Icar, 405, 115719

Hussmann, H., Sohl, F., & Spohn, T. 2006, Icar, 195, 258

Kargel, J., & Croft, S. 1989, 20th Lunar and Planetary Science Conf. (Houston, TX: LPI) 47

Kuntz, M. A., Skipp, B., Champion, D. E., et al. 2007, Geologic map of the State of Hawai'i: U.S. Geological Survey Scientific Investigations Map 3143, U.S. Geological Survey Scientific Investigations

Macdonald, G. A. 1953, AmJS, 251, 169

McKinnon, W. B. 1984, Natur, 311, 355

McKinnon, W. B., Nimmo, F., Wong, T., et al. 2016, Natur, 534, 82

Moore, J., et al. 2017, Icar, 287, 320

Moore, J., Schenk, P., et al. 2016, Sci, 351, 1284

Moore, J. M., Howard, A. D., Schenk, P. M., et al. 2015, Icar, 246, 65

Moore, J. M., & Spencer, J. R. 1990, GeoRL., 17, 1757

National Academy of Sciences 2022, Planetary Science and Astrobiology Decadal Survey 2023–2032, (Washington, DC: United States National Research Council)

Protopapa, S., Grundy, W., Reuter, D., et al. 2017, Icar, 287, 218

Schenk, P. 1990, JGR., 96, 1887

Schenk, P., et al. 2021, RemS, 13, 3476

Schenk, P., & Jackson, M. P. A. 1993, Geo, 21, 299

Schenk, P., & Moore, J. 1995, JGR., 100, 19009

Schenk, P., White, O., Moore, J., & Byrne, P. 2018b, Enceladus and the Icy Moons of Saturn (Tucson, AZ: Univ. Arizona Press) 237

Schenk, P. M., Beyer, R. A., McKinnon, W. B., et al. 2018a, Icar, 314, 400

Schenk, P. M., & Zahnle, K. 2007, Icar, 192, 135

Singer, K., McKinnon, W., & Schenk, P. 2021, Icar, 364, 114465
    Quick, L. C., et al. 2022, Icar, 387, 115185

Singer, K. N., McKinnon, W. B., Gladman, B., et al. 2019, Sci, 363, 955

Singer, K. N., Schenk, P. M., McKinnon, W. B., et al. 2022, NatCo, 13, 1542

Smith, B. A., Soderblom, L. A., Banfield, D., et al. 1989, Sci, 246, 1422

Soderblom, L. A., Kieffer, S. W., Becker, T. L., et al. 1990, Sci, 250, 410

Trowbridge, A., Melosh, H. J., Steckloff, J., & Freed, A. 2016, Natur, 534, 79

Chapter 5

Volatile Cycles and Surface–Atmosphere Interactions

Tanguy Bertrand, Will Grundy, Emmanuel Lellouch, François Forget and Leslie Ann Young

5.1 Introduction: Pluto and Triton as a Distinct Class of Planetary Objects

Triton and Pluto are often seen as siblings, as they both share similar sizes, densities, and atmospheric and surface ice composition. Yet their surface appearance, including their topography, surface albedo and ice distribution strongly differ, suggesting a different geological history. In fact, Triton and Pluto are both thought to have formed beyond Neptune and then to have evolved differently. On the one side, Pluto remained in the Kuiper Belt and was hit by a twin to form the Pluto-Charon moon system. On the other side, Triton was captured by Neptune, as strongly suggested by its retrograde and highly inclined orbit around the ice giant planet, and its interior subsequently experienced intense tidal deformation and heating. Geological activity on Triton may still be powered today by tidal activity.

   In addition to belonging to the same family of Kuiper Belt Objects (KBOs), Triton and Pluto share similar tenuous atmospheres. At their distance from the Sun, their surface ices (nitrogen, N_2, methane, CH_4, and carbon monoxide, CO) are indeed volatile enough to allow for a tenuous and condensable N_2 atmosphere with traces of CH_4 and CO. This is a distinct class of atmosphere, which shares similarities with Io and Mars (tenuous and condensable), Titan (a N_2–CH_4 atmosphere with photolytically-produced hazes) and which is representative of other atmospheres that could be exhibited near perihelion by large volatile-rich KBOs (e.g., Eris, Makemake). As on Mars, this class of atmosphere leads to interactions between the atmosphere and surface, via relaxation to solid-gas equilibrium, showing up in a variety of phenomena ranging from the formation of glaciers and icy dunes to the establishment of complex volatile cycles and climate systems. However, the processes involved on Pluto and Triton are often very

doi:10.1088/2514-3433/ad5278ch5

different from what we can find on other objects. For instance, Pluto and Triton's atmospheres are characterized by long radiative timescales, a unique near-surface stratosphere-like thermal structure and a general atmospheric circulation controlled by the condensation-sublimation flows of nitrogen. In addition, their surfaces display geologic features involving complex mixtures of volatile ices and exchange processes between surface and atmosphere that are unique in the Solar System.

Consequently, Triton and Pluto represent invaluable natural laboratories to study the physics of planetary surface–atmosphere interactions. Comparing these objects together offers numerous insights into the physical mechanisms occurring on large volatile-rich objects of the Kuiper Belt. Above all, Triton and Pluto give us a perspective on the great diversity of planetary surfaces and atmospheres in the outer Solar System. Their peculiar climate and geology is a challenge to test our fundamental understanding of atmospheric and surface physics, usually based on what is known on the Earth.

We review the observations of Triton and Pluto relevant to surface–atmosphere interactions in Section 5.2, and cover the fundamentals of the physics involved in Section 5.3. In Section 5.4, we review various numerical models, and then we detail the volatile cycles and the climate system in Section 5.5. We conclude and cover open questions and future work in Section 5.6.

5.2 Basic Parameters and Observations of the Surface and Atmosphere of Pluto and Triton Relevant to Surface–Atmosphere Interactions

Observing the distant and cold worlds of Pluto and Triton is a challenging task. So far, only flyby missions visited these two objects (respectively, New Horizons in 2015, and Voyager 2 in 1989). The rest of the available observations were obtained through ground-based or space-based telescopes. Although the current telescopes do not provide much spatial resolution of the distant Pluto and Triton (e.g., ~3 pixels across with the Hubble Space Telescope or with the James Webb Space Telescope), they can inform on disk-averaged surface temperatures and volatile ice distribution, atmospheric trace gas abundances, surface pressure, etc. These datasets are crucial to assess the seasonal changes on Pluto and Triton.

5.2.1 Orbits, Seasons and Climate Zones

Triton and Pluto are located at relatively similar distances from the Sun (~30 au for Triton and ~30–50 au for Pluto). Pluto's orbital period and year duration is ~248 Earth years, while a year on Triton lasts ~160 Earth years. However their received insolation varies significantly across latitudes during an annual cycle, due to their orbital properties and axial tilts (i.e., obliquity). In addition, their obliquity and orbital parameters are not constant over time and evolve over "astronomical timescales."

The orbital and obliquity parameters of Pluto evolve on a timescale of thousands of years, similar to Earth where Milankovitch cycles influenced its past climate; the obliquity of Pluto (current: 119.6°) varies between 104° and 127° over 2.8 million years (Myr), the solar longitude of perihelion (current: 3.7°) varies from 0° to 360° with a precession period of 3.7 Myr, and the eccentricity (current: 0.2488) oscillates between 0.222 and 0.266 with a 3.95-Myr period (Dobrovolskis & Harris 1983; Dobrovolskis et al. 1997; Earle et al. 2017). With such high obliquities, Pluto experiences highly pronounced seasons with long polar nights lasting for several Earth-decades and reaching very low latitudes. As a result, a significant portion of Pluto's surface is "tropical" but also experiences "arctic" seasons (Figure 5.1, Binzel et al. 2017; Young et al. 2021). Furthermore, the generally high eccentricity of Pluto results in a season asymmetry, with a more intense and shorter northern summer compared to the southern summer.

The orbital and obliquity parameters and the seasons of Triton are more complex to define because Triton orbits Neptune. Change in Neptune's inclination, precession period and obliquity are significant over a timescale of tens of millions of years and the Triton-Sun distance has not evolved much since its capture by Neptune (Laskar & Robutel, 1993), therefore the seasons on Triton over the last millions of years can be reconstructed from its current orbit around Neptune. Triton has a retrograde and circular orbit around Neptune with an orbital inclination of 157°.

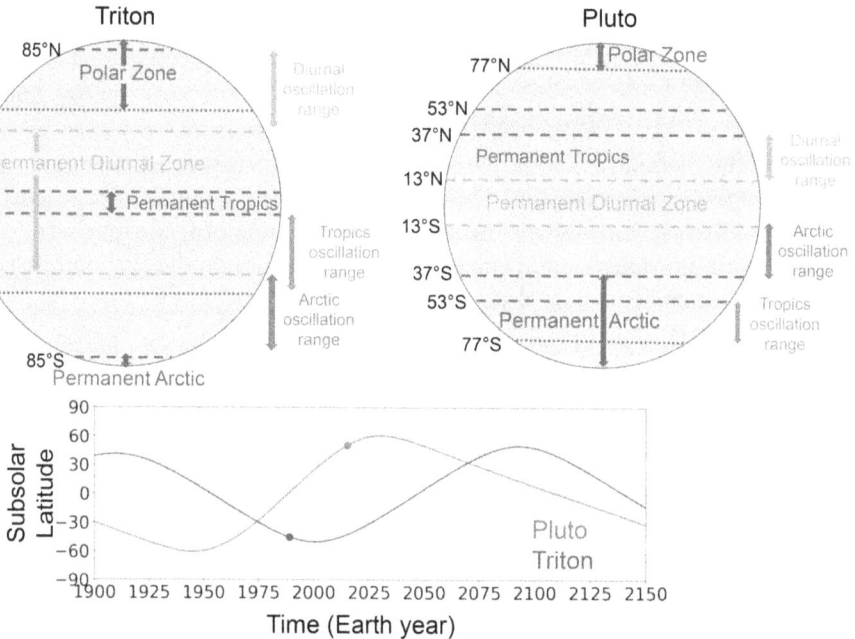

Figure 5.1. Top: Climate zones on Triton and Pluto, inspired by Binzel et al. (2017). Tropics: Latitude range where the Sun reaches zenith during orbital year. Arctic: Experiences periods of continuous Sun in summer, continuous dark in winter, throughout orbital year. Diurnal: Latitude range where a day/night cycle occurs continuously, throughout orbital year. Polar: Sun never reaches the overhead point at any time during orbital year. Bottom: Subsolar latitudes on Pluto and Triton over time and during the Voyager 2 and New Horizons flybys (circles).

The combination of this inclination with Neptune's obliquity of 28° leads to "chaotic seasons", with summer solstices oscillating between low (~5° latitude), moderate (~20° latitude) and extreme (~50° latitude) over a period of 140–180 Earth years (Trafton 1984; Bertrand et al. 2022). Given Neptune's orbital period of 165 years, each season on Triton lasts ~35–45 Earth years. These oscillations are equivalent to a "planet-like obliquity" (i.e., the angle between Triton's rotation axis and Neptune's orbit plane) varying between 5°, 20° and 50° over the same period, which corresponds to an average of ~30° over several Myr. Triton is therefore an object with a relatively low mean obliquity.

These differences are further highlighted by Figure 5.1, which compares Triton and Pluto climate zones. As defined in Binzel et al. (2017), the tropics are the latitude range where the Sun can reach the overhead point (zenith) on at least one date during the orbital year. For instance, in the case of Pluto where the planetary obliquity is > 90°, the latitude limits of the tropics correspond to the 180° complement of the obliquity. The "arctic" zone corresponds to the region where the Sun may never be above the horizon for one or more planetary rotations (in this same region the Sun can also remain completely above the horizon for a full rotation i.e. "midnight Sun."). The "diurnal zone" is the region continuously experiencing daily sunrise and sunset. As a result of its low obliquity, Triton's diurnal zone extends to high latitudes (up to 85°), while the arctic zone (where polar nights occur) can extend from 90° down to 39° in latitude. On Pluto, the diurnal zone remains confined below 37° latitude while the arctic zone can extend from 90° down to 13°. Finally the polar zone is the region where the Sun never reaches the overhead point in the sky at any time during any orbital year. It is more extended on Triton than on Pluto because the maximal obliquity reached on Triton is less than on Pluto.

The recent epoch during which the highest quality observations of Triton were gathered (from 1980 to today) and during which Voyager 2 flew by Triton corresponds to a period of intense summer in Triton's southern hemisphere at the south pole, as the subsolar latitude of Triton reached ~50°S in 2000. This follows a relatively intense southern winter (subsolar latitude ~50°N in 1910). On Pluto, the highest quality observations were gathered from Fall equinox (1988) to mid-northern spring (New Horizons flyby 2015) and northern summer solstice will be reached in 2030.

5.2.2 The Frozen but Active Surfaces of Pluto and Triton: Two Fundamentally Different Landscapes

Despite both sharing seasonally mobile, volatile N_2, CO, and CH_4 ices on their surfaces, Triton and Pluto exhibit remarkably distinct collections of landforms, as described in greater detail in Chapter 4. Perhaps the most striking contrast between the two is the much smaller topographic extremes on Triton, less than ±1 km compared with ±4 km on Pluto (Schenk et al. 2018, 2021). Triton's subdued topography could be related to higher internal heat flow and a thinner, less rigid lithosphere that is less able to support topographic extremes. Various resurfacing processes could also be at work. Triton's surface is uniformly youthful, featuring very few impact craters (e.g., Strom et al. 1990; Schenk & Zahnle 2007). Although a global resurfacing probably occurred in the immediate aftermath of Triton's capture into its retrograde orbit around Neptune (Agnor & Hamilton 2006), that event probably happened early in solar system history, so in itself cannot explain the paucity of craters observed at present. Triton's youthful surface points to processes that are probably still operating today. Pluto, too, has some youthful terrains that are evidently being modified by ongoing activity, but it also has widespread, ancient, heavily-cratered terrains that are inconsistent with any recent global-scale resurfacing process. Triton and Pluto also differ from the location of their main reservoir of volatile ice. On Pluto, the main reservoir of N_2 ice is confined within the massive topographic basin of Sputnik Planitia in the equatorial regions (Stern et al. 2015), while on Triton, N_2 ice seems to form an extended cap in the southern hemisphere, possibly from 90°S to ~15°S (Smith et al. 1989).

A number of distinctive landforms are seen only on Pluto or only on Triton. For instance, the cellular pattern produced by convective overturning in Pluto's giant Sputnik Planitia glacier has no known counterpart on Triton (see McKinnon et al. 2016; Trowbridge et al. 2016; White et al. 2017). Likewise nothing seen on Triton resembles Pluto's network of valley glaciers flowing down from eastern Tombaugh Regio into Sputnik Planitia (e.g., Umurhan et al. 2017) nor the numerous regions showing evidence of past glacial modification (e.g., Howard et al. 2017a). Another feature unique to Pluto is the equatorial belt of dark, reddish region ("Tenebrae"), exemplified by Cthulhu. The coloration of these regions is thought to result from complex organic haze particles that accumulate after settling to the surface (e.g., Grundy et al. 2018). Once established, such regions appear to be able to resist accumulation of volatile ices, at least at low latitudes owing to their low albedos and attendant daytime temperatures (Binzel et al. 2017; Earle et al. 2018a). Other distinctive landforms on Pluto include the "Bladed Terrain", interpreted as analogous to terrestrial penitentes, albeit at a much larger spatial scale (Moores et al. 2017; Moore et al. 2018), and the pitted and mantled uplands suggestive of collapse into subsurface voids (Howard et al. 2017b). Pluto's immense Wright Mons and Piccard Mons structures have been interpreted as cryovolcanic edifices (Singer et al. 2022) with no known analog on Triton.

Likewise, Triton has features not detected on Pluto, such as the plumes erupting ~8 km up into Triton's atmosphere (Soderblom et al. 1990; Kirk et al. 1995) and the numerous dark streaks presumed to have been deposited by plumes that had erupted some time earlier (e.g., Hansen et al. 1990; Hofgartner et al. 2022; see also Chapter 3). An especially distinctive Triton landform is the cantaloupe terrain (Schenk & Jackson 1993; Croft et al. 1995), which appears to be diapiric in nature. Although the horizontal scale of the cells in the cantaloupe terrain are similar to cells in Pluto's Sputnik Planitia, the topography is entirely different, presumably because the two units involve very different materials (Schenk et al. 2021). Other distinctive landforms on Triton include several linear ridges that somewhat resemble Europa's double ridges, and the enigmatic guttae surrounded by brighter aureoles (Smith et al. 1989). Although it appears to lack anything like Pluto's towering Wright Mons, Triton has large regions mapped as volcanic plains, with walled low-lying areas within this region resembling volcanic calderas (Croft et al. 1995), although subsequent glacial modification as seen on Pluto has been proposed (e.g., Sulcanese et al. 2023).

Comparisons between landforms on Pluto and Triton have been hampered by the lack of spatially resolved infrared spectral imaging for Triton. Ground-based observations established the presence of volatile N_2, CO, and CH_4 and non-volatile CO_2 and H_2O ices on Triton (Cruikshank et al. 1993). As on Pluto, longitudinal variation of the volatile ice distribution was evidenced (Grundy et al. 2010, Holler et al. 2016), but such observations do not provide sufficient spatial resolution to confidently assign compositions to specific landforms on Triton. Differences in landform morphology on Pluto and Triton could result from differences in atmospheric composition. For instance, there is much less CH_4 in Triton's N_2-dominated atmosphere relative to Pluto, and less photochemical haze settling to Triton's surface. Another potential difference is less internal heat from Pluto due to its smaller size and rock content. The distinct seasonal and astronomical insolation patterns and timescales (see Sections 5.2.1 and 5.5.1) could also play a role. With little information about the compositions of Triton's landforms, it is difficult to draw such comparisons. Even on Pluto, where compositional mapping was done by New Horizons (e.g., Protopapa et al. 2017; Schmitt et al. 2017) it appears that a volatile seasonal veneer can mask landform compositions in some regions. An example of this is Wright Mons, which exhibits strong CH_4 ice absorptions, although it is thought to be composed of more rigid material such as H_2O ice (Singer et al. 2022).

5.2.3 Characteristics of the Tenuous Condensable Atmospheres of Pluto and Triton

The N_2-dominated sublimation atmospheres of Pluto and Triton significantly evolve over time. This is seen most clearly in the evolution of atmospheric pressure, as monitored by stellar occultations. The body of data is particularly rich for Pluto, where a combined set of twelve occultations showed a threefold monotonic increase in pressure over 1988–2016 (Meza et al. 2019, and references therein). Although Pluto's surface pressure itself is not probed, these data constrain the pressure at a reference level, usually taken at 1215 km from Pluto center, which can be converted to a surface pressure using a reference thermal profile. The approach (e.g., Dias-Oliveira et al. 2015) can be validated against the surface pressure measured by New Horizons Radio Experiment (REX) in 2015, i.e., $P_{surf} = 12.8\pm0.7$ and 10.2 ± 0.7 µbar at entry and exit (Hinson et al. 2017), where the first value, which pertains to Sputnik Planitia, is more representative of global conditions. As discussed below, the pressure increase is related to the progression of spring over the northern hemisphere, exposing the main reservoir of N_2 ice (Sputnik Planitia) to insolation, and can be explained quantitatively by Pluto volatile transport models, which predict a pressure peak in ~2015–2030 (Bertrand & Forget 2016; Johnson et al. 2021a, Young et al. 2021). More recently, three occultations over 2018–2020 have indicated that Pluto's atmospheric pressure has indeed started to plateau since 2015 (Sicardy et al. 2021), with possible but yet unconfirmed signs (Arimatsu et al. 2020,

Young et al. 2021) of the onset of the decline expected to eventually occur due to the combined effects of recession from the Sun and lower insolation at Sputnik Planitia. Ground-based spectroscopy of the same hemisphere of Pluto obtained over a year apart (2014 and 2015) also indicates loss of N_2 ice from the northern polar regions, which may have played a role in the steady atmospheric pressure measured by occultations (Holler et al. 2022).

At Triton, following the detection of the atmosphere by Voyager 2 with $P_{surf} = 14 \pm 2$ µbar (Gurrola 1995), stellar occultations were observed in 1995, 1997 (July and November), 2008, 2017 (and most recently 2022). Based again on the pressure at a reference level (1400 km from Triton's radius) and a consistent interpretation of all curves, the ensemble of data up to 2017 (Marques-Oliveira et al. 2022, and references therein) may indicate (1) a pressure in 2017 back to its 1989 level, and (2) a mild-(~1.5–2) increase over 1989–1997, followed by a similar decrease over 1997–2017, but the error bars are large. Volatile transport models (Bertrand et al. 2022) predict a similar trend for a wide range of model parameters, but with much less amplitude (typically a ~1.2 increase over 1989–1997).

Besides N_2, Pluto's and Triton's atmospheres hold trace amounts of CH_4 and CO gas. Prior to New Horizons, CH_4 had been detected in Pluto's atmosphere from ground-based high-resolution IR spectroscopy (Young et al. 1997; Lellouch et al. 2009, 2015), indicating a mean ~0.5% mixing ratio with at most marginal variations (< 20%) between 2008 and 2012. CH_4 vertical profiles over 80–1200 km were derived from the New Horizons Alice solar occultations (Young et al. 2018), indicating a CH_4/N_2 mixing ratio steadily increasing with altitude due to molecular diffusion; extrapolation to the surface through 1-D transport models (using different eddy diffusivities) indicated a near-surface atmospheric CH_4 volume mixing ratio of 0.28%–0.35% (Young et al. 2018), ~0.4% (Lavvas et al. 2021) or up to 0.84% (Gladstone et al. 2016; see review in Wan et al. 2021).

At Triton, CH_4 was measured only from the Voyager 2 UVS occultation in 1989 (Strobel & Summers 1995, and references therein) and from VLT spectroscopy in 2009 (Lellouch et al. 2010). Although the two techniques and observables are different (CH_4 line-of-sight density from 0 to 50 km, versus vertical column down to the surface, respectively), combining the two datasets indicated a factor of ~4 increase in the CH_4 column over 1989–2009 (Lellouch et al. 2010). Assuming (retrospectively based on Marques-Oliveira et al. 2022) a surface pressure of ~14 µbar for 2009, the latter implies a 0.08% –0.17% mean CH_4 mixing ratio in 2009, while the Voyager 2 results point to a CH_4 mixing ratio of only ~0.03% at the surface in 1989, further decreasing with altitude with a scale height of ~20 km due to photolysis. We note that the CH_4 surface partial pressure measured by Voyager (2.5 nbar) corresponds to solid-gas equilibrium of pure CH_4 at 38.5 K or of $CH_4:N_2$ at a slightly lower temperature (Tan & Kargel 2018; Tan 2022); evidence for small deposits of CH_4-rich ice has been reported by Merlin et al. (2018). CO gas at Triton was first detected in the IR (Lellouch et al. 2010) with a 24 nbar partial pressure, within an uncertainty factor of 4. Assuming $P_{surf} = 14$ µbar, the central value would correspond to an atmospheric CO/N_2 volume mixing ratio of 1.7×10^{-3}. However, based on ALMA mm-wave data, Gurwell et al. (2019) reported $CO/N_2 \sim 10^{-4}$ only.

The discrepancy remains to be understood, all the more that similar IR and mm-observations of CO in Pluto's atmosphere give very consistent results on the CO/N_2 ratio (5–6×10^{-4}; Lellouch et al. 2011, 2017, 2022). Overall, Pluto's atmosphere appears to contain ~10 times more CH_4 than Triton's, while the CO abundances may be similar.

Atmospheric thermal structures for Pluto and Triton are known from stellar, radio and UV occultations and from thermal sounding using CO from ALMA. Pluto's atmosphere exhibits a sharp inversion layer (stratosphere) just above the surface with temperatures rising from ~40 K to ~110 K in 20 km. Although temperature profiles in the first 20 km still differ based on the retrieval technique (see, e.g., discussion in Lellouch et al. 2017 and Meza et al. 2019), the prominent stratosphere has been classically understood as being primarily caused by CH_4 heating/cooling radiative equilibrium, with some additional role due to CO, HCN, and C_2H_2 cooling (Yelle & Lunine 1989; Strobel et al. 1996; Strobel & Zhu 2017). However, it has also been shown that the prominent stratosphere can be primarily caused by a bright ice-dominated haze (Wan et al. 2021), whose cooling and heating rates dominate over gas (see details below). Triton's lower atmosphere is less well characterized, but the high-quality 2017 stellar occultation (90 light curves) along with a reassessment of some of the previous stellar occultations and the Voyager/RSS data (Marques-Oliveira et al. 2022; Elliot et al. 2000; Strobel & Zhu 2017) indicates a mildly positive temperature gradient from ~20 km (where the temperature is ~50 K) to ~200 km (temperature ~70 K); a somewhat steeper gradient must occur below 20 km and above N_2-ice regions to join to the 38 K surface temperature. Strobel & Zhu (2017) demonstrated that the smaller positive gradients in Triton's versus Pluto's stratosphere can be associated to the factor-of-ten smaller CH_4/N_2 near-surface mixing ratio on Triton and its further depletion in Triton's lower atmosphere due to photolysis. Above 200 km altitude, Triton's atmospheric thermal structure is poorly known, except for the fact that Voyager 2/UVS solar occultation data yield tangential N_2 densities consistent with an isothermal $T = 96$ K thermosphere above 400 km (Stevens et al. 1992). Such a temperature cannot be achieved from CH_4 heating only and additional heating due to precipitation of magnetospheric electrons is invoked (Stevens et al. 1992; Strobel & Zhu 2017). The opposite problem occurs on Pluto where both ALMA (from CO profiling; Lellouch et al. 2017) and New Horizons/Alice (from CH_4 and N_2 tangential densities; Gladstone et al. 2016; Young et al. 2018) data show decreasing temperatures above 30 km, with temperatures asymptotically reaching 65–70 K at 400 km and above. This unexpectedly cold upper atmosphere temperature—which also throttles escape rates of CH_4 and N_2 to much smaller values than anticipated prior to the New Horizons flyby—was puzzling in a gas-only context, as cooling by detected (HCN, CO, hydrocarbons) or envisaged (H_2O) gas species appeared insufficient, unless those are enhanced over their measured or estimated abundances by unreasonably large amounts. A drastically different idea calls for the overall control of Pluto's atmospheric thermal state by photochemically-produced organic haze particles, assumed to have Titan-like optical properties, instead of gas (Zhang et al. 2017). However, Lavvas et al. (2021) show that condensation of HCN and various hydrocarbons should occur in Pluto's atmosphere on the organic particles, leading to a haze bulk material dominated by ice components. Although the thermal heating and cooling rates of such an icy haze are expected to be $\sim$20 times smaller at 200 km than that of an organic-only haze, they are still significantly larger than gas rates, and therefore the icy haze radiation could still dominate the energy balance in the atmosphere (Lavvas et al. 2021; Wan et al. 2021).

Near-surface temperature profiles on Pluto and Triton remain poorly characterized and likely vary with location, time of day, and composition of the corresponding surface. At Triton, geysers erupting from the surface up to 8 km altitude where the material starts to be advected by winds, suggests the presence of a cold 8 km deep troposphere (Yelle et al. 1991), but no evidence of it was found by Voyager/RSS (Marques-Oliveira et al. 2022). At Pluto, the entry occultation profile (on the southeast margin of Sputnik Planitia) shows that the inversion layer there ends 3.5 km above the surface, with a 38.9 $\pm$ 2.1 K temperature in the cold boundary layer beneath the inversion, close to the saturation temperature of N_2-rich and CH_4-rich ices. General circulation models (Forget et al. 2017) show that such a cold layer can form in Sputnik Planitia in response to the combined effects of N_2 daytime sublimation, advection of cold air by near-surface katabatic winds and topographic confinement, but fail to reproduce a layer as deep as 3.5 km. At exit, which occurred on the anti-Charon hemisphere, the inversion layer extends down to the (locally warm) surface, with a 57.0 $\pm$ 3.7 K temperature at 1 km, $\sim$18 K warmer than at entry (and thus a smaller temperature gradient exists at exit than at entry).

Above 30 km altitude, the entry and exit Pluto radio-occultation temperatures are very similar, and so are the New Horizons Alice temperature, gas and haze profiles at ingress and egress. Yet, all of these measurements sampled only low latitudes, typically $\sim$16°N and $\sim$16°S. Large-scale (equator to North pole) temperature contrasts at Pluto's stratopause have been tentatively reported by Lellouch et al. (2022) from disk-resolved temperature sounding with ALMA, with summer pole latitudes near 30 km being 7$\pm$3.5 K warmer than low latitudes. This result is surprising because, although the diurnally averaged insolation at the summer pole is $\sim$4 times larger than at the equator, dynamical mixing times associated with the summer to winter flow (e.g., $\sim$1 terrestrial year for 0.1 m s^{-1} meridional velocities; Toigo et al. 2015; Forget et al. 2017) are short compared to radiative timescales ($\sim$10 years in "gas-only" radiative transfer models). If confirmed, this 2-sigma result from ALMA may point to shorter radiative timescales than previously thought, perhaps associated with haze cooling/heating rates 1–2 orders of magnitude stronger than the gas, as advocated by Zhang et al. (2017). Such hemispheric contrasts probably do not occur at Triton, given that Voyager 2/UVS results yielded remarkably similar N_2 and N profiles (over 450–750 and 170–500 km, respectively) at 26°N and 44°S, and that all curves from the 2017 occultation, altogether sampling most Triton latitudes, can be fit with a single temperature profile (Marques-Oliveira et al. 2022).

Extensive wave patterns are observed in Pluto's atmosphere. They were originally detected in stellar occultation data, which revealed perturbations in atmospheric density and temperature profiles (e.g., Sicardy et al. 2003). These oscillations have an increasing vertical wavelength with altitude ($\sim$10 km at 150 km altitude, $\sim$20 km at

400 km altitude), and amplitudes of a few percent in density and a few tens of Kelvins in temperature over a vertical scale of a few kilometers. On a global scale, wave signature is directly seen in New Horizons imaging showing that Pluto's haze up to 200 km is structured in ~1–5 km thick layers over up to 1000 km horizontal lengths (Cheng et al. 2017; Jacobs et al. 2021). About 20 layers are seen, and their properties are consistent with the perturbations seen in stellar occultations. A detailed description of Pluto's haze is presented in Chapter 6. Two main sources of excitation have been proposed to explain these waves: (1) topographic variations, with orographic gravity waves initiated by winds flowing over Pluto's extensive (~4 km high) topography (Cheng et al. 2017), and (2) the diurnal cycle of N_2 ice sublimation–condensation, with thermal tides resulting from the "breathing" of N_2 ice in response to solar forcing (Toigo et al. 2010). The signature of waves is less conspicuous in Triton's atmosphere, though marginally detectable in occultation curves (Elliot et al. 2000, Marques-Oliveira et al. 2022), and waves cannot be distinguished in Triton's haze (too faint and restricted to the near-surface).

5.3 Fundamentals of Surface–Atmosphere Interactions on Objects Such as Pluto and Triton

In this section, we review the fundamental processes of surface–atmosphere interactions specific to cold objects with a tenuous and condensable N_2 atmosphere, such as Pluto, Triton, and maybe other volatile-rich Trans-Neptunian objects during their perihelion passage.

5.3.1 General Properties of the Surface Ices: A Volatile Cocktail

Pluto's and Triton's surfaces have three main volatile ice species: CH_4, CO, and N_2, in order of increasing volatility. All three can sublime into and condense from the atmosphere at prevailing surface temperatures, albeit at very different rates, leading to distinct seasonal distributions. At 40 K, the vapor pressure of CO is about a thousand times higher than that of CH_4, while N_2's is six times that of CO (e.g., Brown & Ziegler 1980; Fray & Schmitt 2009). CH_4 is thus the least mobile of the three, by far.

The sublimation and condensation of these volatile ices and the diverse landforms thus produced depend sensitively on the basic material properties of the ices. These include (1) their heat capacity and thermal conductivity as well as mechanical and rheological properties as reviewed by Umurhan et al. (2021), (2) their optical properties (albedo, emissivity, etc.), which constrain the thermal balance between absorbed sunlight and thermal emission, and (3) their solid phases at low pressures. In brief, all three ices are mechanically weak, being held together with Van der Waals bonds, and are also poor conductors of heat. N_2 is the least optically active of the three, while CH_4 is the most colorful, with numerous overtone and combination vibrational bands throughout the near-infrared wavelength region (e.g., Grundy et al. 2002). Finally, the three ices each have two distinct solid phases at low pressures, with phase changes between the two occurring at 20.4 K for CH_4, 35.6 K for N_2, and 61.6 K for CO.

The volatile ice properties on Pluto and Triton are complicated by the fact that they mix together, thus impacting the temperature and molecular environment of the ice and therefore their behavior (e.g., Quirico & Schmitt 1997a, 1997b; Protopapa et al. 2015). CO and N_2 are fully miscible in one another, with the solid-solid phase transition shifting in temperature depending on the relative abundance of the two (Tegler et al. 2019). N_2 and CH_4 are partly soluble in one another, with a eutectic occurring at about 24% CH_4 and 62.6 K (Prokhvatilov & Yantsevich 1983). CO and CH_4 mixtures probably behave similar to N_2 and CH_4 mixtures, but have been little studied in the laboratory. There is as yet no ternary equation of state for all three ices together that extends to sub-solidus temperatures. In addition, the volatile ice behaviors are presumably influenced by a variety of other minority species, such as the broad array of photochemical products including ices (HCN, C_2H_2, C_4H_2, and C_3H_8) and organic haze materials (Grundy et al. 2018; Lavvas et al. 2021). Laboratory study is needed for the effects of such materials on volatile ice mixtures.

5.3.2 The Important Role of Thermal Inertia on Distant Cold Bodies

Surface temperature evolution on Pluto and Triton is primarily governed by the balance between absorbed solar insolation, thermal emission in the infrared, latent heat exchanges, and thermal conduction in the subsurface (the atmosphere is too tenuous to play a significant role other than through latent heat). On a weakly-irradiated body like Pluto and Triton, where insolation and thermal fluxes of the order of a fraction of one W m^{-2}, the radiative fluxes are small compared to the internal heat stored in the ground (heat capacity does not depend much on temperature). This has two consequences: (1) temperature changes are extremely slow and diurnal temperature variations are of the order of a few Kelvin at most, and (2) the subsurface heat stored during one season can play a major role in the control of the surface temperature at the opposite season. In such conditions, in spite of the century-long fall and winter, regions in the polar night can remain warm (and ice-free e.g., Leyrat et al. 2016). The amount of heat stored in the subsurface is mainly controlled by the thermal inertia of the subsurface (a lower thermal inertia implies a more porous structure and less heat stored), which is therefore a key parameter for the volatile transport models (see review in Young et al. 2021).

5.3.3 Despite the Cold Temperatures, an Active Surface Exists Thanks to the Ice Volatility and Complex Feedbacks

5.3.3.1 Resurfacing through the Continual Cycles of Sublimation and Condensation

The youngest terrains on Triton and Pluto, which are among the youngest in the solar system, involve volatile ices and their continuous cycles of sublimation and condensation. Over the last millions of years, variations of orbital and rotation parameters of Pluto have led to substantial insolation changes, triggering volatile transport and subsequent resurfacing through: (1) local changes in nitrogen ice thickness, with variations of up to 1 km (particularly in Sputnik Planitia), causing glacial flow and erosion of the water-ice bedrock, (2) local changes in methane ice thickness, with variations of tens of meters, and (3) formation of nitrogen and methane ice mantles (e.g., White et al. 2017, 2019; Bertrand et al. 2018; for a review, see Young et al. 2021). Similar resurfacing mechanisms are expected on Triton with large amounts of nitrogen ice transported from one pole to the other (e.g., Moore & Spencer 1990; Bertrand et al. 2022). More details are provided in Section 5.5.

Over several annual timescales, volatile exchanges between the surface and the atmosphere could contribute to the formation of deep sublimation pits (Howard et al. 2017a; Buhler & Ingersoll 2018) and dunes of ice (Telfer et al. 2018), as observed on Pluto (such features were not observed on Triton due to the lack of spatial resolution but could be present there too). Over an annual timescale, these exchanges induce nitrogen and methane frost formation, able to temporarily cover large-scale regions on Pluto and Triton, as predicted by models (Bertrand et al. 2019, 2022).

5.3.3.2 How Topography Locally Controls the Ice Deposition

New Horizons revealed that topography significantly impacts the volatile ice distribution on Pluto, as N_2-rich ice has been mostly detected in the depressions (notably, Sputnik Planitia) whereas CH_4-rich ice was found at higher altitudes, in particular in the equatorial regions (typically, the Bladed Terrains and the summits of Pigafetta Montes; Gabasova et al. 2021. A review of the known mechanisms at play is given in Young et al. (2021) and Forget et al. (2021), and summarized below. We also discuss whether they could apply on Triton too.

N_2 Ice at Low Altitude

Hydrostatic equilibrium dictates that the surface pressure—mostly N_2—is higher at depth and lower at altitude, and so therefore is the equilibrium N_2 ice temperature, as dictated by the solid-gas equilibrium. It follows that the warmer deposits at low altitude emit more heat to space than the colder deposits, which leads to increased N_2 condensation rates at low altitude allowing this difference in thermal radiation to be balanced by latent heat exchange. Consequently, N_2 ice tends to accumulate at low altitude, forming permanent reservoirs there. This atmospheric-topographic process, first described by Trafton et al. (1998) and Trafton (2015), explains why the Sputnik Planitia basin is filled with N_2 ice (Bertrand & Forget 2016; Johnson et al. 2021a).

Young et al. (2021) give estimates of N_2 ice temperature and mass transport flux on Pluto for an idealized 4 km tall mountain and a 4 km deep valley and assuming Pluto conditions at perihelion, in 2015, and at 42 au, for the day and night sides. They found that the mountain ice temperature is typically $\sim$0.6–0.8 K warmer than the valley ice in all the scenarios explored. This leads to a higher condensation rate in the valley by $\sim$2–4 $\times$ 10^{-8} kg m^{-2} s^{-1}, equivalent to a rate of 15–30 cm per Pluto year. Over astronomical timescales, the characteristic timescale for infilling a Sputnik-like basin with N_2 ice is $\sim$7 million years (Young et al. 2021).

This atmospheric-topographic process is expected to operate on Triton as well. One prominent distinction lies in their respective topographies, as Triton's topography appears to be much less pronounced compared to Pluto, as discussed in

Section 5.2.2. The N_2 ice temperature difference between a 1 km tall mountain and 1 km deep valley becomes $\sim$0.2 K only, associated with mountain-to-valley differences in condensation rates 20 times lower than for a $\pm$4 km amplitude. Therefore, although locally (i.e., at a given latitude, where the mean insolation is the same for all longitudes) accumulation of N_2 ice should be favored at low altitude on Triton, the process should be much weaker and much slower than on Pluto.

CH_4 Ice at High Altitude

In the equatorial region of Pluto, methane-rich frost is observed coating mountain tops (Earle et al. 2018b). A first explanation for this phenomenon is related to the sublimation and condensation of volatile ices, causing them to deviate from thermodynamic equilibrium, thus resulting in altitude segregation with N_2-rich ice dominating at low elevations and CH_4-rich ice prevailing at high elevations (Moore et al. 2018; Tan & Kargel 2018; Young et al. 2021). Measurements over Sputnik Planitia at the REX radio ingress have revealed a near-surface temperature of 38.9 $\pm$ 2.1 K (Hinson et al. 2017), and an atmospheric CH_4 mixing ratio of $\sim$0.3% (Young et al. 2018). This is not consistent with a vapor-pressure equilibrium of pure CH_4 at that temperature. Such a disequilibrium is locally driven by net sublimation or condensation of volatile ices, although the exact relaxation timescales remain unknown (Young et al. 2021). Specifically, if we consider a patch of ice composed of N_2-rich and CH_4-rich grains (binary system N_2 + CH_4), preferential sublimation of N_2 ice would lead to the conversion of this ice into a CH_4-rich ice. Given that N_2 sublimation is favored at higher altitude and condensation at lower altitude (see above), CH_4-rich ice would become dominant at high elevations.

A second explanation was proposed by Bertrand et al. (2020b). Using high resolution numerical climate models of Pluto, the authors found that the atmosphere is enriched in gaseous methane at $\sim$4 km altitude and that the highest mountains extend into this CH_4-enriched air and therefore get capped by methane frost. This process of high-altitude CH_4 ice accumulation is further amplified by positive albedo feedback.

These mechanisms are believed to work on Triton as well, although to a lesser degree since the topography is much less pronounced than on Pluto (Schenk et al. 2021). Besides, the "flatness" of Triton may prevent the segregation by altitude of N_2-rich and CH_4-rich ice and contribute to the limited presence of CH_4-rich ice on Triton (Merlin et al. 2018).

5.3.3.3 Complex Feedback Impacts Ices and Sublimation–Condensation Cycles

The climates of Triton and Pluto are controlled by the condensation and sublimation cycles of their volatile ices and are therefore strongly sensitive to the processes impacting condensation and sublimation rates.

First, changes in volatile ice thermophysical properties can modify the surface thermal balance, with albedo and emissivity regulating the sublimation and condensation rates. These properties are very sensitive to the ice grain size, as shown, e.g., on Mars (Langevin et al. 2007). On Triton and Pluto, such changes are also expected for N_2-rich and CH_4-rich ices, due to sublimation and condensation and to the contamination by haze particles accumulating at the surface, thus lowering the surface albedo. Darkening of the ice can also occur directly at the surface through irradiation of CH_4-rich and other hydrocarbon ice by solar UV and cosmic rays (e.g., Johnson 1989; Grundy et al. 2018). In addition, the N_2-rich ice is expected to transition between solid α and β phases over seasonal timescales, which induces drastic changes in N_2 ice emissivity (Stansberry et al. 1996).

Second, complex surface–atmosphere interactions can lead to feedback between the ice properties and the condensation-sublimation rates. On Pluto, runaway forcing (positive feedback) by albedo and composition are efficient mechanisms to maintain albedo contrasts, in particular in the equatorial regions (Earle et al. 2018a; Bertrand et al. 2020b). For instance, non-volatile areas in Cthulhu tend to be covered by dark materials (Bond albedo ~ 0.1, Buratti et al. 2017) and are therefore warmer by 15 K than the bright N_2-rich ice areas of Tombaugh Regio. This allows the dark areas to remain volatile-free, while bright areas stay cold and remain volatile condensation sites. Albedo feedback can also take place within a given patch of volatile ice. Bertrand et al. (2020a) showed that the bright and dark N_2-rich plains of Sputnik Planitia are dominated by net condensation and net sublimation, respectively (at the time of the New Horizons flyby), even if they are located at the same latitude and therefore receive the same insolation on average. This is because a slightly lower albedo of N_2 ice leads to an increase in sunlight absorption which is sufficient to maintain a net diurnal sublimation. Similar mechanisms occur with CH_4-rich ice, with condensation and sublimation respectively maintained on the bright (e.g., East Tombaugh Regio) and dark deposits (e.g., the Bladed Terrain) observed in the equatorial regions (Bertrand et al. 2020b).

Initially, contrasts in albedo and composition observed on Pluto could be locally driven by aeolian activity and then boosted and maintained by runaway mechanisms. The processes may involve the deposition and re-mobilization of ice grains and dark materials by N_2 sublimation winds, and/or the variation in downward night-time sensible heat flux (Telfer et al. 2018; Bertrand et al. 2020a). For the case of Sputnik Planitia, near-surface winds are thought to play a role as suggested by the observation of dark wind streaks (Stern et al. 2015; Telfer et al. 2018). Eventually, as sublimation of volatile ice continues, a lag deposit may form on top of the ice (e.g., of dark materials or CH_4-rich ice as the N_2-rich ice sublimes), which would inhibit further sublimation (e.g., Schmitt et al. 2017), thus leading to a negative feedback.

These key feedback processes, strongly impacting Pluto's climate, are thought to be dominant on Triton too. Although the production of haze particles is much less important on Triton than on Pluto, settling haze particles could contribute to trigger albedo feedback locally. In particular, as on Pluto, the equatorial regions on Triton should be more sensitive to albedo variations than the poles, because they correspond to the diurnal zone with minimal temperature differences over seasonal timescales (despite the fact that the equatorial regions are much warmer than the poles, on average), as shown by Figure 5.2. This could notably explain the diversity of terrains and colors observed at the edge of the southern bright cap (Bertrand et al. 2022). In addition, the geysers observed by Voyager 2 above the

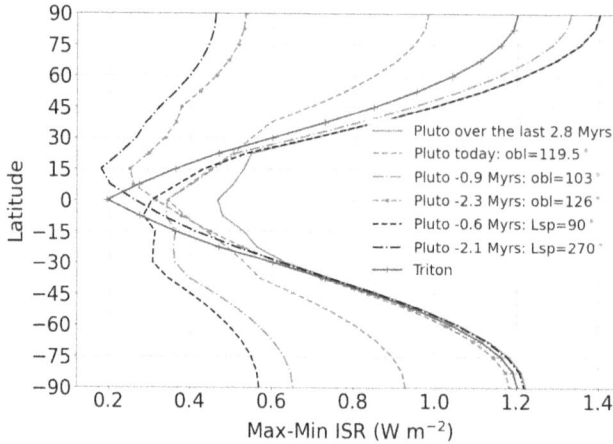

Figure 5.2. Latitude versus the variation between the minimum and maximum incoming solar radiation (ISR) at that latitude over Pluto and Triton. For Pluto we consider the timescale of the last obliquity cycle (i.e. the minimum and maximum ISR are calculated over the last 2.8 Myr, solid red line), and the annual timescale at the current epoch, the epochs with maximal and minimal obliquities (red lines) and the epochs with solar longitude of perihelion of 90° and 270° (black lines). For Triton we consider the timescale of ~4000 years, spanning ~21 Triton-like years. Minimal variations are obtained in the equatorial regions on both objects, suggesting that these regions are the most sensitive to albedo runaway forcings.

southern cap could play a role by depositing darker or brighter particles on top of the ice. For instance, Voyager 2 imaged numerous streaks on Triton's southern cap, possibly related to erupting geysers, that are 10%–20% darker than the surrounding ice (Smith et al. 1989). Finally, pole-to-pole differences in ice contamination by dark material, sensible heat flux from the atmosphere or runaway albedo feedbacks have also been invoked to explain the fact that volatile ices have accumulated in the southern hemisphere of Triton rather than in the northern hemisphere (Brown & Kirk 1994; Moore & Spencer 1990; Bertrand et al. 2022).

5.3.3.4 Small-Scale Surface Processes in Volatile Ices

In addition to large-scale processes directly characterized in observations, smaller spatial scales are fundamentally important to how volatile ices behave, from below the resolution limits of existing imagery all the way down to the molecular level.

Processes at scales just below image resolutions could be important on Triton and Pluto, as suggested by higher resolution imaging of volatile-driven activity on Mars. For instance, spiders or araneiforms are attributed to sublimation at the base of a thin seasonal CO_2 ice sheet leading to build-up of pressure and eventual escape, entraining debris from below the ice sheet (e.g., Piqueux et al. 2003; Attree et al. 2021). On sloped terrain, cycles of seasonal volatile ice deposition followed by sublimation could be an important driver of downslope movement and erosion (e.g., Diniega et al. 2013, 2021). Similar phenomena could occur in seasonal N_2 ice deposits on Pluto and Triton at spatial scales set by the ice thickness, potentially well below the resolution of available images.

In a binary mixture of volatile ices, such as N_2+CH_4, the more volatile component (N_2) will sublime preferentially, as dictated by thermodynamics at the gas-solid interface (e.g., Trafton 2015). But as N_2 is depleted from the surface of the ice, the composition of the remaining ice changes, with the increasing relative abundance of CH_4 gradually choking off the supply of N_2. This is described as a "detailed balancing" scenario (e.g., Trafton 1990; Trafton et al. 1998). However, loss of N_2 from the surface creates a compositional gradient within the ice, leading to diffusion of N_2 from deeper in. This diffusion involves a kinetic factor that depends on the relevant timescales as well as molecular diffusion rates that are as yet poorly known. Additionally, as sublimation and diffusion shift the composition of a mixed ice, solubility limits can be exceeded, triggering exsolution of crystals of a different phase (CH_4-dominated, in the case of an N_2+CH_4 mixture losing N_2), with crystal size and spacing depending on the timescale and diffusion rates (Predel et al. 2013). The nucleation and growth of such crystals can drastically change the ice's mechanical, rheological, and light-scattering properties, but relatively little laboratory work has been done on these phenomena for the volatile ice mixtures and thermal conditions that are relevant to Pluto and Triton.

5.3.4 The Dynamics of Tenuous N_2 Condensable Atmospheres

A review of the knowledge of the atmospheric dynamics on Pluto is given in Forget et al. (2021). In this section, we summarize the main principles and discuss comparisons with Triton.

The general circulation of any planetary atmosphere is fundamentally driven by horizontal gradients of pressure. In the absence of major composition variations, such gradients are primarily induced by temperature gradients resulting from differential solar heating (the atmosphere expands where it is warmer), and, in the case of Pluto and Triton, by condensation-sublimation processes at the surface. The resulting circulation is in turn modified by the rotation of the planet (centrifugal and Coriolis forces), the convergence and divergence of mass, and by the different dynamical waves induced by local or regional perturbations, which can, under some conditions, influence the zonal mean circulation through "wave-mean flow" interaction.

Both surface heating and direct absorption of sunlight can generate atmospheric temperature gradients and dynamically force the atmosphere. Compared to Earth and Mars, the thermal flux from the surface to the atmosphere is relatively small on Pluto and especially on the methane-poor Triton, because the atmosphere is largely transparent in the thermal infrared. That leaves direct absorption of solar radiation by methane (Strobel et al. 1996) and by haze particles (Zhang et al. 2017) as the primary source of atmospheric heating. When hazes and methane are radiatively negligible (e.g., as might be the case on Triton), then the heat conducted from the thermosphere (where EUV radiation can be absorbed by N_2) can be the dominant source of atmospheric heating (Yelle et al. 1991).

Horizontal temperature gradients induced by solar heating have been expected to remain very small in Pluto's and Triton's atmospheres because of the very long radiative timescale of the atmosphere and the slow rotation and small size of these objects (Strobel et al. 1996; Forget et al. 2017, 2021). However hazes particles may act as efficient emitters and absorbers at thermal wavelengths and change this expectation (Zhang et al. 2017). General Circulation Models (GCMs) of Pluto and Triton (see below) do not yet take that into account. Assuming a gaseous atmosphere of N_2 and CH_4, GCMs show that the primary driver of the circulation should be the sublimation–condensation flow from subliming N_2 ice reservoirs to condensing reservoirs. If these reservoirs are in different hemispheres (typically from one polar cap to the other on Triton, or from the spring-summer hemisphere to the fall-winter hemisphere or across Sputnik Planitia on Pluto), the circulation should be characterized by a global "retro-rotation" with westward, retrograde winds at most latitudes. Such winds result from the conservation of angular momentum of the air particles as they flow above the equator, where they are farther from the rotation axis than where they started from (Toigo et al. 2015; Forget et al. 2017; Bertrand et al. 2020a). If the sublimation–condensation is weak (for instance if N_2 does not condense in the winter hemisphere), then any temperature gradient may influence the circulation. For instance, an ice-free warm equatorial region has been proposed to explain the prograde wind (blowing west) at the location of plumes observed by Voyager 2 on Triton (Ingersoll 1990). Indeed, the gradient wind equation predicts the formation of polar to mid-latitude prograde jets if the equator is warmer than the mid-latitudes. In such conditions, some models have even found that planets with a period of rotation of 5–7 Earth days like Pluto or Triton could be in a "super-rotation" regime like Titan (Forget et al. 2017) as a result of wave-mean flow interaction.

Near the surface, in addition to the general circulation patterns described above, the model from Forget et al. (2017) showed that topography should create strong downslope winds ("katabatic" winds). These winds result from the fact that the surface is much colder than the atmosphere. The air close to the slopes is cooled and tends to flow down because it is denser than the air away from the slope at the same altitude level. As Triton is flatter than Pluto and its near-surface atmosphere may not be as warm, these near-surface winds should have less of an impact on the atmospheric dynamics. In addition, the formation of orographic atmospheric waves could be limited on Triton due to its flatness, which is supported by stellar occultations revealing wave-like structures in the density profile (see Section 5.2.3) that appear much more muted on Triton than on Pluto (Elliot et al. 2000; Marques-Oliveira et al. 2022).

5.4 Building Models to Simulate the Volatile Cycles and the Climate System on Different Timescales

In this section, we summarize the principles of numerical models of Pluto's and Triton's volatile cycles and climate and their evolution. More details are given in Young et al. (2021); Forget et al. (2021), and Bertrand et al. (2022).

5.4.1 Energy Balance Models and Volatile Transport Models

Because of their very low densities, the atmospheres of Pluto and Triton have a negligible radiative thermal influence on the energy balance of their surfaces. On this basis, it has been possible to develop "volatile transport models" (VTMs) by simply calculating the local insolation (a simple geometrical calculation if one neglects the atmosphere), the thermal infrared cooling (simply using the Stephan-Boltzmann equation) and the heat storage and conduction in the subsurface (of key importance, as explained in Section 5.3.2). Computing these three processes is sufficient to estimate the evolution of surface temperatures over time. In the presence of nitrogen ice, the surface temperature is forced to remain at the nitrogen frost point (which only depends on pressure), and it is straightforward to calculate the amount of N_2 ice that must be condensed or sublimed to ensure that.

The first numerical VTMs simulating the volatile cycles on Pluto and Triton emerged in the late 1980s, motivated by the detection of volatile ices on their surfaces, the measurements of surface pressure from stellar occultations and the flyby of Triton revealing an active surface (e.g., Trafton 1990; Spencer 1990). During the 1990s, several stellar occultation observations were performed for both Pluto and Triton (see Section 5.2.3), bringing observable constraints (surface pressure) to the models. Modeling efforts were made to explain the observed trends (e.g., Hansen & Paige 1992, 1996; Young 1993) and explore new physical processes, such as the effect of N_2 emissivity and phase change (Stansberry & Yelle 1999) and the North-South albedo asymmetry on Triton (Spencer & Moore 1992; Brown & Kirk 1994). During the period 2010–2015, as the number of stellar occultations on Pluto increased, and as the New Horizons flyby approached, more efforts were made to simulate the nitrogen cycle on Pluto taking into account the effect of diurnal and seasonal thermal inertia and constraining the ice distribution and properties with the stellar occultation measurements (Young 2012, 2013; Olkin et al. 2015; Hansen et al. 2015; Toigo et al. 2010).

In 2015, the flyby of Pluto by New Horizons brought strong constraints on the surface ice distribution (in particular: Sputnik Planitia as the main reservoir of N_2 ice), topography, atmospheric structure, composition, and surface pressure. Post-encounter volatile transport models taking into account these new discoveries were able to (1) explain the seasonal evolution of surface pressure and trace gas abundance (Bertrand & Forget 2016, Johnson et al. 2021a), (2) contribute to the interpretation of geological features involving volatile ices (e.g., White et al. 2017; Moore et al. 2018; Telfer et al. 2018), and (3) explore the formation and evolution of the Sputnik Planitia ice sheet (Bertrand & Forget 2016, Johnson et al. 2021b). and of other seasonal and permanent reservoirs of N_2-rich and CH_4-rich ice (Bertrand et al. 2018, 2019). Finally, models for Triton were also updated based on what has been learned on Pluto, and the volatile cycles on Triton were reanalyzed in light of new observable constraints (Bertrand et al. 2022).

5.4.2 Three-dimensional Global Climate Models

Although energy balance considerations explain a number of global features, such as the location of the main reservoirs of volatile ice and the pressure cycle, they cannot reproduce all the characteristics of glaciers and frosts observed on Pluto and Triton. The complex distribution of volatile ice deposits (e.g., longitudinal distribution of CH_4-rich Bladed Terrain deposits and CH_4-rich frost on mountain tops) must involve additional processes that control the surface–atmosphere interaction.

For instance, the condensation and sublimation fluxes of CH_4 and CO ices depend on the exchange between the surface ices (with vapor pressure affected by the mixing of the ices) and the atmosphere. These exchanges are directly sensitive to (1) the turbulent mixing (and thus to the wind and the atmospheric lapse rate), and (2) the mixing ratios of the CH_4 and CO gases in the atmospheres. The first term depends on the location of nitrogen ice deposits, whose sublimation–condensation flows control the near-surface atmospheric winds (see Section 5.3.4), while the last term depends on the transport from other reservoirs on the planet, the possible condensation in CH_4 or CO clouds, etc.

To model the environment at the surface of Pluto and Triton, it has thus been necessary to develop full 3D global climate models that not only calculate the energy balance of the surface, but also solve the equations to account for atmospheric dynamics, radiative transfer though the atmosphere, mixing by turbulence and convection, condensation of clouds, formation of photochemical haze, formation and propagation of waves and tides, etc.

Historically, two GCMs, taking into account the 3D atmospheric dynamics and nitrogen volatile transport, were developed before the New Horizons encounter of Pluto (PlutoWRF; Toigo et al. 2015; LMD, Vangvichith 2013), following first developments to build a Triton GCM (Vangvichith et al. 2010). After the encounter, the LMD model was updated to take into account the observation from New Horizons and include all physics and key equations to represent the CH_4, CO and N_2 cycles (Forget et al. 2017; Bertrand et al. 2020a, 2020b).

5.5 Volatile Transport on Seasonal and Astronomical Timescales on Pluto and Triton

5.5.1 How Obliquity Changes Everything, From the Surface to the Atmosphere

On Earth, Mars and Titan, the Milankovitch parameters (obliquity, orbital eccentricity, solar longitude of perihelion) vary over time and combine to modulate solar insolation and surface temperatures, thus controlling seasons and forcing the volatiles that form glaciers, frost or lakes to migrate in different regions with time. These Milankovitch cycles are also significant on Pluto, as detailed in Section 5.2.1. Modeling studies showed that Pluto's climate is strongly sensitive to these variations, which drive significant changes in sublimation and condensation rates of the volatile ices and affect the atmospheric pressure and composition, and the surface distribution of volatile ices (Earle et al. 2017; Stern et al. 2017; Bertrand et al. 2018, 2019). On average over one obliquity cycle (2.8 Myr), the poles receive more insolation than the equator, as a result of the relatively high obliquity (mean obliquity of 115°, i.e., 65° in retrograde rotation, see Section 5.2.1), and surface temperatures are predicted to be warmer at the poles than at the equator, if we assume a large thermal inertia ($\geqslant$500 thermal inertia units, tiu= $J\,kg^{-1}\,m^{-2}\,s^{-0.5}$) for the subsurface (Figure 5.3; Bertrand et al. 2018). The observation of massive and permanent volatile deposits in the equatorial regions (the N_2-rich Sputnik Planitia ice sheet and the CH_4-rich Bladed Terrain) seems to be explained by these latitudinal differences in mean insolation, coupled with topographic-atmospheric processes (Bertrand & Forget 2016; Bertrand et al. 2019).

On Triton, the definition of Milankovitch parameters and obliquity is more complex since Triton orbits Neptune. As detailed in Section 5.2.1, Triton can be described as an object with a low mean obliquity (30°). Volatile transport models have shown that in these conditions and on average over several annual cycles (1) Triton's poles tend to remain colder than the equator by 5–6 K (Figure 5.3, Bertrand et al. 2022), and (2) volatile ices accumulate at the poles, forming two polar caps with similar seasonal evolution (both poles on Triton are equivalent in terms of mean received insolation due to the very low eccentricity of Neptune's orbit, e.g., Hansen & Paige 1992; Spencer & Moore 1992; Bertrand et al. 2022). However, if North-South asymmetries in surface or subsurface properties are accounted for in the model (e.g., in internal heat flux, surface albedo, etc., as in Brown & Kirk 1994; Moore & Spencer 1990; Bertrand et al. 2022), then one cap develops over the other and becomes the main cold trap for volatiles. Such North-South asymmetries have been suggested on Triton to explain the presence of a bright southern cap (assumed to be made of N_2-rich ice) extending to the equator and the presumable absence of a

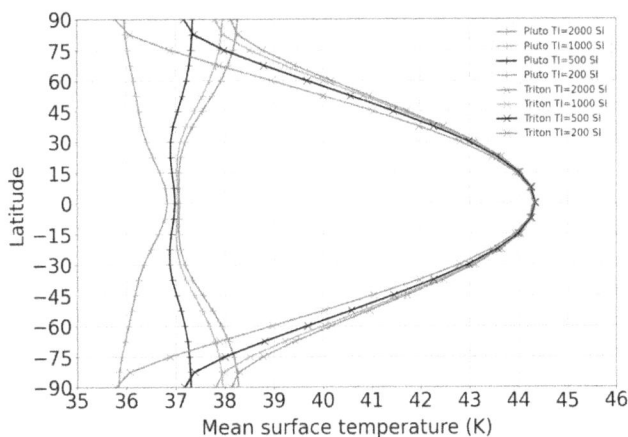

Figure 5.3. Modeled zonal mean surface temperatures on Pluto and Triton, assuming a bedrock uniformly made of water ice (albedo=0.6, emissivity=0.8, no volatile ice) and subsurface thermal inertias of 2000, 1000, 500, 200 SI. The temperatures are averaged over the last 10 Myr (and therefore include the last ~3 obliquity cycles in the case of Pluto) as shown in Bertrand et al. (2018) and Bertrand et al. (2022). On average, the poles are slightly warmer than the equatorial regions on Pluto (except in the case of a low subsurface thermal inertia), and much colder than the equatorial regions on Triton.

northern cap (although latitudes north of 45°N were not probed, being in the polar night at the Voyager 2 epoch).

These modeling results hint at obliquity being the main driver for the ice distribution observed on Pluto and Triton. In order to investigate this in more detail, we performed simulations of Pluto and Triton with the LMD Volatile Transport Model spanning the last 20 Myr, as in Bertrand et al. (2019) and Bertrand et al. (2022). We used a flat topography for Triton, and the New Horizons topography data for Pluto. In addition, we designed two alternative simulations by taking the initial state of the Pluto simulation with the following modifications: (1) the first alternative simulation assumes that Pluto orbits the Sun as the Neptune-Triton system does, i.e., at 30.1 au with an eccentricity of 0.009 (the obliquity remains that of Pluto), and (2) the second one assumes that the subsolar point follows the same variations with latitude and time as on Triton. All simulations start with a global and uniform cover of 200 m of N_2 ice and 4 m of CH_4 ice and share the same initial state for the surface and subsurface temperatures (uniform at 37 K). Only the topography, orbit, and obliquity differ in these simulations.

Figure 5.4 shows the distribution of volatile ice obtained after 20 Myr, with N_2 and CH_4 ice deposits in an equilibrated state. The Triton simulation shows that the volatile ice accumulates at the poles, forming massive polar caps of N_2-rich ice. The Pluto simulation reproduces the formation of perennial deposits of N_2 and CH_4 volatile ices in the equatorial regions, with N_2 ice in the Sputnik Planitia topographic basin and in local depressions outside, in line with New Horizons observations.

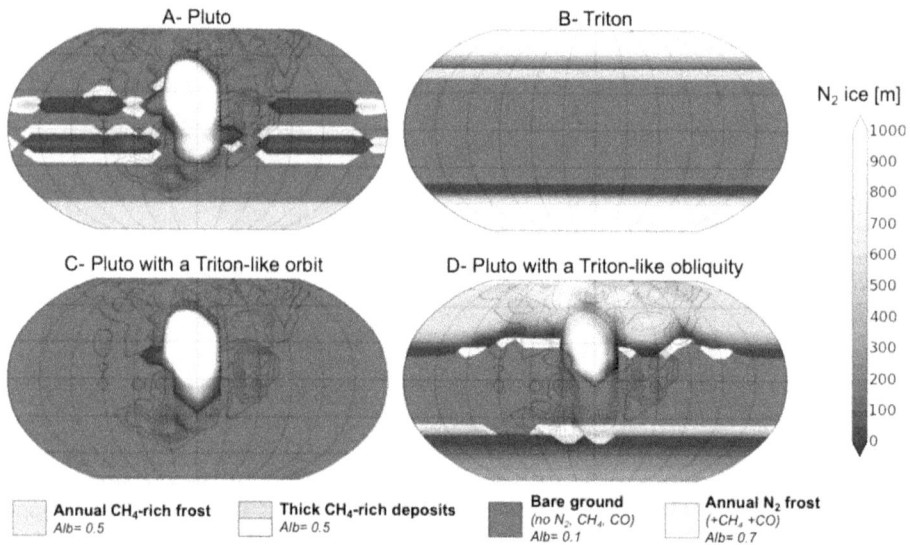

Figure 5.4. Volatile ice distribution obtained after a 20-Myr simulation performed with the LMD volatile transport model for (A) Pluto, (B) Triton, (C) Pluto with a Triton-like orbit around the Sun (eccentricity = 0.009, Sun-Triton distance = 30.1 AU) and with a Pluto-like obliquity (varying over a timescale of 2.8 Myr, see Section 5.2.1), (D) Pluto with a Pluto-like orbit but Triton-like obliquity (i.e, Triton's subsolar latitude variation). Black contours show the bedrock's topography on Pluto (Triton is assumed flat).

This distribution is not strongly impacted by the change in eccentricity; if Pluto was orbiting the Sun in a quasi-circular orbit at the distance of Neptune (alternative simulation 1), the volatile ice would still accumulate in the Sputnik Planitia basin, as shown by Figure 5.4(C). On the other hand, if we consider a Pluto-like orbit with a Triton-like obliquity (alternative simulation 2), then the volatile ice would accumulate at the poles, as shown by Figure 5.4(D). These results demonstrate that obliquity is the main driver explaining the differences in perennial volatile ice distribution observed on Pluto and Triton. Topography also plays an important role to trap the nitrogen ice at the longitudes of the deepest depressions. The fact that we can reproduce, to first order, the observed volatile ice surface distribution on Pluto and Triton by using the same model and initial conditions, tentatively supports the hypothesis that both objects had a common origin (McKinnon 1984), but then their surface ice distribution evolved differently, mostly due to the difference in obliquity.

The obliquity on Pluto and Triton not only impacts the location of their surface volatile ice distribution but also affects their climate at a global scale in two ways. First, the distribution and seasonal evolution of surface N_2 ice reservoirs drive the N_2 condensation-sublimation flows, which strongly controls the winds and the general atmospheric circulation (Toigo et al. 2015; Forget et al. 2017; Bertrand et al. 2020a). Second, the location of the coldest regions may favor (in the case of Pluto) or prevent (in the case of Triton) the formation of extended CH_4-rich ice deposits (see below). On Triton, this absence of massive, perennial and extended CH_4-rich deposits leads to lesser amounts of gaseous CH_4 in the atmosphere than on Pluto, which causes colder atmospheric stratospheric temperatures, less evolved photo-chemical processes and haze formation, and smaller escape rates.

How does the location of the coldest region impact the amount of CH_4-rich ice exposed at the surface? Pluto's low-insolation equatorial regions correspond to an extended area with variegated topography ($\pm$ 5 km) allowing the existence of both perennial N_2-rich and CH_4-rich deposits, as N_2 preferably accumulates in depressions, and CH_4 at altitude. On Triton, however, the low-insolation areas are the poles, which correspond to smaller areas displaying, supposedly, only small variations in topography (at least on the pole imaged by Voyager 2; Schenk et al. 2021). As a result, on Triton, when N_2 condenses at the winter pole, it likely does so covering the entire polar region, leaving no space for exposed CH_4-rich ice deposits over large areas (Bertrand et al. 2022). It could still be possible that a lag deposit of CH_4-rich ice forms at the summer pole as N_2 sublimes, or that previously buried CH_4-rich ice deposits become exposed if the entire polar N_2 ice reservoir sublimed. However N_2 condensation in the winter season would then likely bury CH_4-rich ice again, similarly to the H_2O and CO_2 ice caps on Mars.

The question on the absence of extended CH_4-rich ice deposits on Triton's surface remains open. It is not clear whether it is due to the mechanism described above or if other processes dominate, such as a depletion in the total CH_4 abundance driven by chemistry of the early solar system, internal chemistry or capture history. Current volatile transport models of Triton are not sensitive to the total abundance of CH_4 ice because, in the simulations, any initial CH_4 ice reservoir would migrate to the cold poles and always remain there, buried by N_2 ice deposits and therefore not interacting with the atmosphere. However, these models only treat N_2 and CH_4 as almost pure ice, and do not treat the case of intermediate mixtures, which may play an important role.

5.5.2 Formation of Glaciers and Thick Ice Deposits

New Horizons revealed numerous N_2-rich ice deposits on Pluto, including the prominent ice sheet lying in Sputnik Planitia but also small patches in the eastern side of Tombaugh Regio, in the depressions and valleys between the high altitude "Bladed Terrain" deposits (Moore et al. 2017), and more generally at the bottom of deep craters in the mid-latitudinal and equatorial regions (Grundy et al. 2016, Schmitt et al. 2017). Recent N_2 glacial flow activity is observed, and mostly in Sputnik Planitia; cellular patterns are seen in Sputnik's center and are indicative of active solid-state convection (McKinnon et al. 2016, Trowbridge et al. 2016), and flow advection is seen at the shorelines of Sputnik and glaciers are flowing from the neighboring highlands into eastern Sputnik Planitia (Howard et al. 2017a). Several models have been able to reproduce and explain these observations.

First, in the early-Pluto phase, Sputnik basin, unfilled post-formation, is thought to have represented a locally negative gravity anomaly, which became positive through the combination of N_2 ice infill (Johnson et al. 2021b; see review in Young et al. 2021) and other compensation to the basin structure (such as ocean uplift and/or an ejecta blanket, Keane et al. 2016; Nimmo et al. 2016). This change likely drove the initial position of Sputnik to more closely align with the Pluto-Charon tidal axis, thus inducing a reorientation (true polar wander) of Pluto's surface almost certainly associated with tectonic activity (Keane et al. 2016; Nimmo et al. 2016; McGovern et al. 2021). Johnson et al. (2021a) used a volatile transport model coupled to obliquity cycles, true polar wander, and deformation of the underlying basin and surrounding area as the ice sheet grows in mass, to understand how Sputnik Planitia filled with N_2 ice and reorient to its current position. They show that the basin fills with N_2-rich ice because of the effect of enhanced N_2 condensation in topographic depressions (see Section 5.5.1), but only if the basin is initially located at a low enough latitude ($<50°N$) to overcome the enhanced polar insolation. They estimate the infilling timescale of 8–10 Myr if Sputnik Planitia formed when Pluto was in its current orbit, or much faster ($\sim$2 Myr) if it formed when Pluto's semimajor axis was at 20 au. A timescale of $\sim$10 Myr was also found by Bertrand et al. (2018) for a basin at its current location and for current insolation and atmospheric conditions. Johnson et al. (2021b) also constrain the basin depth to be $\sim$2.5–3 km initially, and 3–4 km after deformation, with a mean N_2 ice thickness of 1–2 km. This is consistent with the modeling of N_2 ice convection cells, which suggests a depth of 3–10 km for the Sputnik basin (McKinnon et al. 2016; Trowbridge et al. 2016). The renewal of the surface by a typical convection cell is estimated to be $\sim$500,000 years ($\pm$ a factor of 2, McKinnon et al. 2016; Buhler & Ingersoll 2018).

Second, in the recent past, Sputnik Planitia has evolved along the Milankovitch cycles. Bertrand et al. (2018) found that it accumulated $\sim$0.3 km of N_2 ice from 1.3 to 0.1 Myr ago, and lost 1 km from its northern edge, and 0.15 km from its southern edge on average over the last 2.8 Myr obliquity cycle. In their model, this net transport of ice is balanced by glacial flow consistent with New Horizons observations. The modeled rise and fall of the Sputnik Planitia shoreline could have contributed to the observed erosion of the water-ice border. Bertrand et al. (2018) estimated it to cover 200 m in altitude over the obliquity cycles (they used a static and 8–10 km deep basin at Sputnik's center) while Johnson et al. (2021b) obtained 10–20 m only, which is sufficient to change the pole position by 0.1°. Finally, the thick N_2-rich ice deposits observed outside Sputnik Planitia in local equatorial depressions are estimated to be very stable over the current obliquity cycles, as the averaged insolation is minimal near the equator (Bertrand et al. 2018).

Overall, based on these modeling studies, the current global inventory of N_2 ice at Pluto's surface (excluding any subsurface reservoirs, and loss from escape rates) could be about 50–500 m in global equivalent layer (Bertrand et al. 2018; Johnson et al. 2021b; Glein & Waite 2018).

On Triton, the relatively coarse spatial resolution of Voyager 2 images and the lack of topography data in the southern hemisphere (assumed to be made of N_2-rich ice) do not allow one to determine with confidence the order of magnitude of the thickness of the bright southern deposits and whether glacial flow of N_2 ice has occurred or not. However, the sharp dichotomy between the southern and northern terrains suggests a topographic barrier (e.g., scarps) of ~100 m to several 100 m (Bertrand et al. 2022), favoring permanent and thick N_2 ice deposits rather than seasonal and thin. In addition, volatile transport simulations performed with low global N_2 reservoirs (< 1 m) show that most of Triton's N_2 inventory migrates and remains confined to the poles; however, the modeled cap in 1989 in these simulations does not extend to the equatorial regions as observed (e.g., Stansberry et al. 1990; Spencer 1990; Hansen & Paige 1992). On the contrary, simulations performed with larger global N_2 reservoirs with viscous flow of N_2 ice (Brown & Kirk 1994; Bertrand et al. 2022) obtain a cap in 1989 extending to the equatorial regions, with N_2 ice flowing from the poles toward low latitudes and balancing the net sublimation–condensation flow. Brown & Kirk (1994) show that an anisotropic internal heat flow could also produce permanent caps of considerable latitudinal extent. Recent volatile transport simulations of Triton, performed over a large range of model parameters, and constrained by several observations, suggest a global N_2 reservoir of 200–400 m, similar to the estimates on Pluto (Bertrand et al. 2022). Their modeled permanent cap quickly reaches a steady state after ~1 Myr and then remains relatively stable over time, with 1 km of N_2 ice at the south pole and 100 m in the equatorial regions. This is consistent with the timescale for viscous flow relaxation over a relatively flat bedrock of a layer of 100 m of N_2 ice, and a characteristic length scale of 1000 km (Umurhan et al. 2017, see their Figure 9(b)).

Voyager 2 images did not reveal terrains resembling a volatile ice cap in the northern hemisphere outside the polar night (southward of 45°N) in 1989. Therefore, it remains unclear whether a northern cap exists on Triton, but if it does, it must be smaller than the southern cap, which extends up to ~15°S. Several mechanisms have been proposed to explain this asymmetry between the northern and southern caps, including differences in internal heat flux (Brown & Kirk 1994), surface ice albedo

(Moore & Spencer 1990; Spencer & Moore 1992; Bertrand et al. 2022), topography (Trafton et al. 1998; Bertrand et al. 2022), and changes in radiative properties of N_2 ice as it goes through the alpha-beta phase transition (Eluszkiewicz 1991 Duxbury & Brown 1993; Tryka et al. 1993). Pole-to-pole difference in ice contamination by dark material and geysers, sensible heat flux from the atmosphere coupled to positive feedback on the surface (shown to have a significant effect on Pluto's ice; Bertrand et al. 2020a) could also help maintaining such an asymmetry.

As on Pluto, the mass of the southern cap of volatile ice could correspond to a positive mass anomaly and cause polar wander on Triton (Rubincam 2003; Hu et al. 2017). However, in this case, models predict that the pole should wander about the tidal bulge (Neptune-Triton) axis, an event expected to make the leading/trailing side crater distribution different. However, no such asymmetry is observed. Reasons could be that (1) Triton's surface volatile inventory is too low to permit wander, (2) the mantle at the southern pole corresponds to a negative mass, thus balancing the mass of the volatile ice cap, (3) the mantle viscosity might be low, so that any uncompensated cap load might be expected to wander toward the tidal bulge axis (Rubincam 2003; Hu et al. 2017).

CH_4-rich glaciers have been observed on Pluto, but seem to be absent on Triton. On Pluto, the major CH_4-rich deposits include the massive Bladed Terrain at the equator and several ice mantles at mid-northern latitudes (Howard et al. 2017a; Moore et al. 2018). Climate models have been able to relate the latitudinal extension of the Bladed Terrain to the Milankovitch parameters history of Pluto (Bertrand et al. 2019), but overall it remains unclear which CH_4 reservoir formed first and how they evolved over longer timescales of several 100 Myr. The ~100 m tall bladed texture of the equatorial CH_4-rich deposits could have formed through sublimation of CH_4 ice over the last 10s of Myr (Moores et al. 2017; Moore et al. 2018), through condensation processes (Bertrand et al. 2020b) or through a coupling of both processes. Finally, the mid-latitude ice mantles display subsurface layering up to several kilometers thick that could be the signatures of past climate processes (Stern et al. 2017).

As reviewed in Section 5.5.1, current volatile transport models do not predict the formation of such broad permanent CH_4-rich deposits on Triton, even when a very large CH_4 reservoir is taken into account (Bertrand et al. 2022).

5.5.3 The Seasonal Cycle of Volatile Ices

The seasonal variations of insolation on Triton and Pluto drive significant changes in sublimation and condensation rates of the volatile ices and affect their atmospheric pressure and composition, as well as their volatile surface distribution. After the unambiguous discovery of Pluto's atmosphere in 1988 and the flyby of Triton by Voyager 2 in 1989, several volatile transport models emerged to simulate the seasonal migration of volatile ice and provide climate predictions on Pluto (Stansberry & Yelle 1999; Hansen & Paige 1996; Young 2013, Hansen et al. 2015; Olkin et al. 2015; Bertrand & Forget 2016; Young 2017; Johnson et al. 2021a; see review in Young et al. 2021) and Triton (Spencer 1990; Spencer & Moore 1992;

Hansen & Paige 1992; Brown & Kirk 1994; see review in Bertrand et al. 2022). These models remain very sensitive to unconstrained parameters, such as the initial ice distribution, bolometric Bond albedo, emissivity, and the seasonal thermal inertia of the subsurface (e.g., Hansen & Paige 1996; Young 2013). Most observational constraints for these free parameters come from stellar occultations (measures of surface pressures), atmospheric spectroscopy (gas abundances), surface spectroscopy (volatile ice surface distribution) and visible and thermal imagery (albedo, emissivity, temperatures). Apart from the flyby measurements, such key observations have been performed on Pluto and Triton over the last 30 Earth years using Earth-based telescopes. However, this period of observation only represents 12% of an entire Pluto year and 16% of a Triton annual-like cycle, and therefore a large degree of uncertainty remains on the volatile cycles and associated free parameters.

The seasonal cycle of nitrogen dominates the climate on Triton and Pluto. In order to illustrate how these cycles behave on Triton and Pluto, results from a volatile transport simulation of Pluto (from Bertrand et al. 2018) and Triton (from Bertrand et al. 2022) are compared in Figure 5.5 (surface pressure) and Figure 5.6 (condensation-sublimation rates) over seasonal timescales. These simulations were chosen because they were performed with the same model and because they match relatively well the available observational constraints on Pluto and Triton. However, note that other models exist and that other combinations of surface properties (albedos, emissivities, thermal inertias, surface ice distribution, etc.) are prone to lead to different results while remaining consistent with available observations (e.g., Johnson et al. 2021a; see review in Young et al. 2021). Nevertheless, to first order the following trends remain valid: (1) Transport of N_2 is predicted from the summer

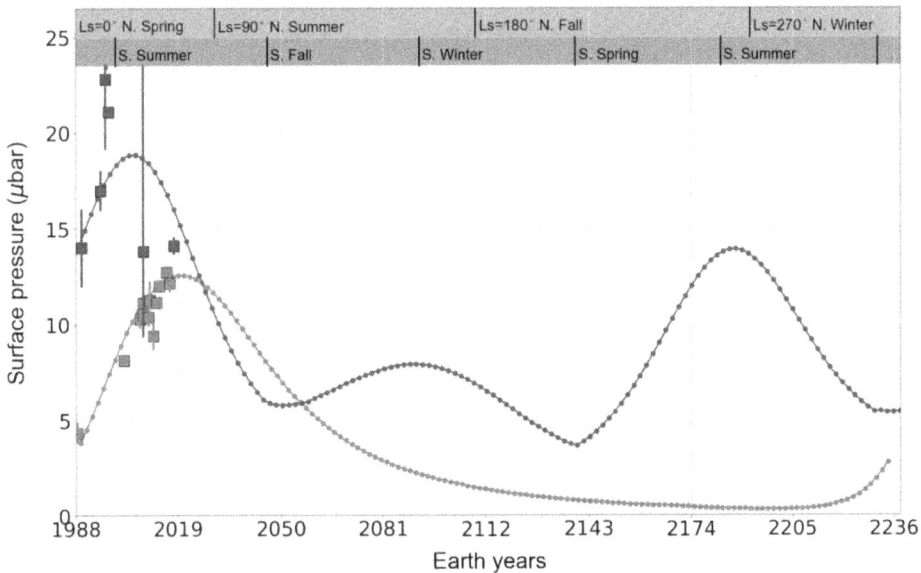

Figure 5.5. Seasonal evolution of surface pressure on Pluto (red) and Triton (blue) as observed by stellar occultation (squares) and modeled (solid line). See text for details.

Daily condensation rates on Pluto

Daily condensation rates on Triton

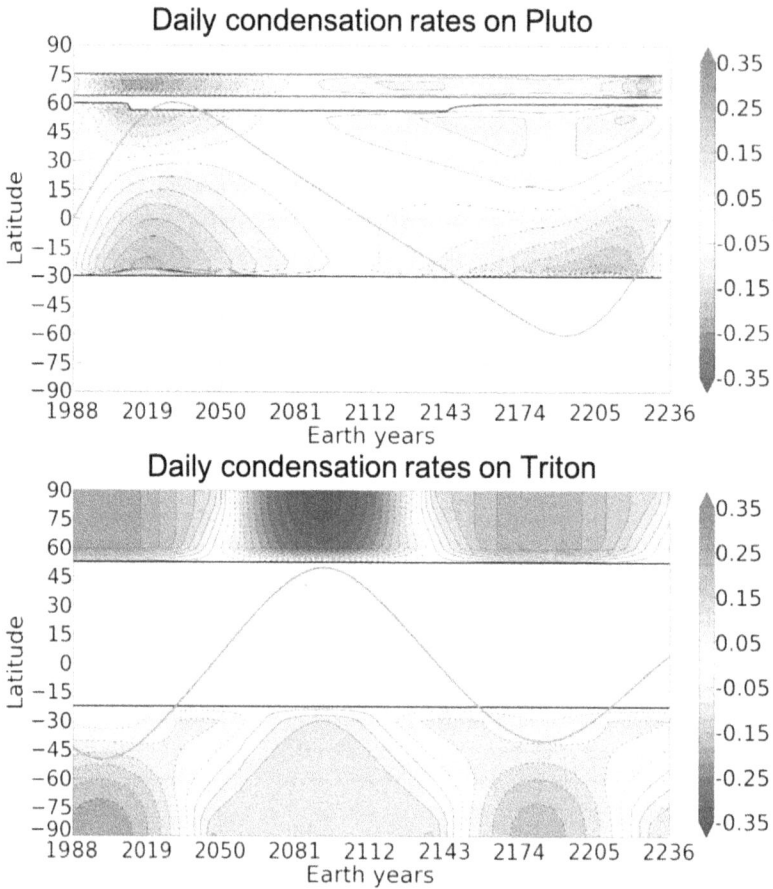

Figure 5.6. N_2 condensation (positive, red) and sublimation (negative, blue) rates on Pluto (top) and Triton (bottom) in mm per Pluto or Triton day as modeled in Bertrand et al. (2018, 2022). In the Pluto simulation, N_2 ice is placed in Sputnik Planitia (30°S-50°N) and at $\sim$70°N. In the Triton simulation, a large southern cap and a smaller northern cap of N_2 ice are permanent. The solid grey line shows the subsolar latitude.

hemisphere (dominated by N_2 ice sublimation) to the winter hemisphere (dominated by N_2 condensation) on Pluto and Triton. (2) Condensation-sublimation rates on Pluto and Triton are similar, typically about 0.05–0.3 mm per Pluto day or Triton day depending on location and season. The net variation of ice thickness is predicted to reach tens of mm over the seasonal timescale, and 100–300 m over 500,000 Earth years. (3) On both Pluto and Triton, the surface pressure is highly variable in time, with a factor of up to 1000 over the present-day seasonal cycles, but the atmosphere never collapses down to the exosphere or even non-global regimes. This is partly due to the effect of the high seasonal thermal inertia suggested for the ice and substrate, which enables storing the heat accumulated during summer and releasing it during winter, thus reducing condensation rates during winter (e.g., Stansberry & Yelle 1999; Olkin et al. 2015; Trafton 1990; Bertrand & Forget 2016). The approximate limit at which point the atmosphere is non-global is 0.06 µbar on Pluto (Johnson et al. 2021a; Spencer et al. 1997), and 0.09 μbar on Triton (owing to the larger size and gravity constant), and modeled surface pressures tend to be higher than this threshold in the best-case simulations. (4) CO abundances tend to follow N_2 as CO condenses where N_2-rich deposits are already present. As a result, the CO fraction in the atmosphere tends to remain constant with time. We detail these points below.

5.5.3.1 The Seasonal Cycles of N_2

On Pluto, stellar occultation measurements acquired between 1988 and 2016 revealed a threefold increase in surface pressure (Figure 5.5). These measurements, along with surface pressure and volatile surface distribution observations by New Horizons, brought strong constraints for the surface properties of N_2-rich ices in volatile transport models, which control the surface pressure. Models with N_2 ice confined to Sputnik Planitia (Bertrand & Forget 2016; Meza et al. 2019; Johnson et al. 2021a) have typical solutions with a subsurface thermal inertia of $\sim$700–800 tiu, a N_2 Bond albedo ~ 0.7, and emissivity $\varepsilon \sim 0.65$–0.9 (or, more generally, $(1 - A)/\varepsilon \sim 0.4$). As shown in Figure 5.5, in Pluto's current orbit, these models show that: (1) there is only one minimum and maximum every Pluto year, (2) the pressure maximum occurs near the time of maximum insolation on Sputnik Planitia, sometime between perihelion and northern solstice (close to the time of the New Horizons flyby in 2015), with a maximum pressure around 11 μbar (with a good match achieved by tuning the N_2 ice albedo and emissivity); (3) the pressure minimum occurs near northern Winter solstice, but is still large enough to maintain a global atmosphere throughout the year in the current orbit. Figure 5.6 shows the corresponding sublimation–condensation rates obtained over time with the model. For instance, in 2015 (northern spring), N_2 sublimation (resp. condensation) is predicted north (resp. south) of 15°N. Over a current Pluto year, a net transport of ice is predicted to reach tens of mm across Sputnik Planitia (see details in Bertrand et al. 2018, their Figure 6) but the sublimation–condensation during this period involves thicker layers of ice (by a factor 10–30, their Figure 9). These modeling results are consistent with the north-to-south increase in abundance and grain size of the N_2-rich ice retrieved from surface spectroscopic measurements of New Horizons (Protopapa et al. 2017; Schmitt et al. 2017).

On Triton, stellar occultation measurements suggest a pressure increase during the 1990–2000 period followed by a decrease, with a surface pressure in 2017 close to the Voyager 2 value (Marques-Oliveira et al. 2022). Volatile transport models have been able to reproduce this trend, and solutions best matching the available observations have a large and several 100 m-thick permanent southern cap of N_2 ice and surface properties close to those found for Pluto, with a soil thermal inertia of $\sim$500–2000 tiu, and $(1 - A)/\varepsilon \sim 0.25$–0.4. These models showed that the pressure maximum occurs near the time of maximum insolation on the southern N_2 ice cap near summer solstice ($\sim$2000–2005). Models by Bertrand et al. (2022) do not suggest a strong surge in surface pressure, as claimed from stellar occultations for the period 1995–1997 (e.g., Elliot et al. 2000), but instead a moderate increase with a peak around 15–19 μbar (Figure 5.5), although ice albedo feedback (not accounted for in their model) may have increased the peak amplitude during this period. Their model predicts a pressure maximum every solstice and a minimum every equinox, due to the presence of significant amounts of N_2 ice at both poles (sublimation during summer solstice, Figure 5.6), even when the northern N_2 ice deposits are seasonal and not permanent (see Figure 22 in Bertrand et al. 2022). In the simulation shown in Figure 5.6, the next maximum is predicted to occur near northern summer solstice ($\sim$2092) and to reach $\sim$8 μbar, which is less than the peak pressure during southern summer solstice because the simulated northern cap is much smaller than the southern cap (despite a similar insolation and a high subsolar latitude of $\sim$50°, Figure 5.6). On average over the period covering the two southern summer solstices of Figure 5.6, the corresponding sublimation–condensation rates are similar to those on Pluto, the thickness of ice involved is of the order 100–400 mm, and net variations of the order of tens of mm. Over the last 30 years, Triton's southern polar cap lost $\sim$0.3 m of N_2 ice by sublimation, while $\sim$0.1 m of ice deposited in the equatorial regions and mid-northern latitudes, according to the model.

Several observations support the fact that the southern cap is currently sublimating away, including (1) the local dark areas on the bright ice seen by Voyager 2, suggestive of erosion due to sublimation (Smith et al. 1989), (2) the tentative seasonal reduction (at 2-σ) seen in Triton's N_2 ice spectral absorption during the 2000–2009 period (Grundy et al. 2010), consistent with a textural reduction in optical path length in N_2 ice, (3) Triton's global color that became redder during the 1992–2004 period, possibly due to plume activity and deposition of dark reddish material over the southern cap, associated with the rapid sublimation of the cap that also concentrates at the surface non-volatile particles included in the ice (McEwen 1990; Buratti et al. 2011). In this context, the statistically significant increase in N_2 band area reported by Holler et al. (2016) is likely due to additional N_2-rich latitudes emerging from polar night as the subsolar latitude migrates northward; this also indicates that the viewing geometry is changing on a faster timescale than sublimation from the southern cap.

5.5.3.2 The Seasonal Cycles of CH_4 and CO

On Pluto, Bertrand & Forget (2016) showed that 0.3% CO in N_2-rich ice filling Sputnik Planitia (as suggested by observations; Merlin et al. 2015) leads to an atmospheric mean gas volume mixing ratio close to 0.04% in 2015 (and throughout the Pluto year), in good agreement with telescopic measurements (Lellouch et al. 2011, 2015). On Triton, Bertrand et al. (2022) used 0.04%–0.08% CO in N_2-rich ice (Quirico et al. 1999) and obtained an atmospheric mean gas volume mixing ratio of 0.006%–0.012% during the period 2000–2020, and slightly lower values for the next decades (0.005%–0.01%), consistent with the 2017 ALMA mm-observations (see Section 5.2.3).

The CH_4 seasonal cycles are more complex (see review in Young et al. 2021). On Pluto, New Horizons revealed that the northern mid-latitude plains (25°N–70° N) are covered in mixtures of N_2-rich and CH_4-rich ices following a latitudinal trend (Schmitt et al. 2017; Protopapa et al. 2017; Earle et al. 2018b) and suggestive of seasonal transport of CH_4 during the last decades (see details in Young et al. 2021). Sublimation of CH_4 in the northern hemisphere and transport toward southern latitudes is supported by simulations (Bertrand & Forget 2016; Bertrand et al. 2019), which reproduce, to first order, the spatial distribution of N_2 and CH_4 ices observed in 2015. These simulations highlight the extreme sensitivity of the CH_4 cycle to the CH_4-rich ice surface properties, such as albedo and emissivity, and to associated positive (or negative) feedback (Earle et al. 2018a; see Section 5.3.3.3). In particular, when the CH_4-rich ice is bright (typically $A > 0.65$), it gets cold enough in autumn and winter that N_2 condenses on it and forms seasonal or perennial N_2-rich deposits at equatorial and mid-latitudes, as observed (Bertrand et al. 2019). Seasonal or perennial CH_4 deposits may even form on the dark, low-latitude terrains (Protopapa et al. 2017; Earle et al. 2018b; Bertrand et al. 2020b). Overall, the high diversity of albedo observed for CH_4-rich terrains on Pluto suggests that numerous surface–atmosphere interactions take place and impact the CH_4-rich ices. As a result, the CH_4 gas volume mixing ratio is very sensitive to the surface properties, surface distribution and the interactions with the N_2 cycle. Nevertheless, Bertrand & Forget (2016) predicted a range in the atmospheric CH_4 mixing ratio from 0.1% to 5% over Pluto's current year. During northern winter, a higher mixing ratio is expected as most post-flyby models anticipate a much thinner atmosphere.

The CH_4 cycle remains to be investigated in detail on Triton. In the simulations of Bertrand et al. (2022), CH_4 ice is trapped into N_2-rich ice and no CH_4-rich ice forms. In two alternative simulations, small CH_4-rich ice patches were artificially placed at the cap edge (i.e., in the equatorial regions) and at the south pole, respectively, to evaluate their impact on the CH_4 atmospheric abundance. While the simulation with CH_4-rich ice at the cap edge produces a CH_4 partial pressure of 1–5×10^{-3} μbar that remains relatively constant during the period 1980–2030, the simulation with CH_4-rich ice at the south pole produces a strong increase in CH_4 partial pressure until 2005, followed by a decrease as the south pole exits polar day. However, as on Pluto, these abundances remain sensitive to the surface properties and as a result both scenarios could reconcile the observations.

5.6 Conclusions and Remaining Questions

Pluto and Triton form a specific class of terrestrial atmospheres. They are both cold objects with a tenuous and condensable N_2 atmosphere, with traces of CH_4 and CO. As a result, they share the same fundamental processes of surface atmosphere interactions (e.g., volatile cycles, complex feedback on ice properties, impact of the topography, etc.), atmosphere dynamics (N_2 condensation-sublimation flows as the major driver of the circulation, katabatic winds, etc.) and climate physics (long radiative timescales, impact of atmospheric waves and of a photochemical haze, etc.).

However, the landscapes of Pluto and Triton are markedly different. Triton is much flatter and much brighter than Pluto and does not display any dark tholin-covered terrains, while its H_2O-CO_2 bedrock is thought to be exposed over a large part of its surface. In addition, the main reservoir of N_2-rich ice is assumed to be located at Triton's south pole (versus Sputnik Planitia in the equatorial regions on

Pluto) and wide CH_4-rich deposits seem to be absent (versus the massive CH_4-rich deposits covering a large part of Pluto's surface).

There are two main reasons thought to combine together to explain these differences. First, Triton is tidally activated by Neptune while Pluto is not. The geological history (and by extension, interior thermal history) of Pluto and Triton differ (e.g., McKinnon 1984, Agnor & Hamilton 2006) and Triton may have undergone and may still undergo subsequent resurfacing processes (e.g., cryovolcanism induced by strong obliquity tides; Nimmo & Spencer 2015). This intense internal activity on Triton explains the flatness of its topography, and could also impact the volatile cycles at the surface, although the exact mechanisms are not well known nor explored in detail yet. Second, as shown in Section 5.5.1, Triton's mean obliquity with respect to the Sun is relatively low, and as a result the poles are on a yearly-averaged basis much colder than the equatorial regions and therefore accumulate volatile ices, while the high obliquity on Pluto leads to volatile ice accumulation in the equatorial regions (Bertrand et al. 2018; Earle et al. 2018a), at low altitude for N_2 ice (i.e., in Sputnik Planitia) and at altitude for CH_4 (e.g., the Bladed Terrains). This difference in obliquity between Pluto and Triton may trigger a runaway positive feedback; the location of the coldest regions may favor (in the case of Pluto) or prevent (in the case of Triton) the formation of extended CH_4-rich ice deposits, which in turn impacts the atmospheric methane abundance, and subsequently, the atmospheric temperatures, photochemical processes and haze formation, escape rates and climate at a global scale.

Many enigmas remain to be solved to further understand how Pluto's and Triton's surface and atmosphere work. Below is a non-exhaustive list of examples.

First, some surface features observed on Pluto could have formed through surface–atmosphere interactions, such as the sublimation pits and the CH_4-rich dunes in Sputnik Planitia and steep ridges in the Bladed Terrain deposits. Although several physical processes have been suggested to explain these landforms (e.g., Telfer et al. 2018; Moore et al. 2017; Howard et al. 2017a), the exact mechanisms and associated timescales need to be explored (and simulated) in detail. While the Bladed Terrain deposits seem to be unique in the Solar System, similarities can be found between the dune-like periodic bedforms and other icy bedforms on Earth and on Mars (e.g., Bordiec et al. 2020). It remains to be explored whether the processes at play on Earth and on Mars are universal and apply on Pluto and Triton.

Second, as Voyager 2 did not carry a near-IR spectrometer, the composition across Triton's surface remains poorly constrained. Energy balance and volatile transport models give hints of where the volatile ices should accumulate, while Earth-based spectroscopic measurements help constrain volatile ice variations over space and time. However, some of these observations suggest that the latitudes poleward of 34°S are depleted in volatile ice (e.g., Grundy et al. 2010; Holler et al. 2016), which is puzzling considering the Voyager 2 images and the fact that the southern pole is a cold trap for volatiles. New observations are therefore clearly needed to decipher the ice distribution on Triton's surface. Another unknown is the surface composition in the polar night of both objects (northern pole on Triton, southern pole on Pluto), which is key as any large-scale N_2 ice deposit present there would significantly contribute to drive the pressure cycle and atmospheric dynamics.

Third, it remains unclear how the complex feedbacks mentioned in Section 5.3.3 impact the volatile cycles of Pluto and Triton. In particular, it is likely that the volatile cycles are impacted by the internal activity and the geysers on Triton, and by ice contamination by haze particles on Pluto, but these effects are not included in models yet and have not been explored in detail.

Finally, Pluto and Triton's atmospheres present many major puzzles: (1) Their atmospheric thermal balance is not well understood yet. In particular, the cooling agent in Pluto's atmosphere (haze particles?) remains to be clearly identified, explored in detail with 3D climate models including haze and hydrocarbon ice microphysics, and tested against thermal emission measurements of the atmosphere (e.g., from Spitzer and JWST data; Zhang et al. 2017; Lavvas et al. 2021; Wan et al. 2023). It also remains unclear how the haze microphysics on Triton, and its radiative impact, compare with those on Pluto. (2) Recent observations of Pluto's atmosphere with ALMA suggest a strong thermal gradient ($\sim$7 K) in latitude (Lellouch et al. 2022), which remains unexplained and inconsistent with the long atmospheric radiative timescale predicted by models. Perhaps this timescale is shortened by the radiative impact of the aerosols, but this remains to be explored in detail. (3) A 3 km deep and 8 km deep cold layer were observed by New Horizons on Pluto and by Voyager 2 on Triton, respectively (Soderblom et al. 1990; Hinson et al. 2017). Although some mechanisms have been identified to explain these cold layers, GCM simulations do not yet reproduce them with the observed atmospheric depth, which remains puzzling (Forget et al. 2017). (4) Orographic waves and sublimation tides are suspected to drive the disturbances observed in the atmosphere of both Pluto and Triton, but these processes remain poorly understood and not modeled in detail. In addition, the impact of these waves on their climate (they can decelerate the large-scale flow and produce local turbulence and mixing when they break; this is well known on the Earth and Mars) has not yet been explored.

Significant research therefore remains to be performed to fully comprehend the surface–atmospheric processes occurring at Pluto and Triton. This calls for (1) new missions toward Triton, already subject of several space mission concepts (e.g., Persephone, Trident) and identified as the highest priority candidate ocean world by the NASA OPAG panel, and toward Pluto to answer the key questions about its atmosphere (mentioned above) and about the processes shaping its surface (see Chapter 9 (2) new observations with Earth-based and space-based telescopes (e.g. JWST, 39-m ELT, etc.) to monitor the atmosphere and surface evolution (see Chapter 9 (3) new experiments (see Chapter 10), and (4) the development of improved numerical models. In particular two kinds of modeling challenges can be identified. On the one hand, it seems that many dynamical processes occur at relatively small scales, creating microclimate on slopes, at the bottom of craters, or at the top of mountains. Much could be learned from very high resolution GCM simulations or limited area mesoscale (kilometers scale) or microscale (meters) simulations. For instance, such high resolution models could simulate orographic waves and sublimation tides with unprecedented details, the formation of icy dunes such as the Bladed Terrains, and the mesoscale effects impacting the near-surface temperature profile, thus shedding light on the puzzles mentioned above for Pluto's and Triton's surface and atmosphere. On the other hand, as detailed below, the timescale that must be represented to model the seasonal cycles (hundreds to thousands of Earth years) and even more to model the formation of perennial glaciers or ice caps (over millions of years) are orders of magnitude too long to be simulated by GCMs, especially if one wishes to simulate the surface at kilometer resolution. Therefore, a new generation of 3D GCM and volatile transport models including surface–atmosphere interactions over thousands to millions of years will have to be developed to fully interpret spectacular glacial landscapes of Pluto and Triton.

References

Agnor, C. B., & Hamilton, D. P. 2006, Natur, 441, 192

Arimatsu, K., Hashimoto, G. L., Kagitani, M., et al. 2020, A&A, 638, L5

Attree, N., Kaufmann, E., & Hagermann, A. 2021, Natur, 540, 86

Bertrand, T., & Forget, F. 2016, Natur, 540, 86

Bertrand, T., Forget, F., Umurhan, O. M., et al. 2018, Icar, 309, 277

Bertrand, T., Forget, F., Umurhan, O. M., et al. 2019, Icar, 329, 148

Bertrand, T., Forget, F., White, O., et al. 2020a, JGRE, 125, e2019JE006120

Bertrand, T., Forget, F., Schmitt, B., White, O. L., & Grundy, W. M. 2020b, NatCo, 11, 5056

Bertrand, T., Lellouch, E., Holler, B. J., et al. 2022, Icar, 373, 114764

Binzel, R. P., Earle, A. M., Buie, M. W., et al. 2017, Icar, 287, 30

Bordiec, M., Carpy, S., Bourgeois, O., et al. 2020, ESRv, 211, 103350

Brown, R. H., & Kirk, R. L. 1994, JGRE, 99, 1965

Brown, G. N., & Ziegler, W. T. 1980, Advances in Cryogenic Engineering, Proc. of the Cryogenic Engineering Conf. Vol 25 (Boston, MA: Springer) 662

Buhler, P. B., & Ingersoll, A. P. 2018, Icar, 300, 327

Buratti, B. J., Bauer, J. M., Hicks, M. D., et al. 2011, Icar, 212, 835

Buratti, B. J., Hofgartner, J. D., Hicks, M. D., et al. 2017, Icar, 287, 207

Cheng, A. F., Summers, M. E., Gladstone, G. R., et al. 2017, Icar, 290, 112

Croft, S. K., Kargel, J. S., Kirk, R. L., et al. 1995, Neptune and Triton (Tucson, AZ: Univ. Arizona Press) 879

Cruikshank, D. P., Roush, T. L., Owen, T. C., et al. 1993, Sci, 261, 742

Dias-Oliveira, A., Sicardy, B., Lellouch, E., et al. 2015, ApJ, 811, 53

Diniega, S., Hansen, C. J., McElwaine, J. N., et al. 2013, Icar, 225, 526

Diniega, S., Bramson, A. M., Buratti, B., et al. 2021, Geomo, 380, 107627

Dobrovolskis, A. R., Peale, S. J., & Harris, A. W. 1997, Pluto and Charon (Tucson, AZ: Univ. Arizona Press) 159

Dobrovolskis, A. R., & Harris, A. W. 1983, Icar, 55, 231

Duxbury, N. S., & Brown, R. H. 1993, Sci, 261, 748

Earle, A. M., Binzel, R. P., Young, L. A., et al. 2017, Icar, 287, 37

Earle, A. M., Binzel, R. P., Young, L. A., et al. 2018a, Icar, 303, 1

Earle, A. M., Grundy, W., Howett, C. J. A., et al. 2018b, Icar, 314, 195

Elliot, J. L., Strobel, D. F., Zhu, X., et al. 2000, Icar, 143, 425

Eluszkiewicz, J. 1991, JGRA, 96, 19217

Fray, N., & Schmitt, B. 2009, P&SS, 57, 2053

Forget, F., Bertrand, T., Vangvichith, M., et al. 2017, Icar, 287, 54

Forget, F., Bertrand, T., Hinson, D., & Toigo, A. 2021, Pluto System After New Horizons (Tucson, AZ: Univ. Arizona Press) 297

Gabasova, L. R., Schmitt, B., Grundy, W., Bertrand, T., Olkin, C. B., & Spencer, J. R.New Horizons Composition Team 2021, Icar, 356, 113833

Gladstone, G. R., Stern, S. A., Ennico, K., et al. 2016, Sci, 351, aad8866

Glein, C. R., & Waite, J. H. 2018, Icar, 313, 79

Grundy, W. M., Binzel, R. P., Buratti, B. J., Cook, J. C., Cruikshank, D. P., & Dalle Ore, C. M. New Horizons Science Team 2016, Sci, 351, aad9189

Grundy, W. M., Schmitt, B., & Quirico, E. 2002, Icar, 155, 486

Grundy, W. M., Young, L. A., Stansberry, J. A., et al. 2010, Icar, 205, 594

Grundy, W. M., Bertrand, T., Binzel, R. P., et al. 2018, Icar, 314, 232

Gurrola, E. M. 1995, PhD thesis, (Stanford Univ.)

Gurwell, M., Lellouch, E., Butler, B., et al. 2019, in EPSC-DPS Joint Meeting, 806

Hansen, C. J., McEwen, A. S., Ingersoll, A. P., & Terrile, R. J. 1990, Sci, 250, 421

Hansen, C. J., & Paige, D. A. 1992, Icar, 99, 273

Hansen, C. J., & Paige, D. A. 1996, Icar, 120, 247

Hansen, C. J., Paige, D. A., & Young, L. A. 2015, Icar, 246, 183

Hinson, D. P., Linscott, I. R., Young, L. A., et al. 2017, Icar, 290, 96

Hofgartner, J. D., Birch, S. P., Castillo, J., et al. 2022, Icar, 375, 114835

Holler, B. J., Young, L. A., Grundy, W. M., & Olkin, C. B. 2016, Icar, 267, 255

Holler, B. J., Yanez, M. D., Protopapa, S., Young, L.A., Verbiscer, A.J., Chanover, N.J., & Grundy, W.M. 2022, Icar, 373, 114729

Howard, A. D., Moore, J. M., Umurhan, O. M., White, O. L., Anderson, R. S., & McKinnon, W. B. New Horizons Science Team 2017a, Icar, 287, 287

Howard, A. D., Moore, J. M., White, O. L., et al. 2017b, Icar, 293, 218

Hu, H., van der Wal, W., & Vermeersen, L. L. A. 2017, JGRE, 122, 228

Ingersoll, A. P. 1990, Natur, 344, 315

Jacobs, A. D., Summers, M. E., Cheng, A. F., et al. 2021, Icar, 356, 113825

Johnson, R. E. 1989, GeoRL, 16, 1233

Johnson, P. E., Young, L. A., Protopapa, S., et al. 2021a, Icar, 356, 114070

Johnson, P. E., Keane, J. T., Young, L. A., & Matsuyama, I. 2021b, PSJ, 2, 194

Keane, J. T., Matsuyama, I., Kamata, S., & Steckloff, J. K. 2016, Natur, 540, 90

Kirk, R. L., Soderblom, L. A., Brown, R. H., Kieffer, S. W., & Kargel, J. S. 1995, Neptune and Triton (Tucson, AZ: Univ. Arizona Press) 949

Langevin, Y., Bibring, J. P., Montmessin, F., et al. 2007, JGRE, 112,

Laskar, J., & Robutel, P. 1993, Natur, 361, 608

Lavvas, P., Lellouch, E., Strobel, D. F., et al. 2021, NatAs, 5, 289

Lellouch, E., Sicardy, B., De Bergh, C., et al. 2009, A&A, 495, L17

Lellouch, E., de Bergh, C., Sicardy, B., Ferron, S., & Käufl, H. U. 2010, A&A, 512, L8

Lellouch, E., de Bergh, C., Sicardy, B., Käufl, H. U., & Smette, A. 2011, A&A, 530, L4

Lellouch, E., De Bergh, C., Sicardy, B., et al. 2015, Icar, 246, 268

Lellouch, E., Gurwell, M., Butler, B., et al. 2017, Icar, 286, 289

Lellouch, E., Butler, B., Moreno, R., et al. 2022, Icar, 372, 114722

Leyrat, C., Lorenz, R. D., & Le Gall, A. 2016, Icar, 268, 50

Marques-Oliveira, J., Sicardy, B., Gomes-Júnior, A. R., Ortiz, J. L., Strobel, D. F., Bertrand, T., & Erpelding, D. 2022, A&A, 659, A136

McEwen, A. S. 1990, GeoRL, 17, 1765

McGovern, P. J., White, O. L., & Schenk, P. M. 2021, JGRE, 126, e2021JE006964

McKinnon, W. B. 1984, Natur, 311, 355

McKinnon, W. B., Nimmo, F., Wong, T., et al. 2016, Natur, 534, 82

Merlin, F. 2015, A&A, 582, A39

Merlin, F., Lellouch, E., Quirico, E., & Schmitt, B. 2018, Icar, 314, 274

Meza, E., Sicardy, B., Assafin, M., et al. 2019, A&A, 625, A42

Moore, J. M., & Spencer, J. R. 1990, GeoRL, 17, 1757

Moore, J. M., Howard, A. D., Umurhan, O. M., et al. 2017, Icar, 287, 320

Moore, J. M., Howard, A. D., Umurhan, O. M., et al. 2018, Icar, 300, 129

Moores, J. E., Smith, C. L., Toigo, A. D., & Guzewich, S. D. 2017, Natur, 541, 188

Nimmo, F., & Spencer, J. R. 2015, Icar, 246, 2

Nimmo, F., Hamilton, D. P., McKinnon, W. B., et al. 2016, Natur, 540, 94

Oliveira, J. M., Sicardy, B., Gomes-Júnior, A. R., et al. 2022, A&A, 659, A136

Olkin, C. B., Young, L. A., Borncamp, D., et al. 2015, Icar, 246, 220

Piqueux, S., Byrne, S., & Richardson, M. I. 2003, JGRE, 108,

Predel, B., Hoch, M., & Pool, M. J. 2013, Phase Diagrams and Heterogeneous Equilibria: A Practical Introduction (Berlin: Springer Science & Business Media)

Prokhvatilov, A. I., & Yantsevich, L. D. 1983, SJLTP, 9, 94

Protopapa, S., Grundy, W. M., Tegler, S. C., & Bergonio, J. M. 2015, Icar, 253, 179

Protopapa, S., Grundy, W. M., Reuter, D. C., et al. 2017, Icar, 287, 218

Quirico, E., & Schmitt, B. 1997a, Icar, 128, 181

Quirico, E., & Schmitt, B. 1997b, Icar, 127, 354

Quirico, E., Doute, S., Schmitt, B., de Bergh, C., Cruikshank, D. P., Owen, T. C., & Roush, T. L. 1999, Icar, 139, 159

Rubincam, D. P. 2003, Icar, 163, 469

Schenk, P., & Jackson, M. P. A. 1993, Geo, 21, 299

Schenk, P. M., & Zahnle, K. 2007, Icar, 192, 135

Schenk, P. M., Beyer, R. A., McKinnon, W. B., et al. 2018, Icar, 314, 400

Schenk, P. M., Beddingfield, C. B., Bertrand, T., et al. 2021, RemS, 13, 3476

Schmitt, B., Philippe, S., Grundy, W. M., et al. 2017, Icar, 287, 229

Sicardy, B., Widemann, T., Lellouch, E., et al. 2003, Natur, 424, 168

Sicardy, B., Ashok, N. M., Tej, A., et al. 2021, ApL, 923, L31

Singer, K. N., White, O. L., Schmitt, B., et al. 2022, NatCo, 13, 1542

Smith, B. A., Soderblom, L. A., Banfield, D., et al. 1989, Sci, 246, 1422

Soderblom, L. A., Kieffer, S. W., Becker, T. L., et al. 1990, Sci, 250, 410

Spencer, J. R. 1990, GeoRL, 17, 1769

Spencer, J. R., & Moore, J. M. 1992, Icar, 99, 261

Stansberry, J. A., Lunine, J. I., Porco, C. C., & McEwen, A. S. 1990, GeoRL, 17, 1773

Stansberry, J. A., Pisano, D. J., & Yelle, R. V. 1996, P&SS, 44, 945

Stansberry, J. A., & Yelle, R. V. 1999, Icar, 141, 299

Stern, S. A., Bagenal, F., Ennico, K., et al. 2015, Sci, 350, aad1815

Stern, S. A., Binzel, R. P., Earle, A. M., et al. 2017, Icar, 287, 47

Stevens, M. H., Strobel, D. F., Summers, M. E., & Yelle, R. V. 1992, GeoRL, 19, 669

Strobel, D. F., & Summers, M. E. 1995, Neptune and Triton (Tucson, AZ: Univ. Arizona Press) 1107

Strobel, D. F., Zhu, X., Summers, M. E., & Stevens, M. H. 1996, Icar, 120, 266

Strobel, D. F., & Zhu, X. 2017, Icar, 291, 55

Strom, R. G., Croft, S. K., & Boyce, J. M. 1990, Sci, 250, 437

Sulcanese, D., Cioria, C., Kokin, O., et al. 2023, Icar, 392, 115368

Tan, S. P., & Kargel, J. S. 2018, MNRAS, 474, 4254

Tan, S. P. 2022, MNRAS, 515, 1690

Telfer, M. W., Parteli, E. J., Radebaugh, J., et al. 2018, Sci, 360, 992

Tegler, S. C., Stufflebeam, T. D., Grundy, W. M., et al. 2019, AJ, 158, 17

Toigo, A. D., Gierasch, P. J., Sicardy, B., & Lellouch, E. 2010, Icar, 208, 402

Toigo, A. D., French, R. G., Gierasch, P. J., et al. 2015, Icar, 254, 306

Trafton, L. M. 1990, ApJ, 359, 512

Trafton, L. M. 1984, Icar, 58, 312

Trafton, L. M., Matson, D. L., & Stansberry, J. A. 1998, in Solar System Ices: Based on Reviews Presented at the International Symposium, (Dordrecht: Springer) 773

Trafton, L. M. 2015, Icar, 246, 197

Trafton, L. M., & Stansberry, J. A. 2015, in AAS/Division for Planetary Sciences Meeting Abstracts 47, 210

Trowbridge, A. J., Melosh, H. J., Steckloff, J. K., & Freed, A. M. 2016, Natur, 534, 79

Tryka, K. A., Brown, R. H., Anicich, V., Cruikshank, D. P., & Owen, T. C. 1993, Sci, 261, 751

Umurhan, O. M., Howard, A. D., Moore, J. M., et al. 2017, Icar, 287, 301

Umurhan, O. M., Ahrens, C. J., & Chevrier, V. F. 2021, The Pluto System After New Horizons (Tucson, AZ: Univ. Arizona Press) 195

Vangvichith, M., Forget, F., Wordsworth, R., & Millour, E. 2010, in AAS/Division for Planetary Sciences Meeting Abstracts 42, 6

Vangvichith, M. 2013, Doctoral dissertation, Ecole Polytechnique

Wan, L., Zhang, X., & Bertrand, T. 2021, ApJ, 922, 244

Wan, L., Zhang, X., & Hofgartner, J. D. 2023, ApJ, 955, 108

White, O. L., Moore, J. M., McKinnon, W. B., et al. 2017, Icar, 287, 261

White, O. L., Moore, J. M., Howard, A. D., et al. 2019, NatAs, 3, 62

Yelle, R. V., & Lunine, J. I. 1989, Natur, 339, 288

Yelle, R. V., Lunine, J. I., & Hunten, D. M. 1991, Icar, 89, 347

Young, E. F. 1993, Doctoral dissertation, Massachusetts Institute of Technology

Young, L. A., Elliot, J. L., Tokunaga, A., de Bergh, C., & Owen, T. 1997, Icar, 127, 258

Young, L. A. 2012, Icar, 221, 80

Young, L. A. 2013, ApL, 766, L22

Young, L. A., Kammer, J. A., Steffl, A. J., et al. 2018, Icar, 300, 174

Young, L. A., Bertrand, T., Trafton, L. M., et al. 2021, Pluto's Volatile and Climate Cycles on Short and Long Timescales (Tucson, AZ: Univ. Arizona Press) 321

Zhang, X., Strobel, D. F., & Imanaka, H. 2017, Natur, 551, 352

Chapter 6

Clouds and Hazes in the Atmospheres of Triton and Pluto

Peter Gao and Kazumasa Ohno

6.1 Introduction

Clouds and hazes are abundant in the thin ($\sim$10 μbar) and cold ($\leqslant$40 K) atmospheres of Triton and Pluto, where they are thought to be produced by interactions between atmospheric gases and ultraviolet (UV) photons from the Sun and those scattered by the local interstellar medium. These interactions lead to a rich network of chemical reactions that produces higher order hydrocarbons and nitriles that condense out to form ice clouds, and ultimately complex haze particles that rain down onto the surface that impact the atmospheric thermal structure, gas chemistry, and surface evolution. In this chapter, we will review the observational evidence for clouds and hazes in the atmospheres of Triton and Pluto and theoretical interpretations thereof, and the emerging set of experiments aiming to produce Triton and Pluto clouds and hazes in the lab to learn about them in detail.

6.1.1 A Note on Nomenclature

It is important to give formal definitions to "clouds" and "hazes" to avoid confusion, as definitions for these terms vary between Earth science and planetary science literature. Here, we will follow the latter and use definitions related to provenance: Cloud particles form through equilibrium condensation reactions, where the ambient temperature and condensate vapor abundance are such that condensation is energetically favorable and occurs spontaneously. In contrast, haze particles form through disequilibrium processes, typically photochemical reactions, where injection of energy (e.g., UV photons) breaks apart atmospheric gas molecules and leads to successive reactions that gradually build the haze particles. As a result, cloud particles are typically volatile, while haze particles are typically not. A complication of these definitions involves particles that form from condensation of gases that were themselves produced photochemically; while we will refer to these

doi:10.1088/2514-3433/ad5278ch6   6-1

structures as clouds here, they have also been referred to as hazes in the literature. Finally, we will use "aerosols" as a catch-all term for any particulates suspended in an atmosphere, i.e., both clouds and hazes,[1] in cases where the provenance is uncertain.

6.2 Observations of Triton and Pluto Aerosols

6.2.1 Voyager 2 Observations of Triton Aerosols

Voyager 2's historic flyby of the Neptune system on August 25, 1989 revealed the existence of a N_2-dominated atmosphere on Triton with surface pressure and temperature of 16 ± 3 μbar and 38^{+3}_{-4} K, respectively, consistent with conditions of vapor equilibrium with surface N_2 ice (Broadfoot et al. 1989; Tyler et al. 1989; Conrath et al. 1989). The observed near-surface mixing ratio of CH_4 (few to a few hundred ppm) was subsaturated by a factor of 30 and spatially variable with a scale height of 7–10 km, which is considerably smaller than expected from hydrostatic equilibrium and thus suggested the action of photochemistry (Broadfoot et al. 1989). Imaging observations uncovered two aerosol components: bright, discrete structures at 8 km and mostly poleward of 30°S (Figure 6.1) and a global aerosol layer that extends to ∼30 km at almost everywhere on Triton (Smith et al. 1989; Pollack et al. 1990).

Voyager 2 revealed the scattering properties of Triton's aerosols with photometric observations at visible wavelengths (Figure 6.2). The upturn in the scattered light brightness at large phase angles strongly indicates the presence of reflective aerosols. From the disk-averaged brightness, which predominantly traces the photometric

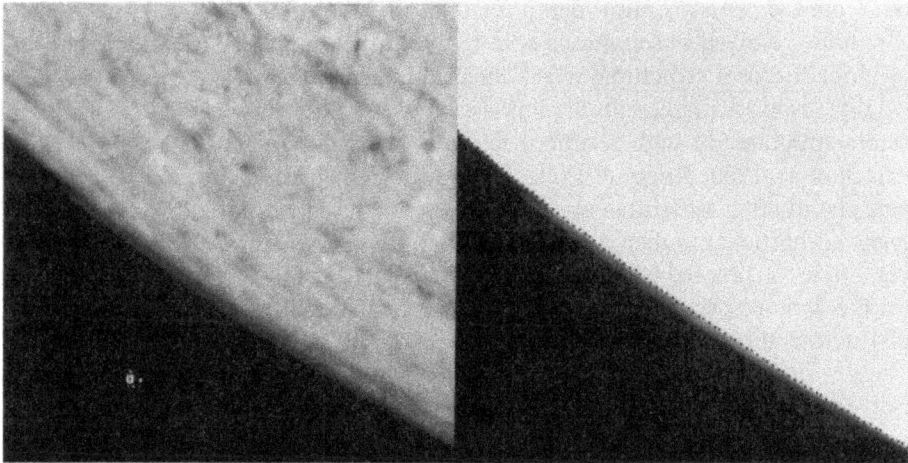

Figure 6.1. Images of aerosols over Triton's south polar cap taken at the west limb. The left panel highlights the surface features of Triton, while the right panel highlights the limb aerosols, with dark pixels marking Triton's limb. Figure taken from Smith et al. (1989). Adapted with permission from AAAS.

---

[1] This is also different from its definition in Earth literature.

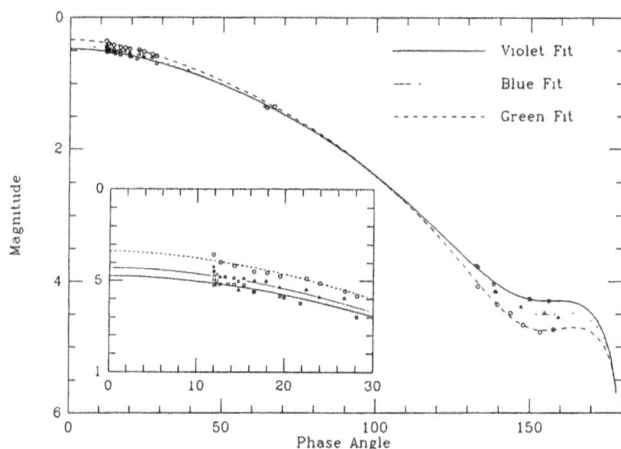

Figure 6.2. Model phase curves of Triton (Hillier et al. 1990) compared with the Voyager 2 violet, blue, and green filter data. The vertical axis is normalized such that the magnitude at the zero degree phase is $-2.5 \log(p_{geo})$, where p_{geo} is the geometric albedo. The disk-averaged image is brighter at smaller phase angles due to surface scattering, while the upturn at large phase angles is due to the scattered light from aerosols. Figure is taken from Hillier et al. (1990). Reprinted with permission from AAAS.

properties of the near-surface discrete structures (Hillier & Veverka 1994), the vertical aerosol optical depth was constrained to ~0.03 at $\lambda = 0.56$ μm and followed a wavelength dependence of $\propto \lambda^{-2}$, with an asymmetry parameter ~0.6 (highly forward scattering) and a single scattering albedo ~0.99 (Hillier et al. 1990, 1991). However, it should be noted that the single scattering albedo cannot be independently determined apart from the aerosol optical depth for optically thin aerosols.

Spatially resolved observations help to avoid the ambiguity between the discrete and global aerosol structures when constraining their scattering properties. From spatially resolved images at high phase angles, the average particle size of the discrete structures at high southern latitudes was constrained to be ~0.2–1.5 μm (Pollack et al. 1990; Rages & Pollack 1992). Meanwhile, at low southern latitudes where the discrete structures are largely absent, the mean particle radius, column-integrated particle number density, and aerosol scale height of the global aerosol layer were estimated to be 0.17 ± 0.012 μm, $2.0 \pm 0.6 \times 10^6$ cm^{-2}, and 11 ± 0.6 km, respectively, yielding vertical scattering optical depths of ~0.002–0.004 across the visible wavelengths (Rages & Pollack 1992). The fact that the derived scale height was considerably smaller than the pressure scale height at the observed region (~16 km) suggests that aerosol formation occurred at altitudes below 20 km, as the aerosol and pressure scale heights would be identical otherwise.

The vertical extinction profile of Triton's global aerosol layer was constrained by stellar occultation observations from Voyager 2's UV spectrometer. The observed extinction at wavelengths of 0.14–0.16 μm at altitudes <20 km had a wavelength dependence suggestive of submicron (<0.3 μm) particles, as the inferred optical depth at UV wavelengths was orders of magnitude higher than that at visible wavelengths derived from the imaging observations (Smith et al. 1989; Pollack et al. 1990). The extinction also differed between ingress and egress (Herbert & Sandel 1991), while the extinction scale height was identical to the pressure scale height, in contrast to the smaller scale height obtained from spatially resolved, visible observations (Rages & Pollack 1992).

6.2.2 Pre-New Horizons Observations of Pluto's Atmosphere

Pluto's atmosphere was detected from ground-based observations of a stellar occultation on June 9, 1988 (Hubbard et al. 1988; Elliot et al. 1989; Millis et al. 1993), which showed a gradual—rather than sudden, in the case of a bare surface—loss of flux from the target star as Pluto occulted it (Figure 6.3). The "knee" in the light curve at ~60 s after 10:35:50 UTC, where the stellar flux drops more precipitously, was the first hint that aerosols may pervade the lower atmosphere of Pluto. This is because extinction by aerosols would reduce the stellar flux more quickly than a clear isothermal atmosphere (Elliot & Young 1992). However, the "knee" can also be explained by a reduction in the atmospheric scale height due to a drastic temperature decrease near the surface (Hubbard et al. 1990). Subsequent stellar occultations from 2002 to 2015 (e.g., Elliot et al. 2003; Pasachoff et al. 2005; Young et al. 2008; Person et al. 2008; Olkin et al. 2014; Gulbis et al. 2015) found evidence for atmospheric waves and showed that both aerosols and a steep temperature gradient were likely present. Evidence for aerosol extinction was provided by the wavelength-dependence of minimum stellar fluxes consistent with a fall off in aerosol extinction caused by the decrease in absorption of submicron aerosol particles at longer wavelengths (Elliot et al. 2003; Gulbis et al. 2015).

The atmospheric composition of Pluto was initially inferred from detections of N_2, CH_4, and CO ices on its surface via infrared spectroscopy (Cruikshank & Silvaggio 1980; Owen et al. 1993), which implied an atmosphere of similar

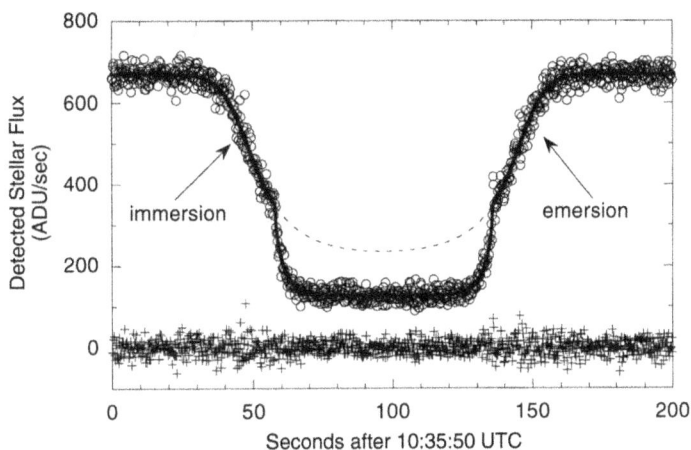

Figure 6.3. Light curve from the Pluto occultation of 1988 June 9 (points) fit with an isothermal atmospheric model that includes refraction and extinction (solid curve) from Elliot & Young (1992). The dashed curve shows the light curve if extinction were not included. The pluses at the bottom show the fit residuals to the well-fit model. Figure reprinted from Elliot & Olkin (1996). Copyright (1989), with permission from Elsevier.

composition in vapor equilibrium. The finding that N_2 ice was more abundant than the others by a factor of 50, and subsequent measurement of the low surface temperature of Pluto of $\sim$40 K (Tryka et al. 1994) all but confirmed N_2 as the dominant atmospheric gas due to its volatility. Direct detections of gaseous CH_4 and CO were subsequently made through high resolution spectroscopy (Young et al. 1997; Lellouch et al. 2011).

6.2.3 New Horizons Observations of Pluto Aerosols

The New Horizons spacecraft flew through the Pluto system on 2015 July 14 and confirmed the existence of aerosols in Pluto's atmosphere (Figure 6.4). The aerosols were found to be optically thin but global, with nadir optical depths of $\sim$0.01 at optical wavelengths and wrapping around the full limb (Gladstone et al. 2016). Unlike Triton, no discrete aerosol structures were detected conclusively (Stern et al. 2017).

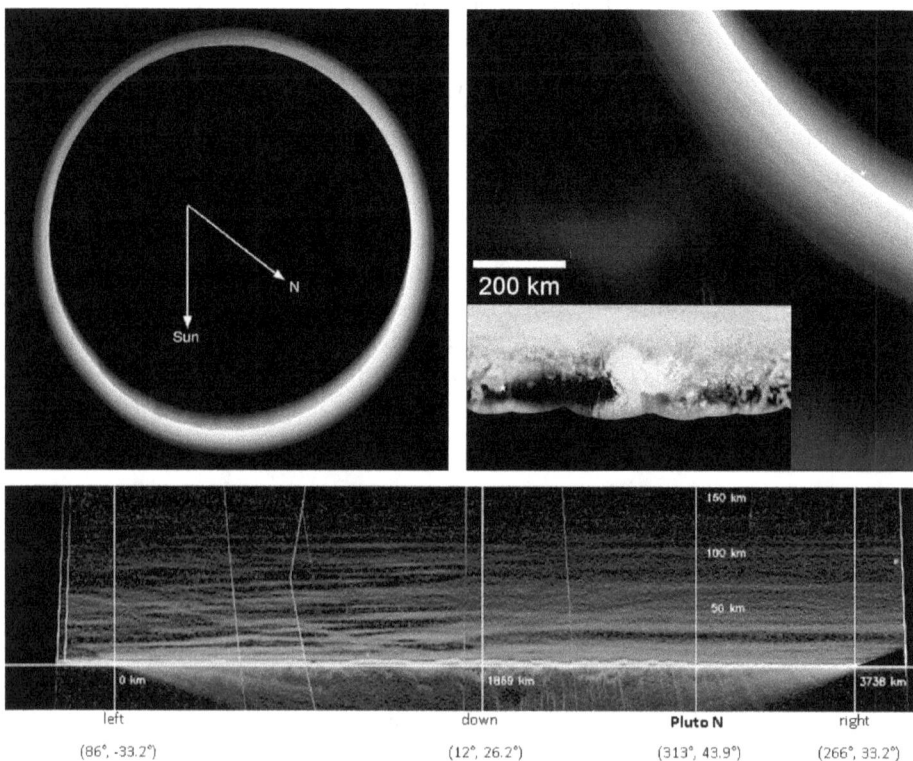

Figure 6.4. Top left: New Horizon's view of Pluto backlit by the Sun, showing the aerosol layers around the full limb. Aerosol scattering is brighter toward the northern (summer) hemisphere rather than the Sun's direction. Image has been stretched and sharpened. Top right: Close-up of the $\sim$20 aerosol layers at Pluto's limb, with the inset map of Pluto showing the location of the observed aerosols. Bottom: Unwrapped mosaic of haze layers around the Pluto limb. Thin, tilted white lines are mosaic seams. Horizontal distance and vertical altitude scales and Pluto longitude/latitude are given. All images are taken by the New Horizons Long Range Reconnaissance Imager (LORRI) and adapted from Cheng et al. (2017). Copyright (2017), with permission from Elsevier.

Aerosol scattering was brighter toward the northern (summer) hemisphere by a factor of 2–3 compared to equatorial regions, suggesting a factor of 2 greater aerosol mass loading. Aerosol scattering at optical wavelengths was observed to 200 km altitude above the surface, while UV extinction by aerosols was seen reaching 500 km. The vertical distribution of aerosols was not smooth, but consisted of ~20 fine distinct layers each with a thickness of a few kilometers separated on average by about 10 km, though in general aerosol scattering increased with decreasing altitude. Contrast between the bright layers and the darker regions in between is a few percent. Each thin aerosol layer was observed to extend for hundreds of kilometers around the limb, and sometimes merging with others or splitting apart (Cheng et al. 2017; Jacobs et al. 2021). Pluto's aerosols appear to be more Titan-like and absorbing than Triton's, with single scattering albedo ranging from 0.9 to 0.95 over wavelengths of 500–900 nm (Hillier et al. 2021).

The wavelength and phase angle dependence of aerosol extinction point to a complex aerosol particle distribution (Figure 6.5). The aerosol scattering intensity in the visible and near-IR showed a gradual decrease with increasing wavelength, suggesting aerosol particle sizes of ~10 nm (Gladstone et al. 2016; Grundy et al. 2018; Kutsop et al. 2021). However, the aerosols were also found to be highly

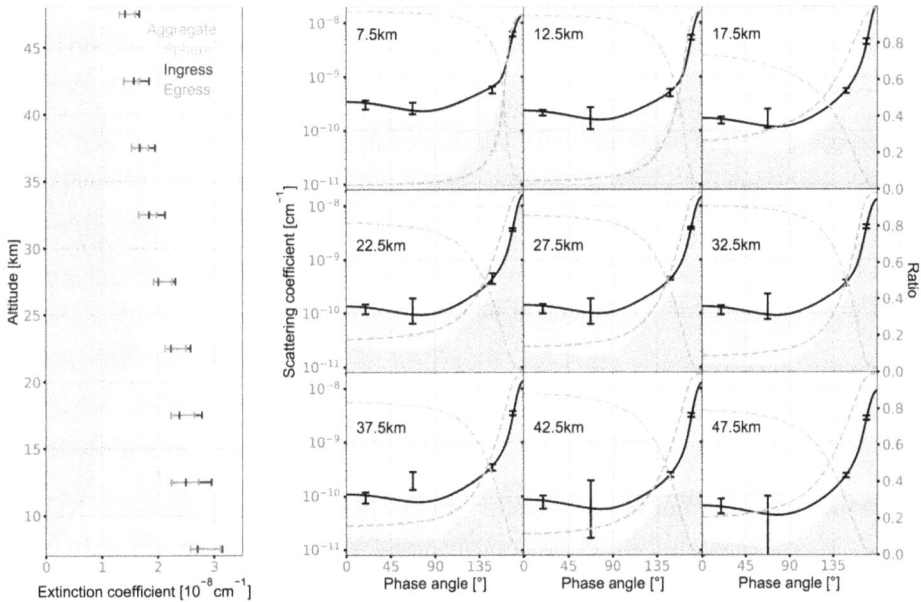

Figure 6.5. Left: Near-surface aerosol extinction at Pluto observed by the Alice spectrograph during ingress (indigo) and egress (red) of solar occultation compared to the modeled contributions from large ~1 μm aggregates (blue) and smaller particles of a few tens of nanometer radius (orange). Right: Near-surface scattered light observations obtained by LORRI (black) compared to the modeled contributions from large aggregates and small particles. Black curves denote the total contributions of aggregates and spheres. Colored dashed curves represent the ratio of contribution from each component. Figures adapted from Fan et al. (2022). CC BY 4.0.

forward scattering in the visible, which is indicative of larger (>0.1 μm) particles. Furthermore, the UV aerosol extinction indicated a large aerosol UV cross section that cannot be produced by even 0.1 μm spherical particles (Young et al. 2018; Kammer et al. 2020). Porous fractal aggregates—irregularly shaped, extended particles composed of loosely bound, small, roughly spherical monomers (see Section 6.3.2)—provide a possible explanation for these seemingly conflicting observations (West & Smith 1991): the wavelength dependence and large UV cross section are due to scattering and extinction by the collection of $\sim$10 nm monomers, while the forward scattering is accomplished by the large (>0.1 μm) extended nature of the particles (Cheng et al. 2017). However, while fractal aggregate particles fit the observations at altitudes above 50 km, an increase in the aerosol backscattering intensity near the surface cannot be reproduced by this population alone. Here, two separate populations of aerosol particles are required: smaller particles (spherical or aggregate) with radii of a few tens of nanometers that are capable of producing the observed backscattering, and $\sim$1 μm fractal aggregates with $\sim$10 nm monomers that can explain the large forward scattering and UV extinction and wavelength dependence (Figure 6.5; Kutsop et al. 2021; Fan et al. 2022).

The abundance profiles of several key gases in Pluto's atmosphere were constrained through solar and stellar UV occultations observed by the New Horizons Alice UV spectrograph (Young et al. 2018; Kammer et al. 2020). N_2 and CH_4 were shown to be the dominant gases in the atmosphere, with mixing ratios of 99% and 1% near the surface, respectively. Diffusive separation at high altitudes results in the mixing ratio of CH_4 approaching 5% at $\sim$500 km. C_2H_2, C_2H_4, and C_2H_6 were also identified, which are likely photochemical in origin (see Section 6.3.1). Interestingly, these species are not well-mixed in the atmosphere below their nominal photochemical production altitude, and instead show depletion below 400 km suggestive of chemical destruction and/or condensation. The mixing ratio profile of HCN, which was measured by ALMA at nearly the same time as the New Horizons flyby, shows similar decreases below 400 km, but also a highly supersaturated mixing ratio in the upper atmosphere (Lellouch et al. 2017).

6.2.4 Summary

Observations of Triton and Pluto's atmospheres show that both worlds possess global aerosol layers with intriguing differences and similarities. The biggest difference is the vertical extent of the aerosols: while Triton's aerosols are confined to the lower $\sim$30 km of the atmosphere, Pluto's extend to >500 km. In addition, Triton possesses two distinct aerosol populations: a low optical depth global layer and higher optical depth near-surface discrete structures. By comparison, Pluto lacks discrete structures (though there are extensive vertical layering, which is not present on Triton), but still appears to possess two particle populations near the surface, with particle sizes similar to the two Triton aerosol populations. The total vertical optical depths of the aerosols on both worlds are also strikingly similar at around 0.01, though Triton's is more variable due to the discrete structures.

6.3 Theory of Triton and Pluto Aerosols

6.3.1 Photochemistry and Aerosol Formation on Triton and Pluto

The origins of aerosols in the cold-reducing atmospheres of Triton and Pluto are inexorably tied to the photochemistry of N_2 and CH_4. N_2 is destroyed by extreme-UV (EUV) solar photons ($\lambda < 100$ nm) and, additionally for Triton, high energy electrons from Neptune's magnetosphere (Majeed et al. 1990, Yung & Lyons 1990; Lara et al. 1997); CH_4 is photolyzed mostly by Lyα photons from the Sun and those scattered by the local interstellar medium (Strobel et al. 1990; Lara et al. 1997). These reactions yield radical species such as N, H, CH, $^3CH_2$, and CH_3, which rapidly react with each other to form more complex hydrocarbons and nitriles like C_2H_2, C_2H_4, C_2H_6, C_4H_2, and HCN (Strobel et al. 1990; Krasnopolsky & Dale 1995; Lara et al. 1997; Wong et al. 2017; Luspay-Kuti & Mandt 2017; Benne et al. 2022).

A key difference in the photochemistry of Triton and Pluto is the near-surface gaseous CH_4 abundance, which is a factor of 10–100 smaller on the former compared to the latter. This results in CH_4 being photolyzed, and thus more complex hydrocarbons and nitriles being produced, at a much lower altitude on Triton—at around 25 km (Krasnopolsky 2012)—in contrast to 400 km on Pluto (Lara et al. 1997; Wong et al. 2017; Krasnopolsky 2020). The rapid destruction of CH_4 near the surface of Triton explains the observed small CH_4 scale height of 7–10 km and the lack of CH_4 at higher altitudes (Broadfoot et al. 1989). In contrast, the abundant CH_4 on Pluto overwhelms any photochemical destruction, allowing it to persist into the upper atmosphere (Figure 6.6; Krasnopolsky 2012; Young 2017).

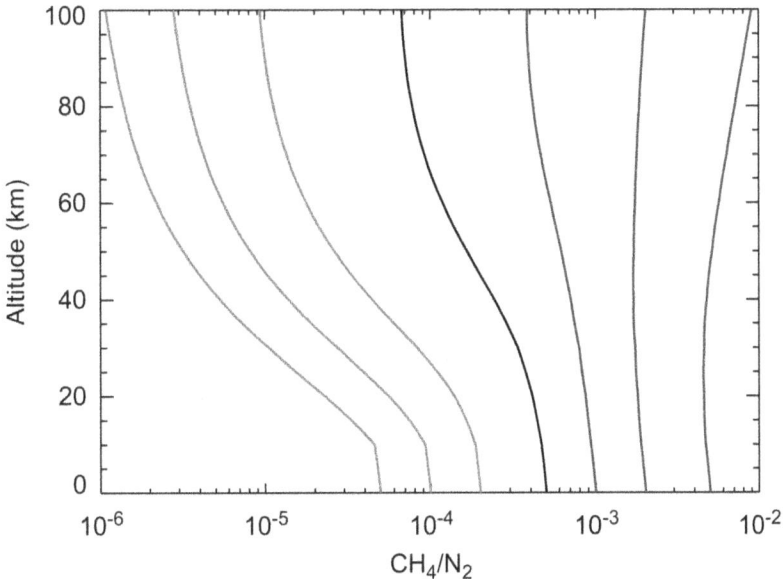

Figure 6.6. CH_4 mixing ratio profiles on Triton as a function of its value near the surface, computed by Krasnopolsky (2012). Photolysis and chemical reactions tend to reduce methane (red) while diffusive separation enhances it (blue). Transition between the two regimes (black) occurs at a surface CH_4 mixing ratio of ~5 ×10^{-4}. Reprinted from Krasnopolsky (2012), Copyright (2012), with permission from Elsevier.

Lessons learned from the analysis, modeling, and experimental interpretation of Cassini observations of Saturn's moon, Titan, can be applied to understand the connection between photochemistry and haze formation on Triton and Pluto. All three bodies possess N_2-dominated atmospheres with trace amounts of CH_4, though Titan's atmosphere is significantly more massive. Similar to Triton and Pluto, N_2 and CH_4 are destroyed in the upper atmosphere and ionosphere of Titan ($\sim$1000 km altitude) through photolysis by EUV and far-UV (FUV) photons and energetic particles from Saturn's magnetosphere (Ågren et al. 2009; Galand et al. 2010), producing radicals and progressively more complex organic molecules and ions (e.g., Yung et al. 1984; Wilson & Atreya 2004; Vuitton et al. 2019). Cassini observations show that this process of atmospheric species gradually increasing in complexity and mass through successive neutral and ion reactions continues smoothly up to heavy negative ions with masses of $\sim 10^4$ Da q^{-1} (Coates et al. 2007; Vuitton et al. 2009). These negative ions attract the less massive (up to 350 Da q^{-1}; Crary et al. 2009) positive ions in Titan's ionosphere and grow rapidly as a result, eventually forming the nanometer-sized "seeds" of Titan's global hazes (Lavvas et al. 2013).

The complex chemistry and formation of hazes on Titan depend heavily on the coupling between neutral and ion chemistry in its ionosphere. Titan's ionosphere exhibits an electron density peaking at $\sim$3000 e^- cm^{-3} (Ågren et al. 2009) and a smorgasbord of organic molecules and ions sourced from N_2 and CH_4. In comparison, New Horizons only placed an upper limit of 1000 e^- cm^{-3} on Pluto's ionosphere, which is otherwise populated by positive molecular ions (Hinson et al. 2018). Triton's ionosphere exhibits an even greater electron density than Titan of 25,000–45,000 e^- cm^{-3} (Tyler et al. 1989) due to the lack of CH_4 allowing the survival of atomic species and ions (Krasnopolsky & Dale 1995), which in turn possess long enough lifetimes to ensure high electron densities (Strobel & Summers 1995). However, the low abundance of CH_4 at high altitudes also makes complex organic chemistry difficult. In addition, the atmospheres of Triton and Pluto both contain about 10 times the abundance of CO as Titan (500 ppm versus 50 ppm; Gurwell 2004; Lellouch et al. 2010, 2017). These differences could result in significant disparities in the haze properties of these atmospheres.

6.3.2 Formation, Evolution, and Dynamics of Triton and Pluto Aerosols

The size and spatial distributions of aerosols are controlled by transport processes such as sedimentation, diffusion, and advection, as well as microphysical processes including nucleation, growth through condensation and coagulation, and loss through evaporation/sublimation. In the following sections, we summarize several key microphysical processes relevant to Triton and Pluto aerosols and what models that consider these processes have shown.

6.3.2.1 Haze Particle Transport

Upon formation, nanometer-sized nascent haze particles will begin settling downwards at their terminal velocities due to gravity. At the low gas densities of Pluto and Triton's atmosphere, the mean free path of gas molecules is much larger than the haze particle radii (large Knudsen number), so that particle transport occurs in the gas kinetic regime. As a result, the terminal velocity is linearly proportional to the particle mass density and radius (for compact particles) and inversely proportional to the ambient gas density.

In addition to sedimentation, aerosol particles are also transported by atmospheric circulation. Unfortunately, only one study (Bertrand & Forget 2017) has investigated the impact of circulation on Pluto's hazes, and none have focused on similar processes on Triton. Bertrand and Forget (2017) simulated the formation, sedimentation, and horizontal advection of haze particles and showed that the haze distribution is strongly tied to whether N_2 is condensing out on the (unobserved) south pole. Without N_2 condensation, the meridional flow is weak and hazes do not experience significant horizontal transport, resulting in a concentration of hazes in the northern (summer) hemisphere where the Lyα flux is the highest. In contrast, N_2 condensation strengthens north-to-south meridional winds due to conveyance of N_2 from Sputnik Planitia to the south pole, resulting in a more homogeneous haze latitudinal distribution. Hazes would also be depleted near the surface over Sputnik Planitia due to the upward branch of the flow, and enriched over the south pole due to the downward branch (Figure 6.7). Current observations cannot discriminate between these two scenarios (Cheng et al. 2017; Chen et al. 2021).

The observed layering of hazes in Pluto's atmosphere (Gladstone et al. 2016; Cheng et al. 2017; Jacobs et al. 2021) may be caused by atmospheric waves, which have been observed as temperature perturbations during groundbased stellar occultations (e.g., Pasachoff et al. 2005; Young et al. 2008; Person et al. 2008). These waves could be caused by thermal tides arising from N_2 sublimation (Toigo et al. 2015; French et al. 2015) and/or gravity waves arising from orographic forcing (Gladstone et al. 2016; Cheng et al. 2017). Such waves could act to compactify and rarefy haze distributions, generating the observed layers. However, neither explanation is perfect: The thermal tide hypothesis does not explain how haze particles are transported by the waves and assumes zero mean flow while the orographic forcing hypothesis involves topography that may be too localized to explain the global haze layers (Jacobs et al. 2021). Followup work is needed to model these processes more rigorously.

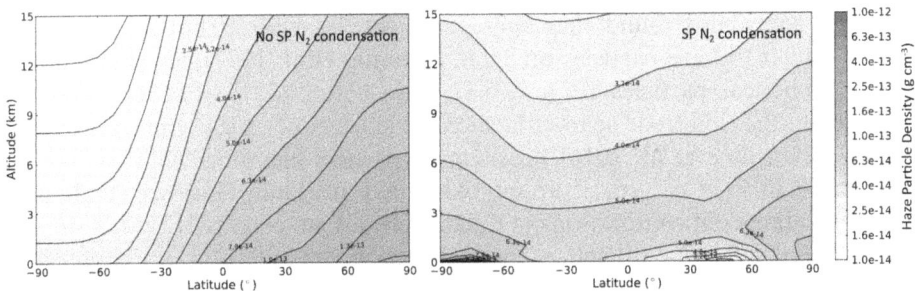

Figure 6.7. Zonal mean Pluto haze particle density without (left) and with (right) N_2 condensation at the south pole, as computed by a 3D model. Adapted from Bertrand and Forget. (2017). Copyright (2017), with permission from Elsevier.

6.3.2.2 Haze Particle Growth

Haze particles can grow as they are transported in the atmosphere. The primary pathway for growth of involatile haze particles is coagulation, when particles collide and stick together. For the submicron particles observed in Pluto's and Triton's atmospheres, thermal Brownian motion is the main driver of relative velocities and collisions. While it is typically assumed that every particle collision leads to sticking, microphysical studies of Titan's hazes have shown that haze particle charging via interactions with ions and electrons may prevent sticking, thereby regulating the coagulation rate and haze particle size distribution (e.g., Lavvas et al. 2010). The impact of particle charge is typically parameterized by a charge-to-radius ratio.

Growth of haze particles via coagulation can lead to non-spherical particle shapes, which drastically alters their aerodynamical and optical properties. Non-spherical haze particles have typically been treated as fractal aggregates composed of smaller spherical monomers. The shape of a fractal aggregate is characterized by the fractal dimension D_f, which is related to the number of monomers in an aggregate N_{mon}, monomer radius r_{mon}, and aggregate characteristic radius r_{agg} (the radius of a sphere that has the same gyration radius) via

$$N_{mon} = k_0 \left(\frac{r_{agg}}{r_{mon}} \right)^{D_f},$$
(6.1)

where k_0 is an order unity prefactor. Particles with $D_f = 1$ are chain-like aggregates, while a $D_f = 2$ particle possesses a mass that is proportional to its cross-sectional area. Equation (6.1) indicates that the internal density of aggregates decreases with addition of monomers, while the terminal velocity of the aggregate is the same as that of its monomers when $D_f \sim 2$, which is typical of aggregate formation via ballistic cluster–cluster collisions (Cabane et al. 1993).

Haze microphysical models of Pluto and Triton typically simulate the vertical transport and growth (by coagulation) of haze particles initially produced at the pressure level of CH_4 photolysis, with the production rate as a free parameter. For Pluto, microphysical models typically require haze production rates on par with the methane photolysis rate due to solar and local interstellar medium-scattered UV (a few times 10^{-14} g cm^{-2} s^{-1}) to match the observed stellar occultation light curves (Stansberry et al. 1989) and UV extinction from New Horizons (Gao et al. 2017). The latter study also found that aggregate particles with 10 nm monomers and $D_f = 2$, similar to haze particles on Titan (Rannou et al. 1997), are preferred over compact spherical particles (Figure 6.8). Gao et al. (2017) further found that reproducing the inferred near-surface particle size of $\sim$0.1 μm observed by LORRI (Gladstone et al. 2016) necessitated particle charging, with a charge-to-radius ratio of $\sim$30 e$^-$ μm^{-1}, about twice that of Titan (Lavvas et al. 2010). However, subsequent comparisons of the models of Gao et al. (2017) to MVIC data suggested a higher value of $\sim$60 e$^-$ μm^{-1} (Kutsop et al. 2021). For Triton, similar modeling has been conducted by Ohno et al. (2021), who in addition explicitly simulated the evolution of aggregate volume through different-sized aggregate collisions in each mass bin. They found that haze particles on Triton can grow

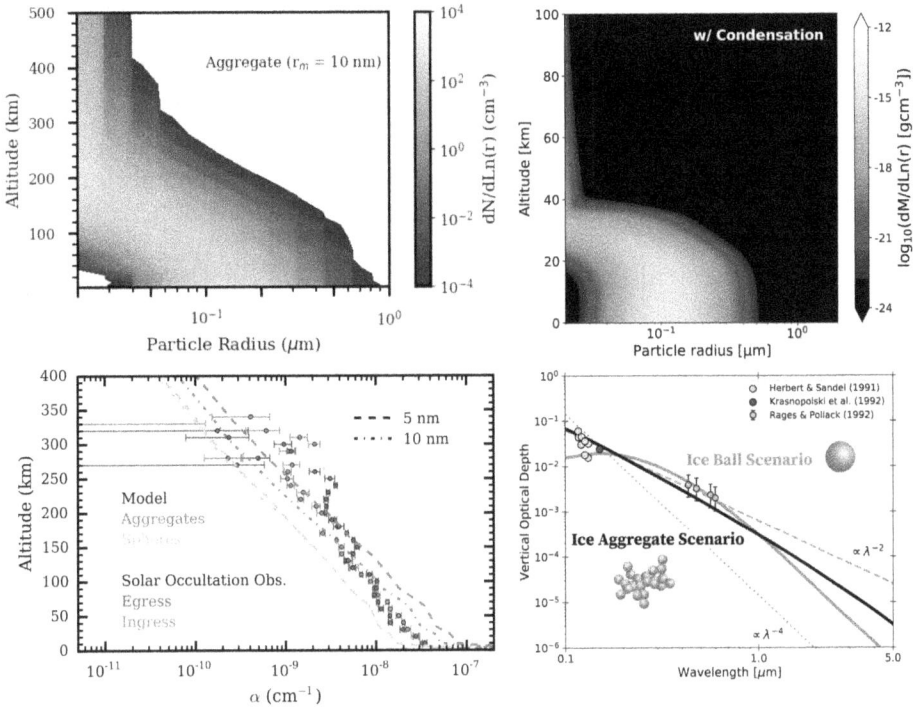

Figure 6.8. Number densities of aerosol particles as a function of altitude and particle radius for Pluto hazes, assuming aggregates with 10 nm monomers (top left) and Triton hazes and ice clouds (top right) from the microphysical models of Gao et al. (2017) and Ohno et al. (2021), respectively. Bottom left: Comparison of extinction coefficient α of spherical (orange) and aggregate (green) haze particles with 5 nm (dashed) and 10 nm monomers (dashed dotted; the 10 nm aggregate model is the same as that presented in the top left panel) with the observed ingress (red) and egress (blue) solar occultation observations at Pluto (Gladstone et al. 2016). Bottom right: Comparison of the vertical optical depth of the ice balls (green line; same model as that plotted top right) and ice aggregates (black line) models for Triton aerosols with the observed optical depths from Herbert & Sandel (1991) (gray), Krasnopolsky et al. (1992) (blue), and Rages & Pollack (1992) (orange). All figures adapted from Gao et al. (2017), Copyright (2017), with permission from Elsevier, and Ohno et al. (2021).

into aggregates with $D_f \sim 1.8$–2.2, though the available observations are insufficient for constraining the actual fractal dimension of Triton aerosols and their charge-to-radius ratios. Critically, Ohno et al. (2021) found that hazes could not simultaneously explain both the observed UV extinction coefficient and visible scattered-light intensity due to absorption by the haze, and that brighter material, such as hydrocarbon ices, must be present.

6.3.2.3 Cloud Formation

The low temperatures of Triton's and Pluto's atmospheres likely lead to ice cloud formation. Cloud particles form through nucleation, when gas molecules cluster together freely (homogeneous nucleation) or around a cloud condensation nucleus (CCN; heterogeneous nucleation) to form a volatile particle. In the latter case, the

CCN is typically another type of particle in the atmosphere. The nucleation rate is dependent on the saturation vapor pressure and the material properties of the condensate, including its surface energy and molecular weight; in the case of heterogeneous nucleation, the rate also depends on the number density, size, and material properties of the CCN, in particular the desorption energy of condensate molecules on the CCN surface and the contact angle between the condensate and the CCN. In general, high surface energies and mean molecular weights, low CCN number densities, small CCN sizes, low desorption energies, and large contact angles lead to lower nucleation rates for a fixed temperature and condensate vapor supersaturation (Pruppacher & Klett 1978). The inclusion of charged particles can increase the nucleation rate through the introduction of electric potential energy, lowering the energy barrier to nucleation (Moses et al. 1992).

Following nucleation, cloud particles can grow through condensation. In the gas kinetics regime, the condensation rate depends on the abundance and thermal velocity of the condensate vapor, as well as the cloud particle size through the Kelvin curvature effect. The Kelvin effect is the increase in the saturation vapor pressure over a curved surface as compared to that over a flat surface owing to the weaker binding energy of molecules on the former (Pruppacher & Klett 1978). As such, smaller particles tend to grow slower and evaporate faster.

Ice clouds as an explanation for the aerosol structures on Triton was a popular hypothesis following the Voyager 2 encounter, owing to their high albedo and supporting evidence from photochemical models (Smith et al. 1989; Strobel et al. 1990; Pollack et al. 1990; Hillier et al. 1991; Rages & Pollack 1992; Strobel & Summers 1995). In particular, Strobel et al. (1990); Strobel & Summers (1995), and Krasnopolsky & Dale (1995) showed that CH_4 photolysis yielded sufficient C_2 hydrocarbons that condensed out in the lower 30 km of Triton's atmosphere (column production rate of $\sim 4 - 8 \times 10^{-15}$ g cm^{-2} s^{-1}) to match the observed optical depth of the global aerosol structure. Meanwhile, the discrete aerosol structures below ~ 8 km altitude were assumed to be N_2 ice clouds, as they were mostly detected over the southern N_2 polar cap where they could form through large scale upwelling from surface N_2 sublimation and/or plume activity, with N_2 vapor possibly nucleating on the C_2 hydrocarbon ice particles (Pollack et al. 1990; Rages & Pollack 1992). More recently, Ohno et al. (2021) considered condensation of C_2H_4 onto haze CCN and found that the resulting ice cloud could explain both the observed UV extinction coefficient and visible scattered-light intensity assuming a C_2H_4 production rate consistent with the photochemical production rate of higher order hydrocarbons computed by Strobel et al. (1990) (Figure 6.8). The ice cloud is present primarily below 30 km, matching the vertical distribution of Triton's global aerosol layer. Under this scenario, haze CCN is produced at >250 km with a rate of only 6×10^{-17} g cm^{-2} s^{-1}, resulting in little observable haze above 30 km. These results were corroborated by Lavvas et al. (2021), who further found through tracking the evolution of D_f from relative contributions of ice condensation and coagulation that $D_f = 3$ above 50 km and transitions to 2 below 25 km as coagulation becomes more important.

Ice cloud nucleation onto haze CCN in Pluto's middle atmosphere may explain the depletion of gaseous HCN and C_2-hydrocarbons below 400 km. Using a photochemical model, Wong et al. (2017) showed that the shapes of their observed mixing ratio profiles can be reproduced if they condensed onto haze CCN, though the C_2H_4 saturation vapor pressure may need to be lower than that extrapolated from laboratory measurements, or else atmospheric mixing may need to be stronger and/or ion chemistry may need to be considered (Krasnopolsky 2020). In contrast, Luspay-Kuti & Mandt (2017) and Mandt et al. (2017) considered adsorption of gas molecules on haze particles and found similar magnitudes of gas depletion. Adsorption is more dependent on the surface chemistry of haze particles than saturation vapor pressures, with Luspay-Kuti & Mandt (2017) finding that Pluto's hazes must become less "sticky" to gas molecules as they age due to chemical evolution, which is similar to what is seen for Titan's hazes (Dimitrov & Bar-Nun 2002). Lavvas et al. (2021) also found that adsorption of C_2-hydrocarbons is possible, but on ice clouds made of HCN, C_3H_4, C_4H_2, and C_6H_6 instead of haze CCN, as they found condensation of these ices to be abundant in their model below 400 km. Lavvas et al. (2021) further found that ice clouds could make up a significant fraction of the mass of Pluto aerosols, and that $D_f = 3$ above 400 km and 2 below 300 km, with a smooth transition in between as sedimenting spherical particles coagulate into aggregates.

Gas condensation also impacts Pluto's upper and near-surface atmosphere. At high altitudes where haze CCN is absent, nucleation rates of ice particles may be extremely slow (Rannou & West 2018), potentially explaining the large ($>10^6$) HCN supersaturation observed by ALMA (Lellouch et al. 2017).[2] Meanwhile, Stern et al. (2017) simulated the condensation of hydrocarbon and nitrile ice clouds within 15 km of Pluto's surface where temperatures reach 40 K and found relatively low cloud optical depths ($\tau \sim 10^{-4}$), though small (~ 1 m s^{-1}) updrafts could promote much more optically thick ($\tau \sim 0.1$) clouds. Such cloud formation could lead to a sharp increase in D_f, as shown by Lavvas et al. (2021). It should be noted, however, that the microphysical parameters used in all of the aforementioned studies, such as the saturation vapor pressure, surface energy, and contact angles between the condensing gases and haze CCN are largely unknown and/or uncertain due to a lack of laboratory measurements at the relevant temperatures.

6.3.3 Summary

The photochemical destruction of N_2 and CH_4 in Triton's and Pluto's atmospheres likely serves as the origins of the clouds and hazes therein. Such photochemical processes produce an array of simple hydrocarbon and nitrile molecules that can react further to form massive charged ions and nanometer-sized organic haze particles. Alternatively, the simple molecules can condense out and form highly reflective ice clouds that likely nucleate on haze CCN. Models of Triton's and

---

[2] In comparison, supersaturation of water vapor on Earth almost never exceeds a few % due to the ample abundance of CCN (Pruppacher & Klett 1978).

Pluto's aerosols predict that the differences in the aerosol composition and distributions of the two bodies are likely caused by the gaseous methane abundance near their surfaces: the relatively high abundance of methane in Pluto's atmosphere allows methane to survive to hundreds of km altitude, where its photolysis leads to the formation of complex haze particles; in contrast, the lower abundance of methane in Triton's atmosphere results in its photolysis within a few tens of kilometers of its surface, leading to the production of hydrocarbon molecules that condense to form ice clouds. In both Triton and Pluto's atmospheres, aerosol particles sediment from their source toward the surface, with little impact from diffusional processes, though global circulation may be important on Pluto. Both spherical and fractal aggregate particles are likely present, and they may transition from one form to another depending on whether condensation or coagulation is dominant. Finally, it is uncertain whether condensation or sticking/adsorption is more important in Pluto's atmosphere due to uncertainties surrounding the microphysical parameters of the relevant gaseous and CCN species such as their saturation vapor pressures and surface energies.

6.4 Laboratory Investigations of Triton and Pluto Aerosols

Laboratory experiments are essential for constraining the material and optical properties of aerosols necessary for modeling aerosol formation and evolution and interpreting observations. To this end, a number of studies have attempted to synthesize photochemical haze analogs by subjecting gas mixtures replicating the compositions of Pluto's and Triton's atmospheres to an energy source. For example, both Thompson et al. (1989) and McDonald et al. (1994) analyzed Triton haze analogs produced from exposing a $N_2/CH_4(0.1\%)$ gas mixture to cold plasma and found remarkable differences in their C/N ratio and spectral features compared to Titan haze analogs, indicating that differences in N_2/CH_4 ratios led to different reaction pathways in haze production, and that electrons coming from Neptune's magnetosphere likely promote the production of absorbing organic hazes. More recently, Moran et al. (2022) performed a cold plasma experiment on a $N_2/CH_4(0.2\%)/CO(0.5\%)$ gas mixture at 90 K and found that the addition of CO results in nitrogen-rich (rather than carbon-rich) Triton haze analogs that incorporate oxygen at $\sim$10 wt%, indicating that the relative abundance of CO also plays key roles in controlling the properties of organic hazes. Meanwhile, Jovanović et al. (2020, 2021) investigated the altitude dependence of haze composition on Pluto by considering a $N_2/CH_4/CO$ gas mixture where the CO abundance is fixed to 500 ppm and the CH_4 abundance is allowed to vary from 0.5 to 5% to simulate high altitude diffusive separation. They found that more N and O atoms were incorporated in the sample with lower CH_4 abundances, showing that haze composition likely varies with altitude. They also found that the Pluto haze analogs were less absorbing at optical wavelengths compared to Titan haze analogs, which are produced without CO gas (Figure 6.9). Finally, He et al. (2017) considered a $N_2/CH_4/CO$ gas mixture with varying CO abundances that is applicable to Titan, Triton, and Pluto, and

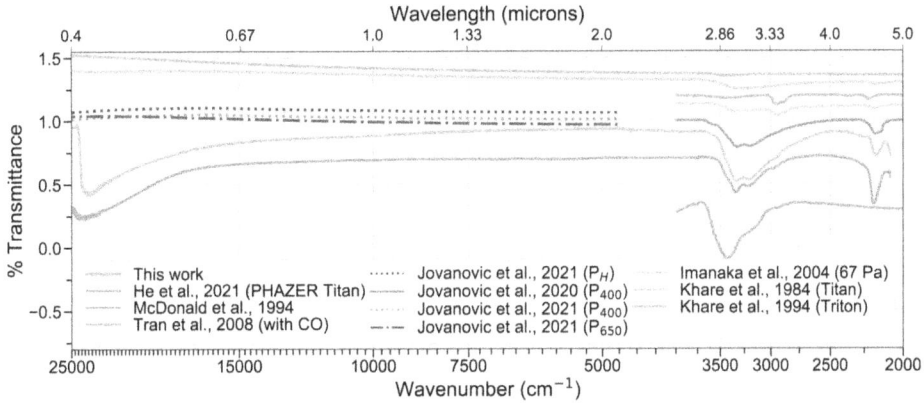

Figure 6.9. Transmittance spectra of organic haze analogs produced in laboratory experiments from mixtures of N_2 and CH_4 (and CO in some cases) replicating the atmospheres of Titan (orange, pink, yellow, and salmon), Pluto (light/dark purple dotted and dashed-dotted), and Triton (cyan, green, and light blue). See Table 3 of Moran et al. (2022) for details on the individual studies. Figure adapted with permission from Moran et al. (2022). CC BY NC. ("This work" in the figure).

found an increase in the density and oxygen content and a decrease in the saturation of the haze analogs with increasing CO.

To aid our understanding of ice cloud formation in Triton's and Pluto's atmosphere, several studies have measured and compiled the saturation vapor pressures and surface energies of relevant hydrocarbon and nitrile ices (Moses et al. 1992; Fray & Schmitt 2009; Yu et al. 2023). However, the temperature ranges for which the measurements are valid often do not contain the low temperatures of Triton and Pluto, and thus caution is required when extrapolating the empirical saturation vapor pressure and surface energy expressions. Measurements of other microphysical parameters, such as the desorption energies and contact angles between condensates and CCN, are relatively rare and confined mostly to methane and ethane condensates and haze CCN (Curtis et al. 2008; Rannou et al. 2019).

6.5 Triton and Pluto Aerosols: Unknowns and Outlook

Despite similarities in atmospheric composition and surface pressures and temperatures, the aerosols of Triton and Pluto are distinct from each other (Table 6.1). Triton hosts a global aerosol layer likely composed of organic ices that is confined to the lower few tens of kilometers of its atmosphere, with discrete clouds below 10 km, while Pluto's aerosols are seemingly dominated by organic hazes extending up hundreds of kilometers with multiple fine layers, though organic ice adsorption and/ or condensation also likely play a role. This difference has been hypothesized as being driven by the difference in CH_4 abundance on the two worlds: the lower CH_4 abundance on Triton leads to CH_4 and N_2 photochemistry in the lower few tens of kilometers, leading to production and condensation of simple hydrocarbon and nitrile ices. Meanwhile, the higher CH_4 abundance on Pluto allows it to survive into the upper atmosphere, resulting in the production of organic haze particles at

Table 6.1. Summary of Triton and Pluto Aerosol Properties (See Sections 6.2 and 6.3 for Details)

	Triton	Pluto
Spatial distribution	Global + discrete structures, up to 30 km[1]	Global, layered, up to 500 km[9]
Production rate	2–8×10^{-15} g cm^{-2} s^{-1} [2]	~10^{-14} g cm^{-2} s^{-1} [10]
Composition	N$_2$ ice (discrete)[3] Organic ices (global)[4]	Organic haze + ices[11]
Size	~0.2–1.5 μm (discrete)[5] ~0.15 μm (global)[6]	~0.1–1 μm[12]
Shape	Unconstrained[2]	Spheres + aggregates[12]
Charge	Unconstrained[2]	30–60 e$^-$ μm^{-1} [13]
Optical depth	~0.03 (discrete)[3] ~0.003 (global)[7]	~0.01[9]
Single scattering albedo	0.99[8]	0.9–0.95[14]

[1] (Smith et al. 1989; Pollack et al. 1990)
[2] (Ohno et al. 2021)
[3] (Pollack et al. 1990)
[4] (Ohno et al. 2021; Lavvas et al. 2021)
[5] (Pollack et al. 1990; Rages & Pollack 1992)
[6] (Krasnopolsky et al. 1992; Rages & Pollack 1992)
[7] (Rages & Pollack 1992)
[8] (Hillier et al. 1990, 1991)
[9] (Gladstone et al. 2016)
[10] (Gao et al. 2017)
[11] (Gao et al. 2017; Lavvas et al. 2021)
[12] (Fan et al. 2022)
[13] (Gao et al. 2017; Kutsop et al. 2021)
[14] (Hillier et al. 2021)

hundreds of kilometer altitude from a rich network of neutral and ion reactions; these particles then sediment downwards and could act as nucleation centers and/or adsorbing surfaces for gaseous hydrocarbon and nitrile species.

Groundbased stellar occultations, the Voyager 2 and New Horizon flybys, and lessons learned from Cassini observations of Titan have increased our knowledge of Triton's and Pluto's aerosols considerably, but many vital unknowns remain. Due to the limitations in wavelength and phase angle of the existing observations, the composition, and therefore formation mechanism of Triton's and Pluto's aerosols remain uncertain. While parallels could be drawn between the organic hazes and ice clouds of Titan and those on Triton and Pluto, the different atmospheric compositions (i.e., higher CO on the latter bodies), energy sources, and ionospheric environments could result in significant differences in the formation and evolution of the aerosols of the three worlds. In addition, the impact of aerosols on the atmospheric radiative transfer and dynamics of Triton and Pluto requires further study. For example, Zhang et al. (2017) explained the observed low escape rate of Pluto's atmosphere by showing that Pluto's aerosols could significantly cool its upper atmosphere. This would indicate a fundamental connection between atmospheric and surface volatile evolution, photochemistry, and aerosol formation in

Pluto's atmosphere that is likely modulated by Pluto's extensive seasonal cycles caused by its eccentric orbit (Johnson et al. 2021). Similar connections between Triton's aerosols and the rest of Triton's atmosphere may also exist and therefore should be investigated.

Furthering our understanding of Triton's and Pluto's aerosols requires synergy between observations, modeling, and laboratory studies. A dedicated orbiter at Triton and/or Pluto would greatly increase our knowledge of their aerosols by observing their atmospheres at a wide range of phase angles and wavelengths, and by monitoring them for temporal and spatial variations in aerosol distributions and optical properties. In particular, spectral features associated with solid organic material are abundant at longer, mid-infrared wavelengths (Zhang et al. 2017). Meanwhile, laboratory measurements of key microphysical parameters at Triton and Pluto temperatures, such as the saturation vapor pressure, surface energy, desorption energy, and contact angles of organic ices and haze analogs would be essential for more accurate modeling of ice cloud formation and how hydrocarbon and nitrile species interact with organic haze particles. Further laboratory work on haze formation is also needed, particularly in how they initially form from gaseous species and how they grow through condensation and/or coagulation into spheres or aggregates. In addition, uncertainties in the aerosols' optical properties remain a major bottleneck in interpreting observations and thus more measurements are needed. At the same time, modeling efforts that take into account the laboratory measurements and additionally focus on how hazes interact with the three-dimensional (3D) radiative environment and dynamics of the atmosphere, along with how they evolve on seasonal timescales would also greatly aid in understanding current and future observations. Finally, given that the differences between Triton's and Pluto's aerosols likely hinge on the surface CH_4 abundance, observing and modeling these surfaces and the near-surface atmosphere through time would help us better understand whether Triton's and Pluto's aerosols are fundamentally different or they are simply different points on the same evolutionary path.

References

Ågren, A., Wahlund, J. E., Garnier, P., et al. 2009, P&SS, 57, 1821

Benne, B., Dobrijevic, M., Cavalié, T., Loison, J. C., & Hickson, K. M. 2022, A&A, 667, A169

Bertrand, T., & Forget, F. 2017, Icar, 287, 72

Broadfoot, A. L., Atreya, S. K., Bertaux, J. L., et al. 1989, Sci, 246, 1459

Cabane, M., Rannou, P., Chassefiere, E., & Israel, G. 1993, P&SS, 41, 257

Chen, S., Young, E. F., Young, L. A., et al. 2021, Icar, 356, 113976

Cheng, A. F., Summers, M. E., Gladstone, G. R., et al. 2017, Icar, 290, 112

Coates, A. J., Crary, F. J., Lewis, G. R., et al. 2007, GeoRL, 34, L22103

Conrath, B., Flasar, F. M., Hanel, R., et al. 1989, Sci, 246, 1454

Crary, F. J., Magee, B. A., Mandt, K., et al. 2009, P&SS, 57, 1847

Cruikshank, D. P., & Silvaggio, P. M. 1980, Icar, 41, 96

Curtis, D. B., Hatch, C. D., Hasenkopf, C. A., et al. 2008, Icar, 195, 792

Dimitrov, V., & Bar-Nun, A. 2002, Icar, 156, 530

Elliot, J. L., Ates, A., Babcock, B. A., et al. 2003, Natur, 424, 165

Elliot, J. L., Dunham, E. W., Bosh, A. S., et al. 1989, Icar, 77, 148

Elliot, J. L., & Olkin, C. B. 1996, AREPS, 24, 89

Elliot, J. L., & Young, L. A. 1992, AJ, 103, 991

Fan, S., Gao, P., Zhang, X., et al. 2022, NatCo, 13, 240

Fray, N., & Schmitt, B. 2009, P&SS, 57, 2053

French, R. G., Toigo, A. D., Gierasch, P. J., et al. 2015, Icar, 246, 247

Galand, M., Yelle, R., Cui, J., et al. 2010, JGRA, 115, A07312

Gao, P., Fan, S., Wong, M. L., et al. 2017, Icar, 287, 116

Gladstone, G.R., Stern, S.A., Ennico, K., et al. 2016, Sci, 351, aad8866

Grundy, W. M., Bertrand, T., Binzel, R. P., et al. 2018, Icar, 314, 232

Gulbis, A. A. S., Emery, J. P., Person, M. J., et al. 2015, Icar, 246, 226

Gurwell, M. A. 2004, ApJL, 616, L7

He, C., Hörst, S. M., Riemer, S., et al. 2017, ApJL, 841, L31

Herbert, F., & Sandel, B. R. 1991, JGR, 96, 19241

Hillier, J., Helfenstein, P., Verbiscer, A., & Veverka, J. 1991, JGR, 96, 19203

Hillier, J., Helfenstein, P., Verbiscer, A., et al. 1990, Sci, 250, 419

Hillier, J., & Veverka, J. 1994, Icar, 109, 284

Hillier, J. H., Buratti, B. J., Hofgartner, J. D., et al. 2021, PSJ, 2, 11

Hinson, D. P., Linscott, I. R., Strobel, D. F., et al. 2018, Icar, 307, 17

Hubbard, W. B., Hunten, D. M., Dieters, S. W., Hill, K. M., & Watson, R. D. 1988, Natur, 336, 452

Hubbard, W. B., Yelle, R. V., & Lunine, J. I. 1990, Icar, 84, 1

Jacobs, A. D., Summers, M. E., Cheng, A. F., et al. 2021, Icar, 356, 113825

Johnson, P. E., Young, L. A., Protopapa, S., et al. 2021, Icar, 356, 114070

Jovanović, L., Gautier, T., Broch, L., et al. 2021, Icar, 362, 114398

Jovanović, L., Gautier, T., Vuitton, V., et al. 2020, Icar, 346, 113774

Kammer, J. A., Gladstone, G. R., Young, L. A., et al. 2020, AJ, 159, 26

Krasnopolsky, V. A., Sandel, B. R., & Herbert, F. 1992, JGR, 97, 11695

Krasnopolsky, V. A. 2012, P&SS, 73, 318

Krasnopolsky, V. A. 2020, Icar, 335, 113374

Krasnopolsky, V. A., & Dale, P. 1995, JGR, 100, 21271

Kutsop, N. W., Hayes, A. G., Buratti, B. J., et al. 2021, PSJ, 2, 91

Lara, L. M., Ip, W. H., & Rodrigo, R. 1997, Icar, 130, 16

Lavvas, P., Lellouch, E., Strobel, D. F., et al. 2021, NatAs, 5, 289

Lavvas, P., Yelle, R. V., & Griffith, C. A. 2010, Icar, 210, 832

Lavvas, P., Yelle, R. V., Koskinen, T., et al. 2013, PNAS, 110, 2729

Lellouch, E., de Bergh, C., Sicardy, B., Ferron, S., & Käufl, H. U. 2010, A&A, 512, L8

Lellouch, E., de Bergh, C., Sicardy, B., Käufl, H. U., & Smette, A. 2011, A&A, 530, L4

Lellouch, E., Gurwell, M., Butler, B., et al. 2017, Icar, 286, 289

Luspay-Kuti, A., & Mandt, K. 2017, MNRAS, 472, 104

Majeed, T., McConnell, J. C., Strobel, D. F., & Summers, M. E. 1990, GeoRL, 17, 1721

Mandt, K., Luspay-Kuti, A., Hamel, M., et al. 2017, MNRAS, 472, 118

McDonald, G. D., Thompson, W. R., Heinrich, M., et al. 1994, Icar, 108, 137

Millis, R. L., Wasserman, L. H., Franz, O. G., et al. 1993, Icar, 105, 282

Moran, S. E., Hörst, S. M., He, C., et al. 2022, JGRE, 127, e06984

Moses, J. I., Allen, M., & Yung, Y. L. 1992, Icar, 99, 318

Ohno, K., Zhang, X., Tazaki, R., & Okuzumi, S. 2021, ApJ, 912, 37

Olkin, C. B., Young, L. A., French, R. G., et al. 2014, Icar, 239, 15

Owen, T. C., Roush, T. L., Cruikshank, D. P., et al. 1993, Sci, 261, 745

Pasachoff, J. M., Souza, S. P., Babcock, B. A., et al. 2005, AJ, 129, 1718

Person, M. J., Elliot, J. L., Gulbis, A. A. S., et al. 2008, AJ, 136, 1510

Pollack, J. B., Schwartz, J. M., & Rages, K. 1990, Sci, 250, 440

Pruppacher, H. R., & Klett, J. D. 1978, Microphysics of Clouds and Precipitation (Dordrecht: Reidel Publishing)

Rages, K., & Pollack, J. B. 1992, Icar, 99, 289

Rannou, P., Curtis, D., & Tolbert, M. A. 2019, A&A, 631, A151

Rannou, P., & West, R. 2018, Icar, 312, 36

Rannou, P., Cabane, M., Botet, R., & Chassefière, E. 1997, JGR, 102, 10997

Smith, B. A., Soderblom, L. A., Banfield, D., et al. 1989, Sci, 246, 1422

Stansberry, J. A., Lunine, J. I., & Tomasko, M. G. 1989, GeoRL, 16, 1221

Stern, S. A., Kammer, J. A., Barth, E. L., et al. 2017, AJ, 154, 43

Strobel, D. F., & Summers, M. E. 1995, Neptune and Triton (Tuscon, AZ: Univ. Arizona Press) 1107

Strobel, D. F., Summers, M. E., Herbert, F., & Sandel, B. R. 1990, GeoRL, 17, 1729

Thompson, W. R., Singh, S. K., Khare, B. N., & Sagan, C. 1989, GeoRL, 16, 981

Toigo, A. D., French, R. G., Gierasch, P. J., et al. 2015, Icar, 254, 306

Tryka, K. A., Brown, R. H., Cruikshank, D. P., et al. 1994, Icar, 112, 513

Tyler, G. L., Sweetnam, D. N., Anderson, J. D., et al. 1989, Sci, 246, 1466

Vuitton, V., Lavvas, P., Yelle, R. V., et al. 2009, P&SS, 57, 1558

Vuitton, V., Yelle, R. V., Klippenstein, S. J., Hörst, S. M., & Lavvas, P. 2019, Icar, 324, 120

West, R. A., & Smith, P. H. 1991, Icar, 90, 330

Wilson, E. H., & Atreya, S. K. 2004, JGRE, 109, E06002

Wong, M. L., Fan, S., Gao, P., Liang, M.-C., et al. 2017, Icar, 287, 110

Young, E. F., French, R. G., Young, L. A., et al. 2008, AJ, 136, 1757

Young, L. A. 2017, Icar, 284, 443

Young, L. A., Elliot, J. L., Tokunaga, A., de Bergh, C., & Owen, T. 1997, Icar, 127, 258

Young, L. A., Kammer, J. A., Steffl, A. J., et al. 2018, Icar, 300, 174

Yu, X., Yu, Y., Garver, J., et al. 2023, ApJS, 266, 30

Yung, Y. L., Allen, M., & Pinto, J. P. 1984, ApJS, 55, 465

Yung, Y. L., & Lyons, J. R. 1990, GeoRL, 17, 1717

Zhang, X., Strobel, D. F., & Imanaka, H. 2017, Natur, 551, 352

Chapter 7

Magnetospheric and Space Environment Interactions with the Upper Atmosphere and Ionosphere

Tom A Nordheim, Adrienn Luspay-Kuti, Lucas Liuzzo,
Peter Gao and G Randy Gladstone

7.1 Introduction

Triton, with its retrograde and highly inclined orbit, is a likely captured Kuiper Belt Object (see Chapter 1). This places the moon in the same family as several dwarf planets in the Outer Solar System, most notably Pluto, whose orbital distance at perihelion is comparable to that of Neptune and Triton. Furthermore, these two icy worlds are of comparable size, mass, and density, and have extremely cold ($\sim$30–40 K) surfaces rich in volatiles, including N_2, CH_4, and CO. Both worlds also possess tenuous but significant collisional atmospheres (Olkin et al. 2015; Strobel & Zhu 2017) sustained by volatile sublimation. However, unlike Pluto, Triton orbits within Neptune's magnetosphere and is exposed to magnetospheric plasma and trapped energetic charged particles that precipitate into its upper atmosphere, providing an additional source of heating as well as ionizing atmospheric neutrals. This is highlighted by the fact that Triton's ionosphere was found by *Voyager 2* to be surprisingly dense, while *New Horizons* failed to detect an ionosphere at Pluto. As the two worlds have similar solar inputs, this suggests that the magnetospheric charged particle input may be highly important at Triton. At the same time, *Voyager 2* detected a significant heavy ion component in Neptune's magnetosphere, indicating that Triton is likely a significant source of plasma to Neptune's magnetosphere. Neptune, its magnetosphere, and Triton's upper atmosphere therefore represent a coupled system, and a holistic system-level approach will be required to further our understanding of the individual components. Furthermore, Neptune's unusual tilted magnetic field leads to a complex and highly variable magnetospheric environment along Triton's orbit, and the moon's magnetospheric interaction was poorly constrained by the Voyager 2 observations. These factors challenge our understanding of the Neptune-Triton system. Here we will provide an overview of the upper atmospheres and ionospheres of Triton and Pluto, including the interaction of these objects with their respective space environments. We further discuss the possible role of magnetospheric particle precipitation as a dominant energy input at Triton, and compare its upper atmosphere and ionosphere to that of Pluto based on the new insights from the recent *New Horizons* flyby of the dwarf planet.

7.2 Neutral Atmospheres

Understanding how an atmosphere interacts with its environment requires knowledge of the neutral composition, surface temperature and pressure, ionospheric density and structure, sources of energy, and thermal structure as a function of distance from the surface. Limited information is available for Pluto and Triton, with the most valuable details provided by spacecraft measurements made during a single flyby of each. *Voyager 2* carried out a flyby of the Neptune system in 1989, measuring the composition and structure of the atmosphere and ionosphere of Triton using the *Voyager 2* Ultraviolet Spectrometer (UVS; Broadfoot et al. 1989) and Radio Science System (RSS; Tyler et al. 1989). *New Horizons* flew by Pluto in 2015 and measured the composition and structure of the atmosphere with the Alice Ultraviolet Spectrograph (Gladstone et al. 2016; Young et al. 2018) and determined an upper limit for the ionospheric density using the Radio Science Experiment (REX) (Hinson et al. 2018).

Both Pluto and Triton have atmospheres that are predominantly (>99%) N_2 with minor amounts of CH_4 and CO. At Triton, the atmosphere was found to have a very low CH_4 abundance of only ~0.01% during the *Voyager 2* flyby (Broadfoot et al. 1989; Herbert & Sandel 1991). Subsequent ground-based observations found that the CH_4 abundance had increased by ~ 4x in the time since the *Voyager 2* flyby, and also made the first measurement of CO in Triton's atmosphere at an abundance of 0.05%, similar to the surface CO ice abundance (Lellouch et al. 2010). At Pluto, the neutral atmosphere was measured during the *New Horizons* flyby to contain ~0.30% CH_4 (Young et al. 2018) and ~0.05% CO (Lellouch et al. 2017). Trace amounts of ethane (C_2H_6), acetylene (C_2H_2) and ethylene (C_2H_4) were also detected at Pluto with abundances on the order of 0.1% in the middle atmosphere but unexpectedly dropping to between 10^{-3} and 10^{-5}% at ~100 km altitude (Young et al. 2018). These hydrocarbon species are likely also present at Triton (e.g., see discussion in Broadfoot et al. 1989), but have as of yet not been detected.

Pluto and Triton had similar surface temperatures during their respective spacecraft flybys, 37±3 K (Gladstone et al. 2016) and 38^{+3}_{-4} K, respectively. Both also had fairly low surface pressures of ~10 μbar for Pluto and ~14 μbar for Triton.

Energy input into an atmosphere determines the temperature profile and drives chemistry. Both Pluto and Triton receive energy from the Sun in the form of photons emitted at a broad range of wavelengths, from infrared through X-ray. They also receive energy from photons at Lyman-alpha wavelengths emitted by interstellar hydrogen emission (Bertaux & Blamont 1971) as well as from galactic cosmic rays. In addition to heating, these energy inputs also lead to dissociation and ionization of the primary molecules in the atmosphere, leading to complex chemistry that eventually produces haze particles. This process is reviewed in detail in Chapter 6 of this book (Chapter 6). What is most notable is that very little haze was observed at Triton, extending only up to 30 km from the surface, while Pluto's atmosphere was observed to have extensive layers of haze extending up to altitudes of at least 500 km.

Precipitating ions and electrons from the surrounding environment are also a source of energy for these atmospheres. Pluto interacts directly with the solar wind, which contains electrons as well as protons, alpha particles, and heavier ions with $\sim$ eV to $\sim$ keV kinetic energies. *New Horizons* observations of pickup ions originating from Pluto showed that methane was the primary molecule lost from the atmosphere at a rate similar to what is expected for Jeans escape (Bagenal et al. 2016). Triton, on the other hand, is located inside Neptune's magnetosphere where magnetospheric ions and electrons with $\sim$eV to MeV energies precipitate into the upper atmosphere and may provide a significant energy source (e.g. Sittler & Hartle 1996).

Although both atmospheres have similar surface pressures and temperatures, the difference in methane abundance, a powerful greenhouse gas, and the additional source of energy from magnetospheric particle precipitation at Triton, leads to very different structures in the temperature profile of the two atmospheres. We compare Pluto and Triton based on the *New Horizons* (Gladstone et al. 2016) and *Voyager 2* (Krasnopolsky & Cruikshank 1995) observations in Figure 7.1, where the exobase is illustrated for both atmospheres using a dashed line. It is first notable that Pluto's collisional atmosphere extends twice as far from the surface as Triton's as a result of more efficient heating due to methane and other greenhouse gases, which is most effective at the lowest altitudes. By comparison, Triton's atmosphere experiences the greatest heating in the upper atmosphere where magnetospheric particle precipitation provides an additional energy input.

7.3 Ionospheres

The only constraints available for the ionospheres of Triton and Pluto are from radio occultation measurements taken during the respective *Voyager 2* and *New Horizons* flybys. For both Triton and Pluto, near-terminator radio occultation measurements were taken on ingress and egress. While a very robust ionosphere was detected at Triton, *New Horizons* failed to detect any signature of an ionosphere at Pluto (Hinson et al. 2018). We illustrate these observations in Figure 7.2 and compare them with the predicted total electron density for Pluto's atmosphere from modelling by Luspay-Kuti et al. (2017). The blue curve represents the predicted densities for an ionosphere at Pluto, which fall well below the lower detection limit of $\sim$1000 cm^{-3} for the *New Horizons* radio occultation experiment (blue shaded region). For comparison, the two ionospheric electron profiles returned by the *Voyager 2* radio science occultations of Triton are also included as the black curves. Finally, measured total ion densities for Titan from Mandt et al. (2012) are included as the yellow points. While Titan differs from Pluto and Triton in several key ways, it nonetheless represents another example of a cold upper atmosphere and

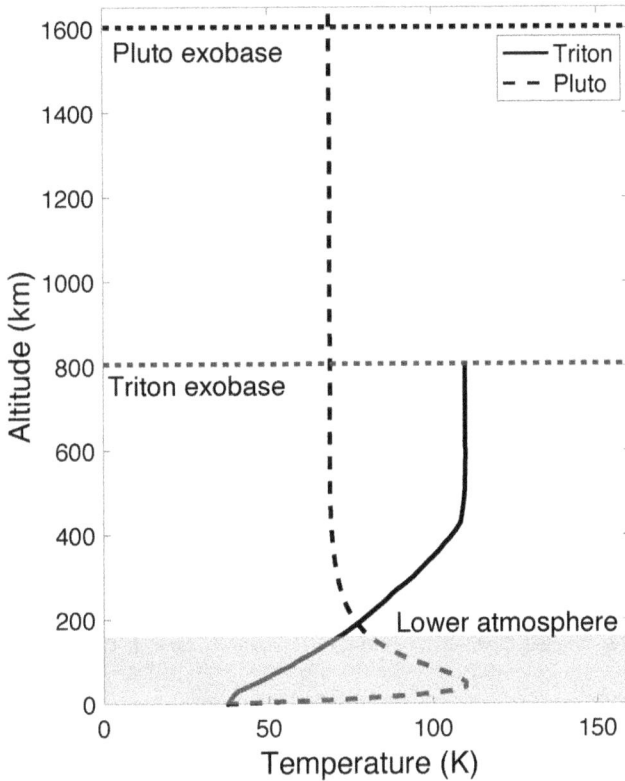

Figure 7.1. Temperature as a function of altitude for the atmospheres of Pluto (Gladstone et al. 2016; dashed) and Triton (Tyler et al. 1989; solid). The gray shaded area is the lower atmosphere. The upper atmosphere consists of the thermosphere and exosphere, which are separated by the exobase (thick dashed horizontal lines).

ionosphere dominated by nitrogen and hydrocarbon chemistry, with input from both solar photoionization and charged particle precipitation. From this comparison it becomes immediately obvious that Triton's ionosphere stands out, with 1–2 orders of magnitude higher ionospheric electron densities than those at Pluto and Titan.

New Horizons carried out radio occultation measurements of Pluto's atmosphere, at solar zenith angles of 90.2° (ingress, sunset) and 89.8° (sunrise, egress). No signature of an ionosphere was detected during these occultation measurements (Hinson et al. 2017, 2018). The instrument 1-sigma sensitivity was estimated to correspond to an integrated electron content (IEC) of 2.3×10^{11} cm^{-2}, corresponding to a peak of ionospheric density in Pluto's terminator region of no more than $\sim$1000 cm^{-3}. Hinson et al. (2018) constructed a simple ionospheric model for Pluto, predicting a peak IEC of 1.8×10^{11} cm^{-2}, which is seemingly consistent with the non-detection of an ionosphere by the *New Horizons* REX. These authors argue that a key reason for the comparatively low ionospheric densities at Pluto is the higher abundance of CH$_4$ at ionospheric altitudes, leading to rapid removal of atomic ions through ion-neutral interactions with CH$_4$. The resulting molecular ions are readily

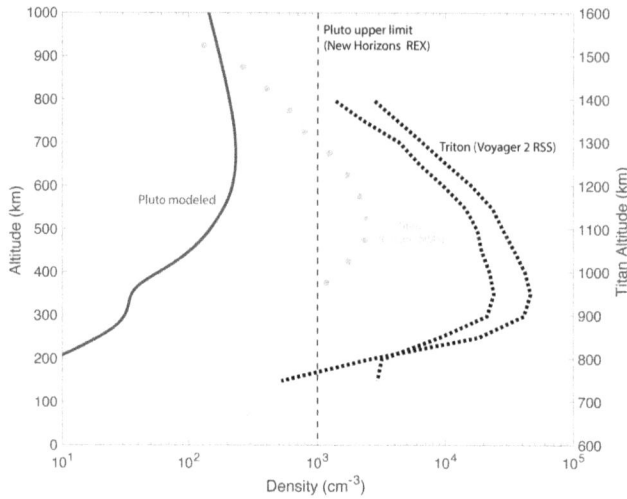

Figure 7.2. Altitude profiles of the electron densities observed at Triton by *Voyager 2* (black dashed lines; Tyler et al. 1989), the upper limit for Pluto's ionosphere based on *New Horizons* (Dashed vertical line: Hinson et al. 2018). Predicted ionospheric electron densities at Pluto from Luspay-Kuti et al. (2017) are shown with the solid blue line to be well below the upper limit for ionospheric electron density determined by *New Horizons*. These are compared to the peak daytime ion densities measured by *Cassini* at Titan (Mandt et al. 2012, yellow points).

lost through dissociative recombination. Subsequent ionospheric modelling by Krasnapolsky (2020) predicted a maximum peak ionospheric density of ~ 800 cm^{-3}, again consistent with the non-detection of an ionosphere at Pluto by *New Horizons*.

The radio occultation observations of Triton's atmosphere and ionosphere by the *Voyager 2* RSS revealed that the Tritonian ionosphere is unusually dense ($n_e \sim 10^4$ cm^{-3}) with an electron density peak located at ~ 350 km altitude (Tyler et al. 1989). The electron peak densities were measured as $\sim 2.3 \times 10^4$ cm^{-3} (350 km) and $\sim 4.6 \times 10^4$ cm^{-3} (340 km) during the RSS ingress and egress occultations, respectively (Figure 7.3). For comparison, this is comparable to, but somewhat larger than the electron densities in Callisto's sunlit ionosphere ($\sim 1.5 \times 10^4$ cm^{-3}) (Kliore et al. 2002) and much larger than those found in Titan's ionosphere (~ 1–4×10^3 cm) (Kliore et al. 2008). This is surprising, as Triton receives a significantly lower flux of ionizing solar photons than these moons, e.g., a factor of ~ 600x and ~ 400x, respectively.

Initial attempts at ionospheric modelling determined that the dominant ion in Triton's ionosphere is N$^+$, and that the ionospheric electron densities observed by *Voyager 2* RSS could only be reproduced when including magnetospheric electron precipitation as the major source of ionization (Ip, 1990; Lellouch et al. 1992; Majeed et al. 1990; Strobel et al. 1990; Yung & Lyons 1990). Weak EUV emissions from the sunlit hemisphere detected by *Voyager 2* UVS may also be consistent with significant magnetospheric electron precipitation (Broadfoot et al. 1989). However, subsequent modeling that included more coupled C-N chemistry proposed that it was possible to reproduce the magnitude and general structure of the Tritonian

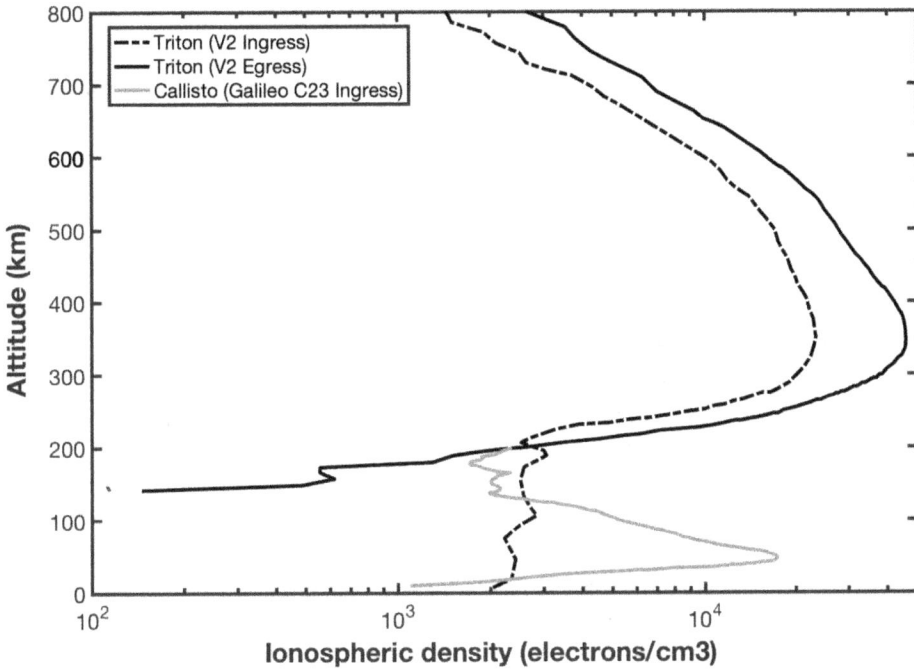

Figure 7.3. Total ionospheric electron number density at Triton as derived from Voyager 2 RSS occultations taken during the Voyager 2 flyby. Shown for comparison is the most intense ionospheric profile (C23 Ingress) observed by *Galileo* at Callisto in terms of peak electron densities. Adapted from Tyler et al. (1989) and Kliore et al. (2002).

ionosphere with solar EUV as the only major ionization source (Lyons et al. 1992). These authors suggested that C^+, rather than N^+, is the major ion in Triton's ionosphere assuming that charge exchange reaction $N_2^+ + C \rightarrow C^+ + N_2$ was taking place at very high reaction rates compared to similar type reactions. Krasnopolsky & Cruikshank (1995) used a lower rate more in line with similar type reactions and also found that C+ was the dominant ionospheric species. However, they concluded that ionospheric chemistry at altitudes above 200 km is driven by solar EUV **and** magnetospheric electrons, requiring an energy input from magnetospheric electrons that is twice that from the solar EUV.

It was also noted that the observed thermospheric temperatures could not be produced by solar EUV alone, perhaps requiring additional input from magnetospheric electrons to explain the heating (e.g., Elliot et al. 2000; Stevens et al. 1992). More recently, Strobel & Zhu (2017) found that roughly equal amounts of solar EUV and magnetospheric particle heating are required to explain the neutral densities in Triton's atmosphere observed by *Voyager 2* UVS (Broadfoot et al. 1989), again indicating that magnetospheric particle precipitation plays an important role. Lellouch et al. (1992) determined that for Triton's ionosphere to be driven primarily by magnetospheric ion precipitation, the precipitating distribution would have to have a peak at unrealistically high energies (e.g., ~1.2 MeV). Thus, the modelling studies in the literature have focused primarily on magnetospheric electron precipitation as an alternative (or complement) to solar photoionization.

It should be noted, however, that the reaction rate coefficient for the charge exchange reaction employed by Lyons et al. (1992) and Krasnopolsky & Cruikshank (1995) is highly uncertain. In order to generate the ionospheric densities observed by *Voyager 2* without magnetospheric input, the models by Lyons et al. (1992) required an unrealistically high rate coefficient for the charge exchange reaction producing C^+. Using that coefficient, this model was able to reproduce the dense ionosphere by producing large amounts of C^+, but failed to reproduce the profile of atomic N measured by the *Voyager 2* UVS between 170 km and 570 km (Krasnopolsky et al. 1993). Krasnopolsky & Cruikshank (1995) considered a total electron energy input based on thermal balance calculations, and concluded that an energy input from magnetospheric electrons that is 2–3 times higher than that from solar UV radiation is necessary to reproduce the measured atomic N profile.

Our understanding of ion-neutral chemistry has increased significantly since the *Voyager 2* era, and recent modeling for Pluto has indicated that the dominant ionospheric ions at Pluto are molecular ions: $HCNH^+$ above and $C_9H_{11}^+$ below the altitude of 600 km (Krasnopolsky 2020), with no indication that the $N_2^+ + C \rightarrow C^+ + N_2$ and other charge exchange reactions are important in Pluto's atmosphere. However, the relative importance of atomic and molecular ions in these atmospheres is predicted to depend strongly on the CH_4 surface mole fraction (Krasnopolsky 2012). These results suggest that at higher surface CH_4 mole fractions (e.g., induced by seasonal variations) Triton's ionosphere would transition to a different chemistry dominated by molecular rather than atomic ions (e.g. "*New Horizons* type" as termed by Krasnoposly 2012). However, with only one ionospheric occultation measurement and no information on the actual ionospheric composition at Triton it is currently not possible to assess the validity of these predictions. An important clue would be if the change in seasons and increasing surface CH_4 mole fraction at Triton would also lead to a drop in the total ionospheric density, which would be expected from the hypothesis of Krasnopolsky (2012) if the chemistry were indeed mainly driven by photoionization.

Most recently, Benne et al. (2022) used the post-*Cassini* chemical scheme of Dobrijevic et al. (2016) developed for Titan and modified it by including association and charge exchange reactions from the early, *Voyager 2*-era studies. They also implemented the approach of Krasnopolsky & Cruikshank (1995) for the treatment of electron impact ionization, using the speculated ionization profiles from Strobel et al. (1990) and arbitrarily moving them up by two scale heights (e.g. as in Summers & Strobel 1991). Similarly, to Krasnopolsky & Cruikshank (1995), they also find that ionization by magnetospheric electrons is ~2.5 times more important than photoionization in producing Triton's ionosphere. However, the predicted C^+ densities in their model is overproduced compared to the total electron density (or ion density, assuming charge neutrality) observed by *Voyager 2* radio occultations.

There is currently no clear consensus on whether Triton's ionosphere is predominantly driven by solar photoionization, magnetospheric charged particle precipitation, or some combination of the two. The large discrepancy between model results highlights the many unknowns or poorly constrained parameters for Triton's neutral atmosphere, ionosphere, and magnetospheric charged particle input. Specifically, the model results are particularly sensitive to the assumed rate coefficient for the $N_2^+ + C \rightarrow C^+ + N_2$ charge exchange reaction. Reaction rate coefficients are generally poorly constrained at the low temperatures relevant to these outer planetary bodies, but the key charge exchange reaction that current photochemical models heavily rely on is almost completely unconstrained.

Another key aspect and major source of uncertainty is our understanding of precipitating magnetospheric charged particles to Triton's upper atmosphere and ionosphere. Some of the published modelling studies (e.g., Yung & Lyons 1990) approximated the magnetospheric energy contribution as a monoenergetic beam of electrons at the top of Triton's atmosphere. This approach does not consider the electron input as a function of energy, which is crucial because electrons of different energies lead to ionization at different altitudes in the atmosphere. Other studies (Ip, 1990; Lellouch et al. 1992; Sittler & Hartle 1996; Strobel et al. 1990) implemented an energy spectrum for the incident magnetospheric electrons at the top of the atmosphere. However, the overall flux and shape of this spectrum was poorly constrained by the *Voyager 2* measurements and was not consistent between the different studies, again likely contributing to the different model outcomes.

7.4 Pluto's Solar Wind Environment and Atmospheric Ion Escape

Pluto's interaction with its space environment is likely that of an unmagnetized body embedded within the solar wind. Thus, prior to the *New Horizons* flyby, it was predicted that Pluto may have a comet-syle interaction, exemplifying a more extended obstacle, or a more compact Venus-like interaction (Bagenal et al. 2015, 2021; Bagenal & McNutt 1989). Firstly, the size of the interaction region depends strongly on the atmospheric escape flux. Secondly, it was realized that due to the low interplanetary magnetic field (IMF) strength at Pluto's heliospheric distance ($\sim$0.1 nT, Table 7.1), the gyroradii of heavy atmospheric pickup ions such as CH_4^+ and N_2^+ would be $\sim 10^2$–10^3 Pluto radii, many times larger than the region defined by Pluto's interaction with the Solar Wind. Thus, it is clear that kinetic effects are important, and that the Pluto interaction cannot be explained by a purely fluid-like (e.g. magnetohydrodynamic) picture (Delamere & Bagenal 2004). Prior to the *New Horizons* flyby, it was expected that, due to Pluto' low surface gravity, the neutral escape flux at the top of its atmosphere would be large, on the order of 10^{28}–10^{29} amu s^{-1} (e.g., Zhu et al. 2014).

Pluto's upper atmosphere was found by *New Horizons* to be cooler and thus significantly more compact than expected, leading to an estimated atmospheric neutral escape rate of only $\sim 10^{27}$ amu s^{-1} (Strobel & Zhu 2017; Gladstone et al. 2016; Hinson et al. 2017; Young et al. 2018). In addition to this substantially lower-than-expected atmospheric escape rate, the Solar Wind ram pressure was found to be significantly enhanced (by a factor of $\sim$3$\times$) during the *New Horizons* encounter

Table 7.1. Representative Background (Upstream) Solar Wind Parameters Applicable to Pluto at a Heliospheric Distance of ~33 AU

	Predicted value at ~33 AU	Observed by *New Horizons*
Interplanetary magnetic field magnitude	0.08–0.28 nT	0.1–0.3 nT
Solar Wind proton density	0.0015–0.01 cm^{-3}	0.025 cm^{-3}
Solar wind proton temperature	0.16–1.5 eV	0.66 eV
Solar wind flow speed	340–480 km s^{-1}	403 km s^{-1}
Solar wind proton ram pressure	0.32–4.0 pPa	6.0 pPa
Alfven speed	22–96 km s^{-1}	14–41 km s^{-1}
Alfvenic Mach number	4.6–20	9.8–29
Magnetosonic Mach number	5–17	9.5–23

The second column shows predicted values based on Voyager 2 heliospheric measurements and appropriate scalings from Bagenal et al. (2015). The third and last column lists the measurements made by the Solar Wind Around Pluto (SWAP) instrument onboard *New Horizons* at the time of the Pluto flyby (Bagenal et al. 2021; McComas et al. 2016). Note that the IMF strength was not directly measured as *New Horizons* did not carry a magnetometer, but rather Inferred from SWAP observations and modelling.

with Pluto, likely due to the arrival of a strong interplanetary shock shortly before the encounter (see Table 7.1) (McComas et al. 2016). Thus, *New Horizons* and its Solar Wind at Pluto (SWAP) instrument observed the Pluto interaction region in a compressed state, which may not be representative of its state during typical quiescent solar wind conditions. McComas et al. (2016) estimated a solar wind standoff distance of only ~2.5 Rp during the encounter, with an estimated plutopause boundary thickness of ~0.9 Rp (see Figure 7.4). Given the observed Solar Wind standoff distance and interaction region seen by *New Horizons*, it was possible to set an upper limit on Pluto's surface dipole magnetic field of < 30 nT, confirming that Pluto is an unmagnetized or only weakly magnetized body (McComas et al. 2016). SWAP observed Pluto to have a heavy ion tail that extended to at least 100 Rp downstream, with an estimated circular diameter of ~30 Rp at that distance. Furthermore, the SWAP observations indicated that the heavy ion tail is structurally variable with significant North-South asymmetries present. Zirnstein et al. (2016) argued that the detected plutogenic heavy ions were likely CH_4^+ and McComas et al. (2016) estimated a CH_4^+ escape rate of ~8 × 10^{24} amu s^{-1}, roughly 1% of the estimated neutral escape rate. Kollmann et al. (2019) reported on observations by the *New Horizons* Pluto Energetic Particle Spectrometer Science Investigation (PEPSSI), which indicated a surprisingly significant interaction between the Pluto obstacle and Interstellar Pickup Ions (IPUIs; pick-up ions generated by interstellar neutrals that penetrate the solar system), including an IPUI wake region that extended to a distance of ~190 Rp downstream of Pluto. Based on the *New Horizons* observations, it was concluded that Pluto's interaction with the

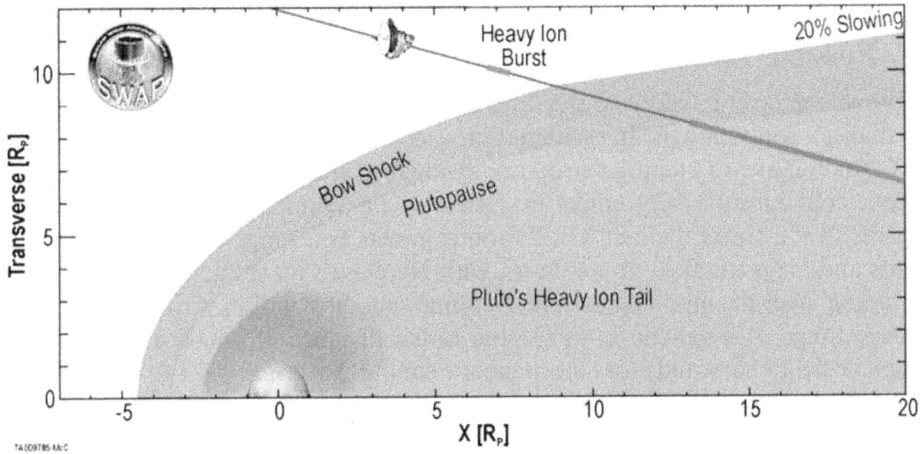

Figure 7.4. Schematic of the Pluto–Solar Wind interaction region as observed by *New Horizons* during its flyby. Reprinted with permission from Barnes et al. (2019).

solar wind is a hybrid of the predicted comet-like and Venus-like interaction modes, with the upstream bow shock generated by mass loading as is the case with comets, but the pressure balance diverting the shocked solar wind flow is sustained by atmospheric thermal pressure like it is at Venus (e.g., Bagenal et al. 2021 and references therein). Based on hybrid modeling Feyerabend et al. (2017) were able to reproduce the main characteristic features (e.g., location and thickness of the bow shock and plutopause, heavy ion tail with significant N–S asymmetry) seen by *New Horizons* SWAP at Pluto. Hale & Paty (2017) carried out MHD simulations and found that the presence of Charon can significantly affect the morphology of the solar wind interaction region, and also partially shield Pluto from the solar wind when located upstream.

Barnes et al. (2019) compared hybrid simulations to SWAP observations during the *New Horizons* encounter and found that while in many respects, the solar wind interaction is similar to that of a weak comet, Pluto's broad heavy ion tail is unique and shaped by the combination of the very low IMF and non-negligible thermal pressure provided by IPUIs.

The observations briefly summarized above indicate that Pluto displays a unique interaction with the solar wind, although possibly shared with other Kuiper Belt Objects, and many questions remain unanswered after the brief snapshot provided by the *New Horizons* flyby. In particular, the fact that the encounter appeared to coincide with significantly enhanced solar wind pressure indicates that the Pluto interaction region would be much larger in scale during quiescent solar wind conditions if the neutral escape rate stayed constant. Furthermore, it is expected that Pluto's atmosphere varies significantly over seasonal and orbital timescales (Bertrand et al. 2019; Krasnopolsky 2020), likely also driving large changes in the neutral atmospheric escape rate and modifying Pluto's interaction with the Solar Wind.

7.5 Neptune's Magnetospheric Environment and Triton's Role as a Possible Plasma Source

Neptune's magnetic field is highly tilted ($\sim$47° for the dipole term) with respect to the planet's rotation axis. In combination with the large ($\sim$156°) orbital inclination of Triton relative to Neptune's rotational equator, this leads to a unique magnetospheric configuration very unlike that seen at, e.g., Jupiter and Saturn's satellites. Figure 7.5 illustrates the fact that Triton experiences a range of magnetic environments and magnetic field orientations, with Neptune's magnetic field lines in effect 'tumbling' over the moon across one Neptune synodic rotation. Triton orbits within a large range of magnetic L shells (the radial distance at which a given dipolar magnetic field line would cross the magnetic equator), from $L=14.3$ to 40 (Ness et al. 1989; Strobel et al. 1990). This also implies that Triton samples a diverse range of magnetospheric charged particle environments as it traverses the different regions of the Neptunian magnetosphere. Shown in Figure 7.6 is a plot of the magnetospheric plasma densities measured by *Voyager 2* during the inbound and outbound legs of its Neptune flyby organized by L-shell. While measurements at Triton's largest L-shell

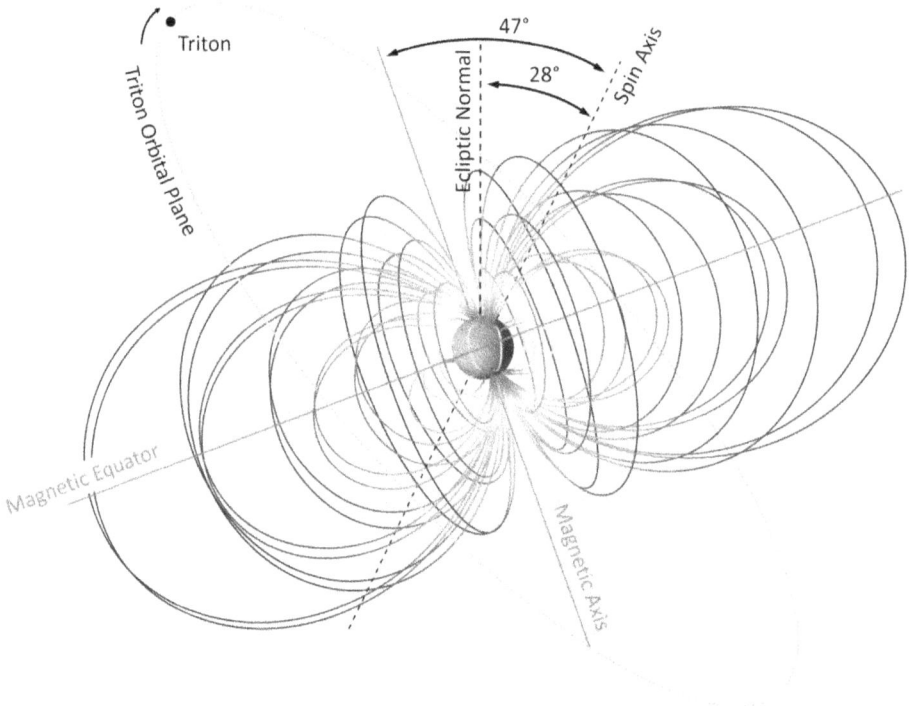

Figure 7.5. Illustration of the Neptune-Triton system, showing the tilt of Neptune's spin axis, the tilt of Neptune's magnetic axis with respect to its spin axis, the orbit of Triton and the magnetic equator populated with corotating plasma particles. Adapted with permission from Cochrane et al. (2022). CC BY-NC. See also Chapter 6 for a further discussion regarding Triton's interaction with Neptune's magnetic field.

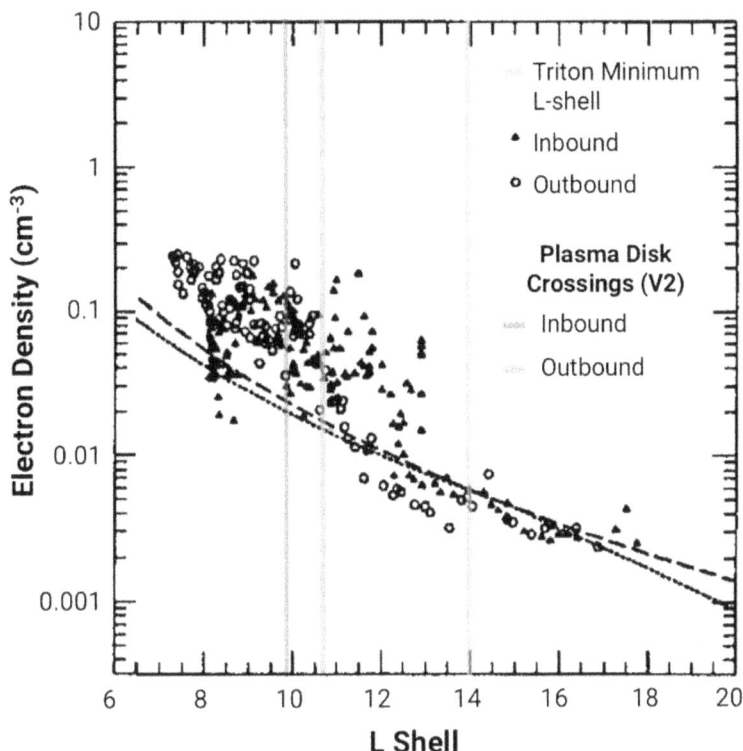

Figure 7.6. The magnetospheric electron density observed on the inbound (triangles) and outbound (circles) segments of the Voyager 2 flyby of Neptune. Observations are organized by magnetic L-shell from 6 to 20. The magnetic equator/plasma disk crossings are indicated for inbound and outbound. The dashed and dotted black lines are theoretically predicted plasma densities from two simple models (a constant flux shell model and a diffusion model), neither of which predict the actual densities well. Adapted with permission from Zhang et al. (1991).

excursions are lacking, this nonetheless demonstrates the inherent variability of the moon's ambient magnetospheric charged particle environment over the course of even a single synodic rotation. In addition to variations in the overall electron number density it is also conceivable, but currently not constrained by the observations, that the shape of the magnetospheric electron spectrum is also variable, e.g., as observed by *Cassini* at Titan (Rymer et al. 2009).

Our current best understanding of Triton's magnetospheric environment is based on measurements taken by *Voyager 2* around the time of its Triton flyby and summarized in Table 7.2 below. We emphasize that this was a distant flyby, where the closest approach to the moon was 39,800 km, and therefore the spacecraft never directly measured Triton's **local** magnetospheric environment. Furthermore, these numbers are representative of the ambient magnetospheric conditions during the flyby, when Triton was located near its **minimum** L-shell (~14) and are not necessarily applicable to Triton at other magnetic configurations. Therefore,

Table 7.2. Representative Background (Upstream) Magnetospheric Parameters Applicable to Triton During the *Voyager 2* Flyby, when the Moon was Located near L~14 in Neptune's Magnetosphere

Parameter	Value	Source
Background magnetic field magnitude	8 nT	Ness et al. (1989); Strobel et al. (1990)
Magnetospheric ion density	Total: 3×10^{-3} cm^{-3} H$^+$: 1.5×10^{-3} cm^{-3} N$^+$: 1.5×10^{-3} cm^{-3}	Sittler & Hartle (1996); Zhang et al. (1991)
Magnetospheric ion temperature	H$^+$: 16 eV N$^+$: 100 eV	Richardson et al. (1991); Sittler & Hartle (1996)
Magnetospheric electron temperature	300 eV	Zhang et al. (1991)
Plasma flow speed relative to Triton	43 km s^{-1}	Strobel et al. (1990)
Plasma β	0.086	Sittler & Hartle (1996)
Alfven speed	1160 km s^{-1}	
Sound speed	95 km s^{-1}	
Alfvenic Mach number	0.037	
Sonic Mach number	0.45	

significant uncertainties exist regarding Triton's magnetospheric environment and the nature and strength of the magnetospheric electron precipitation source along Triton's full orbit.

Our best available estimate for the magnetospheric electron spectrum at Triton during the *Voyager 2* flyby is shown in Figure 7.7 (Sittler & Hartle 1996). This spectrum is derived from a Maxwellian fit to the *Voyager 2* Plasma Science Experiment (PLS; Zhang et al. 1991) and a power-law fit to measurements by the Low Energy Charged Particle Experiment (LECP; Krimigis et al. 1989; Mauk et al. 1991; Strobel et al. 1990) taken near Triton's minimum L-shell (L~14). Notably, there was a gap in measurement capability between the PLS and LECP instruments in the 6 to 20 keV range, so the magnetospheric electron spectrum is unconstrained in this region and populated by extrapolation.

Sittler and Hartle (1996) studied Triton's magnetospheric interaction and, based on analogies to Io and Venus, concluded that Triton's highly conductive ionosphere (Strobel et al. 1990) would likely stand off and divert the magnetospheric plasma flow, preventing a large fraction of magnetospheric electrons from reaching Triton's upper atmosphere and ionosphere. They estimated that a majority of $E < 2$ keV electrons would be diverted around Triton due to their $E \times B$ drifts and therefore not have access to the ionosphere, while only 30% of electrons with $E > 20$ keV would reach the ionosphere. This emphasizes that merely knowing the magnetospheric

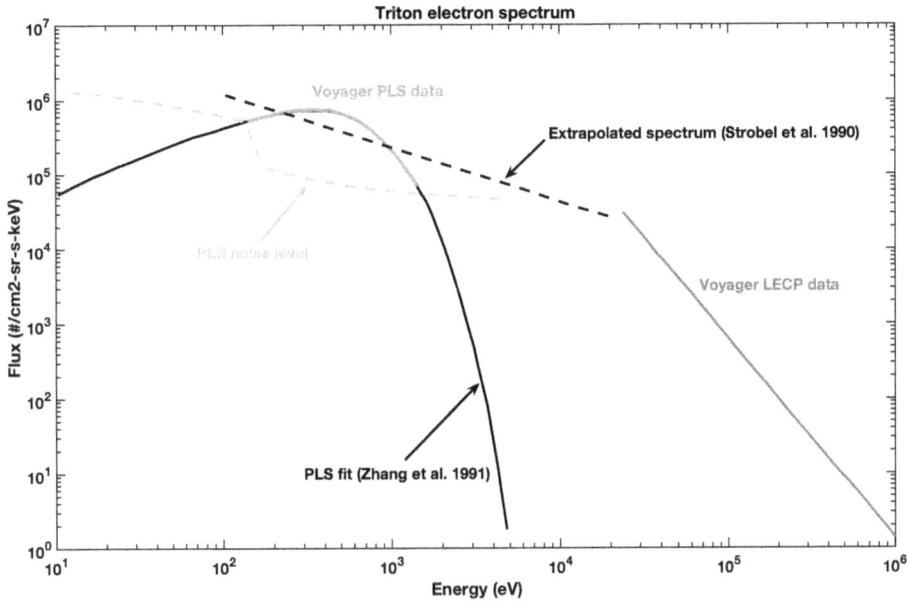

Figure 7.7. Current best estimate for the ambient magnetospheric electron spectrum at Triton during the *Voyager 2* flyby when Triton was located at $L\sim14$. The green curve shows the part of the spectrum populated by a Maxwellian fit to the PLS measurements by Zhang et al. 1991 and the red curve shows a power-law fit derived from LECP measurements (Krimigis et al. 1989; Strobel et al. 1990). The two parts of the spectrum are separated by a measurement gap (blue shaded region) and joined by a simple straight-line extrapolation. Adapted with permission from Sittler & Hartle (1996).

electron spectrum near Triton is insufficient to understand the energy input to Triton's upper atmosphere. The spectrum of magnetospheric electrons precipitating at Triton will be determined by the moon's interaction with Neptune's magneto-sphere and the perturbed local electromagnetic fields created as a response to this interaction.

Pre-encounter models predicted that Triton would be associated with a neutral/ plasma torus (Delitsky et al. 1989). However, although the density of heavy ions in the Neptunian magnetosphere agrees reasonably well with the earlier model predictions, these ions were only observed during the plasma sheet crossings and not when *Voyager 2* crossed Triton's orbit (Belcher et al. 1989; Zhang et al. 1992). On the other hand, the inferred mass (10–40 amu, consistent with N^+) and average temperature (60–100 eV, consistent with pick-up at Triton's minimum L-shell) of the observed heavy ions is consistent with a Triton source. Belcher et al. (1989) calculated a pick-up energy of 120 eV for N^+ at rigid corotation. Cheng (1990) argued that Triton is the dominant source of plasma outside of $L\sim7$ in the Neptunian magnetosphere and Sandel et al. (1990) suggested that a neutral source of ~1 kg s^{-1} from Triton's atmosphere could explain the observed power of the Neptunian aurora. Richardson et al. (1990) found that sputtering from Triton's atmosphere could produce the observed heavy ion densities at Triton's orbit if the ion residence time is ~30 days. However, these authors also found that the Triton neutral source rate of ~1 kg s^{-1} from Sandel et al. (1990) is inconsistent with the *Voyager 2* PLS and UVS measurements. A subsequent study by Decker & Cheng (1994) found similar inconsistencies between the inferred plasma source rates and the predicted neutral escape rates from Triton's atmosphere. Measurements of hot plasma and energetic charged particles by the Voyager LECP instrument found that these populations appear to be strongly affected by Triton and/or interactions with the hypothetical Triton neutral torus (Mauk et al. 1991) and that a distinct trans-Triton heavy ion population exists outside the minimum L-shell of Triton. While Yung & Lyons (1990) concluded that the majority of nitrogen loss from Triton's upper atmosphere was in the form of direct ion escape, subsequent analyses by Richardson et al. (1990) and Summers & Strobel (1991) indicate that most of the nitrogen escaping Triton is in the form of neutrals. In this scenario, the magnetospheric N$^+$ ions would then be sourced from subsequent ionization of the torus neutrals and not locally at Triton.

7.6 Triton's Interaction with its Magnetospheric Plasma Environment

The time variability in the magnetospheric field near Triton's orbit drives currents in conducting layers at the moon that manifest as an induced magnetic field. These time-variable currents may be induced within a conductive, subsurface ocean, if it exists. Based on observations of the geologically young surface, Nimmo & Spencer (2015) have postulated that a global ocean could exist at depths below 200 km at Triton, sustained by tidal heating. However, the conductivity of such an ocean is poorly constrained. Besides a possible subsurface ocean, an additional conducting layer is Triton's ionosphere which is likely generated (at least in part) via precipitation of magnetospheric particles along the perturbed electromagnetic environment onto the moon's atmosphere. Given its high expected conductivities (e.g., Strobel et al. 1990), it is plausible that the time variability of the Neptunian magnetospheric field along the moon's orbit would drive currents within the ionosphere, which would manifest as an induced magnetic field detectable outside of the moon.

The combination of Triton's atmosphere, ionosphere, and induced magnetic field act as an obstacle to the impinging magnetospheric flow. Due to Triton's retrograde orbit ($\boldsymbol{u}_T \approx 4.4$ km s^{-1}), the Neptunian magnetospheric plasma continually encounters the moon's orbital leading hemisphere at a relative velocity of $\boldsymbol{u}_0 \approx 43$ km s^{-1} (Strobel et al. 1990). As the plasma approaches Triton, it is diverted around the moon, generating perturbations in the plasma flow and electromagnetic fields near the moon. Figure 7.8 displays a schematic of Triton's electromagnetic environment for the case where the magnetospheric flow $\boldsymbol{u}_0$ is perpendicular to the background magnetic field $\boldsymbol{B}_0$. Panel (a) is viewed against the direction of the convective electric field ($\boldsymbol{E}_0 = -\boldsymbol{u}_0 \times \boldsymbol{B}_0$), whereas panel (b) is viewed along the direction of the magnetospheric background field. Because the magnetospheric plasma along Triton's orbit is sub-Alfvénic (i.e., the Alfvénic Mach number $M_A = \frac{|v_A|}{|\boldsymbol{u}_0|} < 1$,

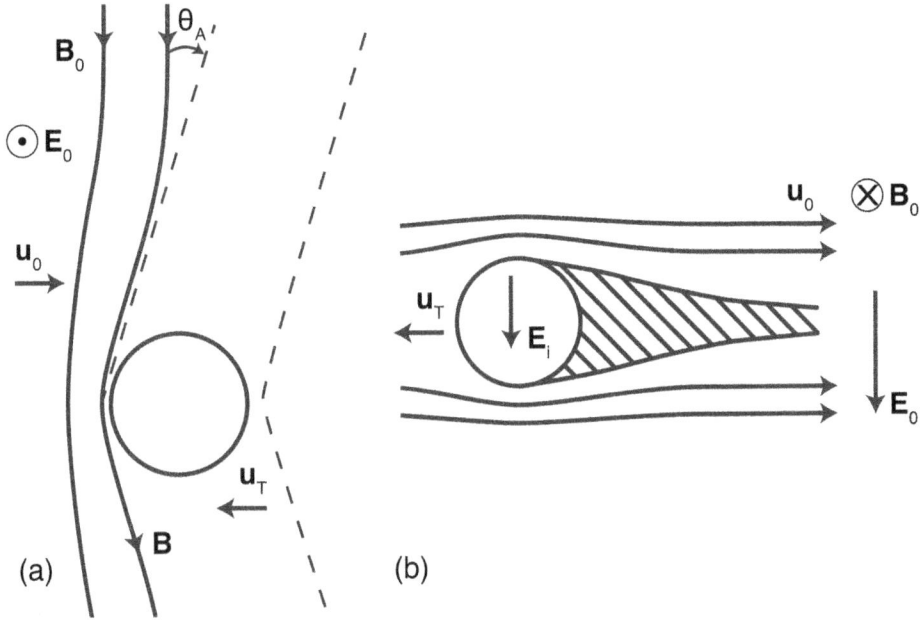

Figure 7.8. Schematic of an analytical model of Triton's plasma interaction. Views are (a) looking against the convective electric field E_0 and (b) along the background magnetic field B_0. Triton's orbital motion is given by vector u_T, magnetic field and flow streamlines are denoted by solid lines, and dashed lines illustrate the Alfvén characteristics. Adapted with permission from Strobel et al. (1990).

with Alfvén velocity v_A and magnetospheric plasma flow velocity u_0; see Table 7.1), the moon's interaction generates Alfvén wings that are inclined at an angle θ_A against the direction of the background magnetic field. These Alfvén wings—a system of nonlinear, standing Alfvén waves—connect Triton to Neptune's iono-sphere. The visual signature where these wings connect to Neptune's ionosphere may be detected during future spacecraft missions to the ice giant, similar to the auroral footprints of the Galilean moons at Jupiter, or of Enceladus at Saturn. Shaded lines in Figure 7.8 denote the location of the moon's plasma wake. Within Triton's ionosphere, the electric field E_i is reduced below E_0.

From an observational standpoint, the geometry of the Voyager 2 encounter was not ideal to detect signatures of Triton's interaction with its magnetospheric environment. Closest approach of the spacecraft occurred nearly 40,000 km from the moon's surface, and Neubauer et al. (1991) presented Voyager 2 magnetic field measurements during closest approach and noted an absence of any perturbations associated with Triton. Despite the lack of far-field observations, we can estimate how strongly the local convective electric field is reduced due to Triton's interaction with the Neptunian magnetospheric plasma. By following the approach of Strobel et al. (1990) and Saur et al. (2013), we define an "interaction strength" parameter $\alpha = 1 - E_i/E_0$, where E_i represents the electric field within Triton's ionosphere. For the case of no plasma interaction, this parameter reaches a value of $\alpha = 0$, while for a strong interaction, α approaches a value of unity. This interaction strength parameter can be estimated via the equation:

$$\alpha = \frac{\Sigma_P}{\Sigma_P + 2\Sigma_A},$$

where Σ_P and Σ_A are the Pedersen and Alfvénic conductances, respectively (see also Neubauer 1980, 1998; Saur et al. 1999). At Triton, Σ_P exceeds 10^4 S (Strobel et al. 1990) while Σ_A is on the order of 1 S. Hence, Triton's interaction is "saturated" as the interaction strength reaches a value of $\alpha \approx 1$, and the electric field within the conducting ionosphere approaches $E_i \approx 0$ V m^{-1}. As a result, Strobel et al. (1990) estimated that streamlines of the magnetospheric plasma would be largely diverted around the moon (see discussion in 7.5 regarding consequences for magnetospheric particle precipitation).

However, this approach is only an estimate of how Triton's interaction perturbs its local environment: due to the low magnetic field along Triton's orbit, the typical magnetospheric and ionospheric ion gyroradii near Triton can approach the size of the moon. Hence, it is important to represent ion dynamics near the moon using a kinetic approach. One such study by Liuzzo et al. (2021) modeled Triton's interaction with the Neptunian plasma using a hybrid model, which accurately represents ion dynamics. Their goal was to determine how strongly Triton's plasma interaction could obscure the signature of an induced magnetic field at the moon. Liuzzo et al. (2021) found that for times during Triton's orbit when the magnetospheric field is perpendicular to the plasma flow direction, signatures of a magnetic field induced at Triton are likely detectable close to the moon's surface, even over the perturbations associated with the moon's plasma interaction. The weak electromagnetic fields associated with the moon's interaction for this scenario resemble those at the icy Galilean moons of Jupiter, where the magnetospheric field piles up and drapes around Triton's ionosphere and induced field, and generates a magnetic field enhancement upstream of the moon and an associated reduction in the total field magnitude within the wake. At greater distances from the moon, Alfvén wings form.

However, at times when the magnetospheric field forms an oblique angle to the flow direction, Liuzzo et al. (2021) found that Triton's interaction generates plasma interaction signatures that are unlike those observed at any other moon in the solar system. These authors identified a region of reduced magnetospheric plasma located downstream of Triton that resembled a typical absorption wake feature. However, the depletion was rotated out of the location of the expected wake (i.e., out of the geometric plasma shadow) by an angle similar to the angle formed between the magnetospheric field and plasma flow vectors (Figure 7.8). Liuzzo et al. (2021) illustrated that, for those times along Triton's orbit where u_0 and B_0 form an oblique angle to one another, one Alfvén wing characteristic (defined by $\mathcal{Z}_{0,\pm} = u_0 \pm v_{A,0}$; see Neubauer 1980) is located *upstream* of Triton. This causes the magnetospheric plasma to "ride" along the Alfvén wing located upstream and be diverted toward Triton. The resulting absorption of the plasma as it impinges onto Triton thereby causes an Alfvén wing absorption feature that is displaced out of the geometric plasma shadow. Such a signature has never been identified at any other solar system moon but may be a key feature of Triton's interaction.

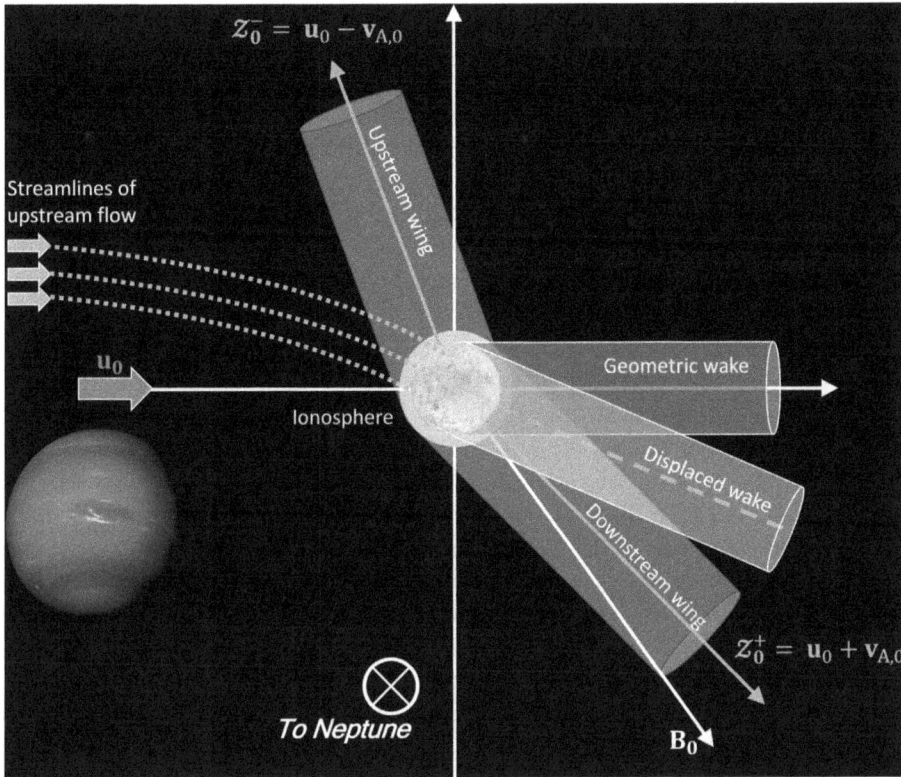

Figure 7.9. Schematic illustrating the geometry of Alfvén wings at Triton, for a case where the magnetospheric flow vector $\boldsymbol{u}_0$ forms an angle to the magnetospheric field $\boldsymbol{B}_0$. When one wing is located upstream of the moon, the plasma flow is diverted toward Triton, generating a displaced plasma wake downstream of the moon that is tilted out of the geometric wake. Adapted with permission from Simon et al. (2022).

Recently, Simon et al. (2022) applied an analytical model to further understand these unique Alfvén wing absorption features associated with Triton's plasma interaction. With their model, they placed constraints on the properties of the magnetospheric plasma required to generate such a feature at any moon. These authors identify a "critical angle" for flow deflection, whereby such a displaced wake would be formed by an Alfvén wing located upstream of a moon (Figure 7.9). They show that the "critical angle," θ_c, satisfies $\sin \theta_c = M_A$. Simon et al. (2022) suggest that a similar feature could likely form at, e.g., exoplanets (see also Saur et al. 2013), the Uranian moons, or even at Saturn's moon Titan (although the Cassini spacecraft did not observe this scenario during any of its encounters).

7.7 Conclusions

Triton and Pluto exhibit several strong similarities, with a likely shared origin in the Kuiper Belt. While the two objects both have cold and relatively dense N_2–CH_4 atmospheres with almost identical surface pressures, there are several key differences between these twin worlds. Firstly, the neutral atmosphere of Pluto is much more extended than that of Triton due to the much larger fraction of CH_4 and other greenhouse gasses leading to low-altitude heating, and comparative studies of Pluto and Triton have yielded new insights into the role of atmospheric CH_4 at this class of object. Secondly, the two bodies reside in very different space environments, with Triton orbiting inside Neptune's magnetosphere while Pluto is almost certainly an unmagnetized body interacting directly with the solar wind and other heliospheric particle populations. Thus, respective flybys by *Voyager 2* and *New Horizons* found that the atmospheres, ionospheres, and space environment interactions of these two bodies are strikingly dissimilar.

Triton was found by *Voyager 2* to have a very dense ionosphere, with peak electron densities over an order of magnitude higher than those found at Saturn's moon Titan. This is also in strong contrast to Pluto, which hosts a very similar cold N_2–CH_4 atmosphere. At the time of the *New Horizons* flyby, Pluto was located at a similar solar distance as Triton, but radio occultations failed to detect any sign of an ionosphere above the *New Horizons* REX detection limit. This indicates that an additional energy source is at work in creating and maintaining Triton's unexpectedly dense ionosphere, likely precipitation of electrons from Neptune's magnetosphere. However, due to the limitations of the *Voyager 2* flyby dataset, it has not been possible to conclusively determine if Triton's ionosphere is primarily solar-driven or if magnetospheric electron precipitation plays a dominant role. A key aspect of this is the determination of the properties of magnetospheric electrons (energy spectrum, flux) incident on Triton, which has not been well constrained from the *Voyager 2* measurements. Furthermore, recent modelling has shown that Triton's plasma interaction with Neptune's magnetosphere may be complex and possibly of a type that is unique among solar system objects. This makes it more difficult to relate the properties of ambient (upstream) magnetospheric plasma to the population of electrons that precipitate at the top of Triton's atmosphere as these particles have first been altered by the perturbed local electromagnetic fields near the moon. Measurements by *Voyager 2* also indicate that Triton has an important role as a plasma source, providing a supply of heavy ions to Neptune's magnetosphere. However, the exact mechanism for atmospheric loss at Triton has not yet been conclusively determined, and while likely present, the long-hypothesized Triton neutral torus has not yet been confirmed by measurements.

The dwarf planet Pluto does not orbit within a magnetosphere and therefore interacts directly with the heliospheric environment determined by the interplanetary magnetic field, solar wind plasma, and local fluxes of interstellar pickup ions. *New Horizons* found that Pluto's solar wind interaction at the time of the flyby was a hybrid between that of a weak comet and an unmagnetized planet (e.g., Venus-like). In particular, it was found that while the overall large-scale interaction was comet-like, Pluto possessed a broad heavy ion tail that appeared to be a consequence of the low IMF and thermal pressure from IPUIs. The size of the Pluto interaction region was found to be much smaller than expected, primarily due to much lower rates of atmospheric neutral and ion escape than what was predicted before the flyby. Importantly, however, the *New Horizons* flyby coincided with a brief period of enhanced solar wind pressure, likely leading to a more compressed solar wind interaction region than what would be observed under quiescent conditions. We now also understand that the Pluto–solar wind interaction is strongly controlled by the rate of neutral escape from its atmosphere. The neutral escape rate is tightly coupled to the state (e.g., temperature, structure, composition) of the atmosphere, which is expected to vary significantly over seasonal and orbital timescales. Thus, it is likely that the nature of the Pluto–solar wind interaction also varies over both short (e.g., solar wind variability) and long (orbital, seasonal) timescales. While *New Horizons* did not detect a robust ionosphere at Pluto during its encounter, it is plausible that the ionospheric state likewise also varies over time due to, e.g., changing atmospheric CH_4 mole fraction and possibly also differing contributions from heliospheric charged particle precipitation due to the changing nature of the solar wind interaction.

While we have individual snapshots in time from both the Neptune-Triton and Pluto systems, many questions remain unanswered, and it is clear that these worlds are highly dynamic, particularly with regards to their upper atmospheres and space environment interactions. Future missions to both systems, as well as to destinations within the Kuiper Belt are therefore warranted.

A New Frontiers class mission concept, the Triton Ocean World Surveyor, was highlighted by the recent 2023–2032 Decadal Strategy for Planetary Science and Astrobiology. If such a mission eventually goes forward, we emphasize that the exploration strategy should consider the fact that Neptune, its magnetosphere, and Triton's upper atmosphere and ionosphere represent an inherently coupled system. For example, such a mission could determine the dominant energy input to Triton's upper atmosphere and ionosphere, and investigate the mechanism and loss rates of neutrals and ions from Triton to Neptune's magnetosphere by carrying out a combination of remote sensing and in-situ measurements in the Neptune-Triton system. Triton's plasma interaction and magnetic induction response are also affected by the properties of the moon's ionosphere and potential subsurface ocean, and therefore, an improved understanding of the upper atmosphere and its coupling to the magnetospheric environment will also aid in the future detection and characterization of a putative subsurface ocean on Triton.

A future mission to Pluto could investigate seasonal and orbital changes to its atmosphere, ionosphere, and surface volatiles that have occurred since the *New Horizons* flyby and determine whether a liquid subsurface ocean persists to present day. A dedicated Pluto orbiter could also investigate how Pluto's solar wind interaction changes in response to solar wind and atmospheric variability, thus addressing one of the major open questions remaining after *New Horizons*. We should highlight that given the long orbital and seasonal timescales of these systems, further missions to both objects would be highly synergistic, even if they were somewhat spaced out in time.

References

Bagenal, F., Delamere, P. A., Elliott, H. A., et al. 2015, JGRE, 120, 1497

Bagenal, F., Horányi, M., McComas, D. J., et al. 2016, Sci, 351, aad9045

Bagenal, F., McComas, D. J., Elliott, H. A., et al. 2021, The Pluto System After New Horizons, ed. S. A. Stern, et al. (Tuscon, AZ: Univ. Arizona Press) 379

Bagenal, F., & McNutt, R. L. 1989, GeoRL, 16, 1229

Barnes, N. P., Delamere, P. A., Strobel, D. F., et al. 2019, JGRA, 124, 1568

Belcher, J. W., Bridge, H. S., Bagenal, F., et al. 1989, Sci, 246, 1478

Benne, B., Dobrijevic, M., Cavalié, T., Loison, J. C., & Hickson, K. M. 2022, A&A, 667, A169

Bertaux, J., & Blamont, J. 1971, A&A, 11, 200

Bertrand, T., Forget, F., Umurhan, O. M., et al. 2019, Icar, 329, 148

Broadfoot, A. L., Atreya, S. K., Bertaux, J. L., et al. 1989, Sci, 246, 1459

Cheng, A. F. 1990, GeoRL, 17, 1669

Cochrane, C. J., Persinger, R. R., Vance, S. D., et al. 2022, E&SS, 9, e02034

Decker, R. B., & Cheng, A. F. 1994, JGR, 99, 19027

Delamere, P. A., & Bagenal, F. 2004, GeoRL, 31, 1

Delitsky, M. L., Eviatar, A., & Richardson, J. D. 1989, GeoRL, 16, 215

Dobrijevic, M., Loison, J. C., Hickson, K. M., & Gronoff, G. 2016, Icar, 268, 313

Elliot, J. L., Strobel, D. F., Zhu, X., et al. 2000, Icar, 143, 425

Feyerabend, M., Liuzzo, L., Simon, S., & Motschmann, U. 2017, JGRSP, 122, 10356

Gladstone, G. R., Stern, S. A., Ennico, K., et al. 2016, Sci, 351, aad8866

Hale, J. P. M., & Paty, C. S. 2017, Icar, 287, 131

Herbert, F., & Sandel, B. R. 1991, JGRA, 96, 19241

Hinson, D. P., Linscott, I. R., et al. 2017, Icar, 290, 96

Hinson, D. P., Linscott, I. R., Strobel, D. F., et al. 2018, Icar, 307, 17

Ip, W.-H. 1990, GeoRL, 17, 1713

Kliore, A. J., Anabtawi, A., Herrera, R. G., et al. 2002, JGRA, 107, 1407

Kliore, A. J., Nagy, A. F., Marouf, E. A., et al. 2008, JGRA, 113, A09317

Kollmann, P., Hill, M. E., Allen, R. C., et al. 2019, JGRA, 124, 7413

Krasnopolsky, V. A. 2012, P&SS, 73, 318

Krasnopolsky, V. A. 2020, Icar, 335, 113374

Krasnopolsky, V. A., Sandel, B. R., Herbert, F., & Vervack, R. J. 1993, JGR, 98, 3065

Krasnopolsky, V. A., & Cruikshank, D. P. 1995, JGR, 100, 21271

Krimigis, S. M., Armstrong, T. P., Axford, W. I., et al. 1989, Sci, 246, 1483

Lellouch, E., Blanc, M., Oukbir, J., & Longaretti, P. Y. 1992, AdSpR, 12, 113

Lellouch, E., de Bergh, C., Sicardy, B., Ferron, S., & Käufl, H.-U. 2010, A&A, 512, L8

Lellouch, E., Gurwell, M., Butler, B., et al. 2017, Icar, 286, 289

Liuzzo, L., Paty, C., Cochrane, C., et al. 2021, JGRA, 126, e29740

Luspay-Kuti, A., Mandt, K., Jessup, K. L., et al. 2017, MNRAS, 472, 104

Lyons, J. R., Yung, Y. L., & Allen, M. 1992, Sci, 256, 204

Majeed, T., McConnell, J. C., Strobel, D. P., & Summers, M. E. 1990, GeoRL, 17, 1721

Mandt, K. E., Gell, D. A., Perry, M., et al. 2012, JGRE, 117, 1

Mauk, B. H., Keath, E. P., Kane, M., et al. 1991, JGRA, 96, 19061

McComas, D. J., Elliott, H. A., Weidner, S., et al. 2016, JGRA, 121, 4232

Ness, N. F., Acuna, M. H., Burlaga, L. F., et al. 1989, Sci, 246, 1473

Neubauer, F. M., Lüttgen, A., & Ness, N. F. 1991, JGR, 96, 19171

Neubauer, F. M. 1980, JGRSP, 85, 1171

Neubauer, F. M. 1998, JGRP, 103, 19843

Nimmo, F., & Spencer, J. R. 2015, Icar, 246, 2

Olkin, C. B., Young, L. A., Borncamp, D., et al. 2015, Icar, 246, 220

Richardson, J. D., Belcher, J. W., Zhang, M., & McNutt, R. L. 1991, JGR, 96, 18993

Richardson, J. D., Eviatar, A., & Delitsky, M. L. 1990, GeoRL, 17, 1673

Rymer, A. M., Smith, H. T., Wellbrock, A., Coates, A. J., & Young, D. T. 2009, GeoRL, 36, L15109

Sandel, B. R., Herbert, F., Dessler, A. J., & Hill, T. W. 1990, GeoRL, 17, 1693

Saur, J., Neubauer, F. M., Strobel, D. F., & Summers, M. E. 1999, JGRSP, 104, 25105

Saur, J., Grambusch, T., Duling, S., Neubauer, F. M., & Simon, S. 2013, A&A, 552, 1

Simon, S., Addison, P., & Liuzzo, L. 2022, JGRA, 127, e29958

Sittler, E. C., & Hartle, R. E. 1996, JGR, 101, 10863

Stevens, M. H., Strobel, D. F., Summers, M. E., & Yelle, R. V. 1992, GeoRL, 19, 669

Strobel, D. F., Cheng, A. F., Summers, M. E., & Strickland, D. J. 1990, GeoRL, 17, 1661

Strobel, D. F., & Zhu, X. 2017, Icar, 291, 55

Summers, M. E., & Strobel, D. F. 1991, GeoRL, 18, 2309

Tyler, G. L., Sweetnam, D. N., Anderson, J. D., et al. 1989, Sci, 246, 1466

Young, L. A., Kammer, J. A., Steffl, A. J., et al. 2018, Icar, 300, 174

Yung, Y. L., & Lyons, J. R. 1990, GeoRL, 17, 1717

Zhang, M., Belcher, J. W., & McNutt, R. L. 1992, AdSpR, 12, 37

Zhang, M., Richardson, J. D., & Sittler, E. C. 1991, JGR, 96, 19085

Zhu, X., Strobel, D. F., & Erwin, J. T. 2014, Icar, 228, 301

Zirnstein, E. J., McComas, D. J., Elliott, H. A., et al. 2016, ApJ, 823, L30

Chapter 8

On the Detection of Subsurface Oceans within Triton and Pluto

Corey J Cochrane, Julie C Castillo-Rogez, Steven D Vance and
Benjamin P Weiss

8.1 Introduction

Although they are two of the most distant planetary bodies in the solar system from the Sun, Pluto and Neptune's moon of Triton may harbor liquid-water subsurface oceans. These bodies are similar in size (radius of Pluto, $R_P = 1188$ km and radius of Triton, $R_T = 1353$ km), and exhibit relatively young surfaces, or at least locally young geologic surface features, potentially as young as a few My for Triton (Schenk & Zahnle 2007) and 1 Gy for Pluto (Singer et al. 2021). These bodies formed in the Kuiper belt, and Triton was likely captured by Neptune's gravity (e.g., Agnor & Hamilton 2006).

In the case of Triton in particular, recent activity likely points at the presence of a deep ocean and significant tidal heating (e.g., Ruiz 2003; Nimmo & Spencer 2015). Several geological features have been interpreted as evidence of high heat flow and cryovolcanic activity (Ivanov et al. 1991; Croft et al. 1995; Nimmo & Pappalardo 2016; Schenk et al. 2021; Singer et al. 2022). Endogenic activity has also been suggested to drive the plumes, although alternative explanations have been proposed (e.g., the occurrence of solid-state greenhousing, Kirk et al. 1995). With its highly inclined retrograde orbit, a deep ocean could be subject to obliquity-driven tides (Chen et al. 2014; Nimmo & Spencer 2015). NASA's Roadmap to Ocean Worlds (Hendrix et al. 2019) identifies Triton as the highest priority ocean world to target with a spacecraft in the near future. Triton's geology and its prospect for significant tidal heating are so intriguing that the recent *Origins, Worlds, Life* Decadal Survey features a Triton orbiter (Ocean World Surveyor) in the New Frontiers 7 mission themes. A recent review of Triton's properties can be found in Hansen et al. (2021).

Pluto is also considered a candidate ocean world (Hendrix et al. 2019) based on images returned by the New Horizons spacecraft that made a close flyby (12,500 km) on 2015 July 14. The signature of a possible deep ocean in the geology is more subtle than at Triton. This hypothesis is based in particular on the interpretation of the equatorial location of Sputnik Planitia that potentially involved reorientation via true polar wander (Keane et al. 2016; Nimmo and Pappalardo 2016). Possible uplift of a thick ocean below the Sputnik Planitia basin is one of several possible explanations (Hamilton 2016). In contrast to Triton, Pluto is not subject to significant tidal forcing at present. Hence, the preservation of a deep ocean until present would require a combination of circumstances: the presence of antifreeze (ammonia being likely, Kimura & Kamata 2020; Singer et al. 2022e) and a layer of clathrate hydrate at the top of the ocean acting as an insulant (Kamata et al. 2019). However, the prospect for buoyant clathrate in Pluto is debated (Courville et al. 2023) and liquid water in Pluto might take the form of a residual ocean sustained by antifreeze (ammonia and salts).

Detecting and characterizing these potential oceans within Kuiper belt planetary bodies is of keen interest to the scientific community and the public. One promising and proven technique to detect subsurface water and hidden oceans is the use of electromagnetic (EM) sounding, which relies either on active or passive probing methods. Active methods involve transmission of pulsed or chirp radio waveforms as the source (e.g., ice penetrating radar, see Schroeder et al. 2020; Picardi et al. 2005 for sounding of ice masses on Earth and Mars respectively) while passive methods leverage naturally occurring phenomenon as the source to passively sound, some including the very slow ($\ll 1$ Hz) magnetic field variations exhibited by planetary magnetospheres arising from relative planetary motion (Kivelson et al. 1999; Khurana et al. 2011) and the higher frequency EM radiation radio emissions resulting from magnetic-field-accelerated electrons from the Sun (Peters et al. 2021) or even large planets like Jupiter (Grimm 2009; Romero-Wolf et al. 2015) and Uranus (Romero-Wolf et al. 2023).

Pluto is a very challenging target for ocean detection. The absence of active tidal forcing and a planetary magnetic field make geodetic and magnetic sounding approaches quite difficult. Magnetic sounding using the solar wind as the source has potential, but would be very challenging as the interplanetary magnetic field (IMF) is only on the order of 0.2 nT, with a factor of 2 variability, through the duration of its eccentric orbit (30–50 AU) around the Sun (Bagenal et al. 1997). The New Horizons spacecraft was unfortunately not equipped with either a radar or a magnetometer, so passively or actively sounding the interior of the dwarf planet was not possible during its single flyby. A magnetometer could have tested whether Pluto has or had a magnetic dynamo (through crustal magnetization fields), the prospect of which depends on many parameters (see Chapter 2). Aside from that, Pluto is not an ideal target for passive sounding. Low-frequency (corresponding to periods of 10–100 s of h) magnetic induction investigation would not be possible as the dwarf planet does not reside in a strong and periodically varying magnetic environment, which is needed to facilitate the phenomenon. Additionally, higher frequency passive EM sounding methods using naturally occurring radio emissions, say from the Sun or Jupiter, are not yet well understood at Pluto and would be very weak at its orbital distance.

A recent concept study for a Pluto orbiter (Persephone, Howett et al. 2021) included a magnetometer, as well as a radar sounder with a requirement to probe 10-km deep to search for evidence of a deep ocean under Sputnik Planitia. The average thickness of Pluto's icy shell is debated. Kamata et al. (2019) have suggested that the shell is thin (~100 km) and convection is impeded by the presence a 10–20 km layer of clathrate hydrate at its base. On the other hand, Courville et al. (2023) find that Pluto's shell could be almost entirely frozen, with only a residual ocean a few 10s of kilometers thick. Looking at the capabilities of the Radar for Icy Moon Exploration (RIME) on-board the Jupiter Icy Moon Explorer (JUICE) mission (9 MHz center frequency, dual bandwidth of 2.8 and 1 MHz) and the Radar for Europa Assessment and Sounding: Ocean to Near-surface (REASON) on Europa Clipper (central frequencies of 9 and 60 MHz, bandwidths of 10 and 1 MHz, respectively) (Bruzzonne & Croci 2019; Chan et al. 2023), the probing depths predicted at Europa and Ganymede are only a few tens of km. Hence, it is unlikely that a radar sounder would be able to probe the shell down to the interface with a deep ocean. In the context of an orbiter mission, gravity mapping might provide constraints on the density and viscoelastic structure of the shell (see description of the methodology in Fu et al. 2015). In particular, a positive gravity anomaly associated with the large Sputnik Planitia could confirm the presence of an ocean uplift under the large basin, as suggested by (Keane et al. 2016).

Similarly, a radar sounder might not be able to peek deep into Triton's ice shell depending on its temperature profile. Triton's ice shell thickness is poorly constrained, where estimates provided in the literature, inferred from forward thermal modeling, suggest an ice shell at least 150 km thick (e.g., Hussmann et al. 2006; Nimmo & Spencer 2015; Hammond et al. 2018). On the other hand, heat flow inferred from geological feature indicate the shell might be only a few 10 s of kilometers thick (e.g., Ruiz 2003; Prockter et al. 2005). Hence it is unlikely that radar sounding would be able to probe Triton's ice shell down to the top of the ocean.

Geodetic methods may possibly be used to search for an ocean inside Triton, though would be difficult. Triton's eccentricity is very small (1.6×10^{-5}), hence, its response to eccentricity tides is also small, such that its Love numbers is not measurable via gravity science (i.e., from tracking the acceleration of a spacecraft flying by the Moon). The forced physical libration in longitude is another way to detect the presence of a deep ocean (Nimmo & Pappalardo 2016). However, it is also a function of eccentricity and thus its amplitude would likely be negligible. Nimmo & Pappalardo (2016) also suggest there could be the signature of a deep ocean in the Moon's obliquity (the tilt of the satellite's rotation axis from the normal to its orbit plane), but this difference would likely be very small and challenging to measure, since the Moon's obliquity is estimated at 0.35° (Chen et al. 2014). As noted earlier, constraints on the density and the viscoelastic structure of the Moon may be obtained from combining topography and gravity measurements, but it requires extensive mapping, which may not be feasible in the case of a multi-flyby mission, i.e., an orbiter around Neptune.

Seismometry from a lander could obtain critical constraints on the deep interior depending on the amplitude of possible seismic sources: cryovolcanic eruptions, tectonic activity, convection in the ice shell; tidally-driven currents in the deep ocean, and large impacts (Vance et al. 2018). The quantification of the seismic energy for each of these processes should be undertaken in order to assess the viability of seismometry at Triton.

Magnetometry using natural magnetic sources probably serves as the most robust method to search for an ocean within Triton. In contrast to Pluto, Triton is immersed in Neptune's time-varying magnetospheric field, which has multiple periodic (low-frequency) variations due to Neptune's multipolar rotating magnetic field geometry and Triton's highly inclined orbit, making it well-suited for magnetic induction investigation of its interior. If present, a conductive ocean would react to this time-varying driving field and generate an induced magnetic field that could be sensed remotely by a magnetometer-equipped spacecraft, revealing much about its interior composition and dynamics. Voyager 2 was Triton's only visitor, flying by at an altitude 40,000 km on 1989 August 25. Although it carried two tri-axial fluxgate magnetometers, the spacecraft did not fly close enough to sense a magnetic induction signal from the ocean. Unfortunately, the highly conductive nature of Triton's ionosphere makes passive sounding using radio bursts from Neptune prohibitive. The peak free electron density of Triton is $N_e = 4.6 \times 10^4$ cm^{-3} (Tyler et al. 1989) which results in an ionospheric cutoff frequency of $9\sqrt{N_e}$, roughly 1.9 MHz, which is just above the end of Neptune's kilometric radio emissions. However, the very long periods associated with the anticipated magnetic waves at Triton from Neptune's rotation (14 hr) and Triton's orbit (141 hr) have the ability to penetrate the highly conductive ionosphere to magnetically sound deep within its interior. For these reasons, the remainder of the chapter will focus on passive magnetic sounding of Triton's interior using magnetometry with Neptune's driving magnetic field as a source. It should be noted that this method of passive detection requires a global scale ocean to exist. Although spherical symmetry of the ocean is not necessary, an asymmetrical ocean slightly complicates the analysis (Styczinski et al. 2022), but if small enough, can simply be treated as an error source in the spherically symmetric case. Unfortunately, localized liquid-water reservoirs below the ice layer are likely not possible to detect with a magnetometer-equipped orbiter (at least based on recent work on Europa by Winkenstern & Saur 2023), but may be detected a dual magnetometer approach on the surface, which stresses the need for a magnetometer (ideally paired with a seismometer) for any future proposed lander mission to any icy body.

8.2 Neptune's Magnetic Environment

Neptune's magnetic field can be approximated by a magnetic dipole offset from the interior of the planet by 0.55 R_T and its axis is tilted 47° with respect to its spin axis (Ness, Connerney et al. 1991). This tilt causes a magnetic variation in the field at Triton at the apparent rotation rate of Neptune, roughly 14 hr in the fixed IAU reference frame of Triton, defined by Archinal (2018). The synodic period is shorter than Neptune's rotation period (16.1 hr) due to Triton's retrograde orbit. Additionally, Triton has an orbital inclination of 129.8° (>90° for retrograde orbit), which causes it to traverse a wide range of magnetic latitudes, resulting in time variations of the magnetic field at the 141 hr orbital period of the Moon. Figure 8.1 illustrates the magnetic field lines of Neptune, developed using the Connerney et al. (1991) field model, and also the magnetic field time series and frequency spectrum evaluated in the fixed frame of Triton. In the Triton reference frame, the Z-axis is aligned with Triton's spin axis, the X-axis is in the direction of Neptune, and the Y-axis completes the right-handed orthogonal system and is roughly in the direction

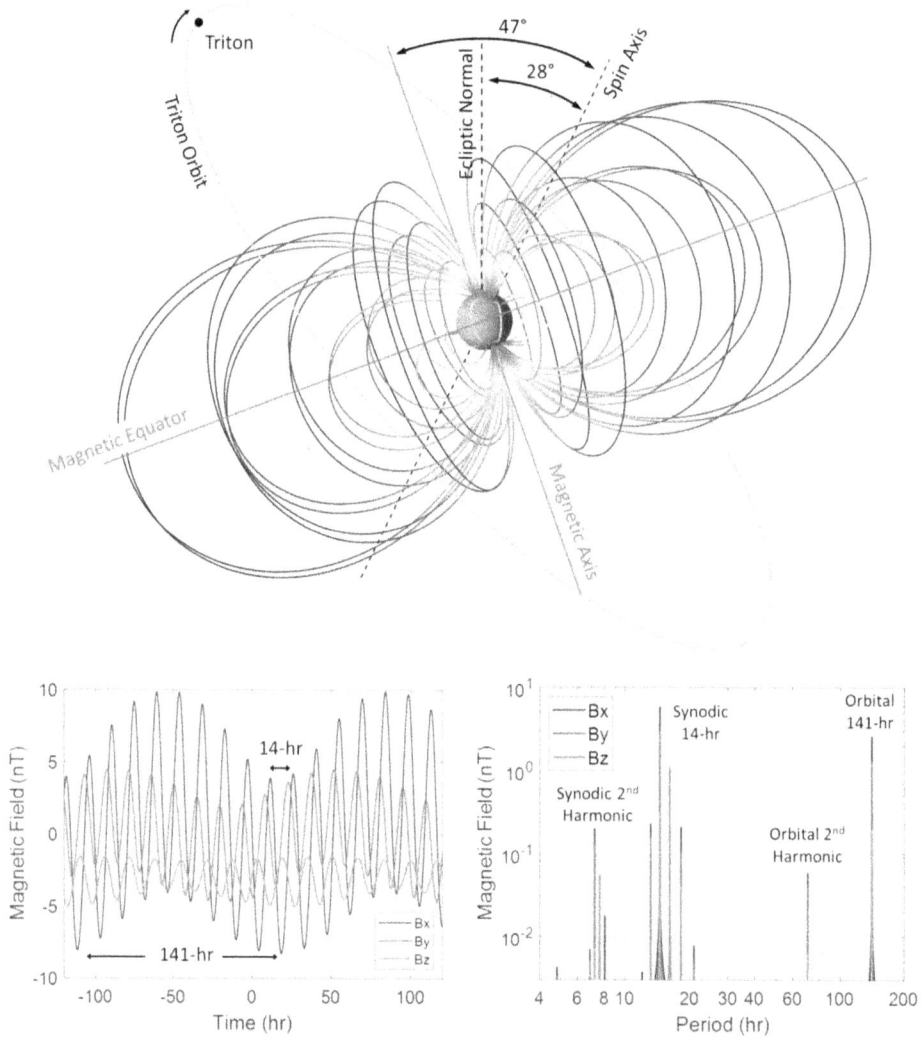

Figure 8.1. Neptune's magnetic field at Triton. (top) Magnetic field line geometry of the Neptune-Triton system. (bottom left) Magnetic field time series of Neptune evaluated at Triton in its fixed frame. (bottom right) Fourier transform of the magnetic field time series showing the 14 hr synodic wave and 141 hr orbital wave of Triton. Also observed are the weaker higher order harmonics and beat frequencies of these two fundamental periods. Adapted with permission from Cochrane et al. (2022). CC BY-NC

of the Moon's orbital motion. As illustrated in the figure, distinct beats and harmonics of the synodic and orbital fundamental periods are also clearly observed, which makes a superposition of sinusoids a good model for the response. The precise tonal response at various frequencies allows the time-varying portion of Neptune's field, $\boldsymbol{B}_N$, in Triton's frame to be modeled by a superposition of sinusoids,

$$\boldsymbol{B}_N(t) = \sum_k \boldsymbol{B}_{N,k}(t) = \sum_k \boldsymbol{B}_k e^{i(2\pi f_k t + \theta_k)}, \tag{8.1}$$

where $\boldsymbol{B}_{N,k}(t)$ is the kth discrete magnetic wave of Neptune at frequency, f_k, $\boldsymbol{B}_k = [B_{x,k} \quad B_{y,k} \quad B_{z,k}]$ are their amplitudes and $\boldsymbol{\theta}_k = [\theta_{x,k} \quad \theta_{y,k} \quad \theta_{z,k}]$ are their phases. The real component of the modeled field in Equation 8.1 represents the physical field at Triton.

Although magnetic induction investigation of the Jovian moons is better due to the large driving fields of Jupiter, the magnetic environment at Triton is especially appealing because the two dominant waves are comparable in amplitude but very far (an order of magnitude) apart in frequency, which enables them to be more easily distinguished. From the magnetic field model used, the synodic wave has an amplitude of 6.7 nT and the orbital wave has an amplitude of 2.6 nT, thus a factor of 2.6 difference. In comparison to the four major moons of the Jovian system, the amplitude of the synodic magnetic wave of Jupiter is more than one order of magnitude larger in magnitude than orbital field variations (Seufert et al. 2011; Vance et al. 2021). Additionally, the ratio of synodic period T_s to orbital period T_o of Triton is 0.1, which in comparison to the Jovian moons is 0.31 for Io, 0.13 for Europa, 0.06 for Ganymede, and 0.025 for Callisto. For the Uranian moons, the orbits are nearly circular, so very little variation in the magnetic field occurs at the orbital period, aside from a small variation at Miranda due to its small orbital inclination (Cochrane et al. 2021; Weiss et al. 2021; Arridge and Eggington 2021). The lack of tilt between the magnetic axis of Saturn with respect to its spin axis makes magnetic induction investigation at Saturn's moons via the synodic period nearly impossible. However, time variations in the magnetic field are expected at the orbital period of Mimas ($\sim$ 30 nT) and Enceladus ($\sim$ 5 nT), due to the distance experienced between the two bodies resulting from the eccentric orbits of the two moons, roughly 0.0196 and 0.0047, respectively (Vance et al. 2021). Close observations from Cassini are not available at Mimas, and the magnetometer measurements from the close flybys of Enceladus reveal an overwhelming field signature from plume-magnetospheric-plasma interactions (Dougherty et al. 2006), which likely mask any induction signal originating from the ocean. Fortunately, the magnetic environment at Triton is unlike the Saturnian system so ocean detection within its moon using magnetic induction is more favorable.

Another advantage of searching for oceans in the Neptunian system over the Jovian system is that Triton spends much of its time orbiting away from Neptune's magnetic equator, where plasma currents are anticipated to be strongest. Figure 8.2 indicates that Triton can be as high as 13 R_N above Neptune's magnetic equator and as low as -7 R_N below it. For this single rotation period in the orbit, Triton spends 93% of the time outside ± 1 R_N of Neptune's magnetic equator. (Note that this will

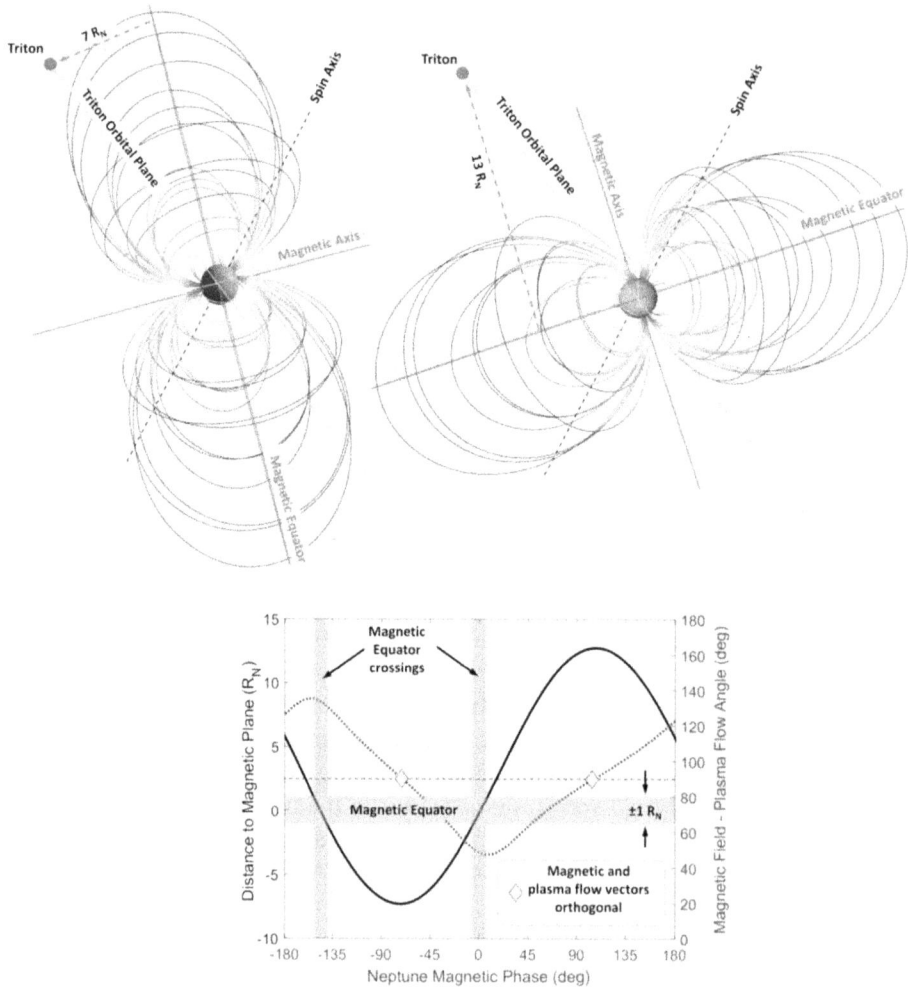

Figure 8.2. Favorable magnetic geometry of the Neptune-Triton system. For a given position of Triton along its orbit, illustrations showing (top left) −75° magnetic phase where Triton lies maximally below the magnetic equator and (top right) 105° magnetic phase where Triton lies maximally above the magnetic equator. (bottom) Triton's distance to the plane of the magnetic equator and magnetic field—plasma flow angle versus Neptune's magnetic phase. Plasma currents are contained near the magnetic equator and rotate with the 14 hr synodic wave yielding large distances between the plasma disk and Triton for the majority of the 14 hr magnetic phase 0°–360° range. Also, the field-flow angle is favorably the worst (e.g., perpendicular) when Triton's distance from plasma sheet (where the densest plasma resides) is the farthest. Adapted with permission from Cochrane et al. (2022). CC BY-NC

vary slightly for different segments of Triton's orbit due to its high orbital inclination). This configuration happens due to the significant tilt in the magnetic axis with respect to the spin axis of the planet. A similar configuration occurs with the Uranian moons (Cochrane et al. 2021). The Jovian moons almost always orbit within the current sheet of Jupiter, complicating interpretation of the induction signal (Schilling 2007; Kivelson 2009). When the plasma disk does approach Trident at $0°$ and $-145°$ phases, the angle between the plasma flow and magnetic field direction is at near maximum, thereby minimizing the effect that plasma has on ocean detection. Strong plasma currents have the ability to perturb the local induction field, especially when the flow-field angle is perpendicular, so making measurements away from these currents is desired. Additionally, measurements made from Voyager 2 over the entire Neptune flyby indicate a three order of magnitude variation in the magnetospheric plasma density, n_0, across a wide range of magnetic latitudes (maximum value on the order of 0.1 cm^{-3} closer to Neptune) due to the variation in the distance of Triton with respect to Neptune's central plasma sheet (e.g., Belcher et al. 1989; Richardson & McNutt 1990; Richardson 1993; Strobel et al. 1990; Sittler & Hartle 1996; Zhang et al. 1991). This suggests, that Triton will experience a maximum density within the plasma sheet of 0.1 cm^{-3}, making detecting an ocean using magnetometry more favorable even when crossing the plasma sheet crossing. Liuzzo et al. (2021) modeled four different field-flow-density scenarios, with ion (50% N+ and 50% H+) density of either 0.003 cm^{-3} (nominal) or 0.11 cm^{-3} (enhanced), the latter value producing perturbing magnetic field signatures on the order of only a few tenths of a nT encountered for the baseline trajectory designed under the Trident Discovery Mission concept. These perturbations can be accommodated when it comes to ocean detection; however, the currents introduce error when attempting to characterize the ocean properties (e.g., thickness and conductivity), at least for single-flyby mission concepts (Cochrane et al. 2021).

The most significant complication that sets Triton apart for detecting hidden oceans relative to other moons is the presence of a highly conductive ionosphere around the Moon. Measurements from Voyager 2 lead to an estimate of ionosphere conductance in the range of 10,000–20,000 Siemens (S), but based on expected local magnetic field variations at Triton, the range could likely extend to 4000–36,000 S (Strobel et al. 1990). A second complication is that the total magnetic driving field at the orbit of Triton is never greater than 12 nT (Saur et al. 2010; Cochrane et al. 2022), meaning that the induction signal at the surface is less than or equal to the same value. Although most space-based magnetometers are sensitive enough to observe this field variation, various sources of uncertainty like spacecraft fields and time-variable fields associated with magnetospheric plasma, both on order of ~ 1 nT could affect accurate measurement of these small magnetic fields.

8.3 Modeling of Induced Magnetic Field

One way to infer what a given magnetic induction measurement indicates about a moon's interior is to compare it with the magnetic responses associated with a representative sampling of interior models. Triton will respond to Neptune's driving field, based on what is known as the interior complex response function. This function comprises the response amplitude A and phase delay ϕ of the induction response with respect to that which would be induced by a perfectly conducting sphere ($A = 1$, $\phi = 0°$) of radius R_T, from various conductive layers for frequency f (e.g., ionosphere and ocean layers), $Q(f) = A(f)e^{-i\phi(f)}$. Magnetic induction

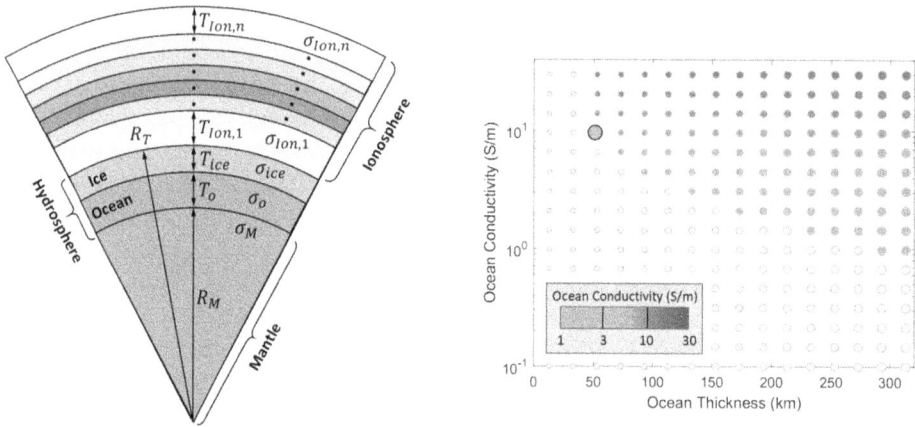

Figure 8.3. Interior model of Triton. (Left) The interior model of Triton consists of a mantle, an ocean, an ice shell, and an ionosphere, each with configurable thickness and conductivity. The mantle and ice are assumed to be non-conductive. (right) The ocean parameter space modeled by Cochrane et al. 2022. The ocean models are colored based on conductivity and sized by thickness. The blue shaded dots represent the most likely oceans as assessed by evolutionary models. The darker the blue shade, the higher the conductivity and larger the dot size, the thicker the ocean. The gray dots represent less likely ocean scenarios included in the analysis for completeness. The single green dot represents the least favorable (hardest-to-detect) of the likely oceans, which defines the minimum boundary for detection. These oceans are convolved with 51 different ionosphere models to account for the potential of increased electron concentrations and electron temperatures, with height integrated conductivity (HIC) ranging from 0 to 100,000 S in log-linear increments, yielding a total of 13, 056 interior models. Adapted with permission from Cochrane et al. (2022). CC BY-NC

measurements are sensitive to ice-shell thickness and ocean depth as the strength of an induced dipolar field degrades per $(R_T/r)^3$ where r is the distance from the point of observation to the center of Triton. Algorithms for computing the magnetic induction response of concentrically stratified conductors, pioneered over 50 years ago (Eckhardt 1963) and recently matured (Vance et al. 2021), determine the specific induction response for each model via a radially-integrated approach as shown in Figure 8.3. Each shell has a specific conductivity and thickness. The ionosphere can be modeled with changing conductivity versus radial distance to more accurately represent ionospheres with electron density gradients.

The inverse problem of inferring interior structure from induction measurements is fundamentally nonunique (Zimmer et al. 2000, Cochrane et al. 2021; Cochrane et al. 2022; Biersteker et al. 2023). In particular, there are degeneracies between ocean and ionosphere thickness, ocean and ionosphere conductivity, and ice thickness. For example, the response of highly conductive and thick ionosphere will approach that of a perfect conductor, masking the response of an interior ocean. Alternatively, in the absence of a significant ionosphere, thick, (modestly or less) conductive oceans will produce similar induction amplitudes and phases as thinner, more conductive oceans. Thicker, more conductive oceans with thick ice shells will also have similar induction amplitudes as thinner, less conductive oceans and thin ice shells.

To illustrate this further, Cochrane et al. (2022) simulated the parameters for numerous Triton interior models, which have ocean conductivity values ranging from 0.1 to 30 S m^{-1} (in log-linear increments) and thicknesses from 10 to 310 km (in increments of 20 km), representing 256 different models as illustrated in the right panel of Figure 8.3. The blue shaded dots, color coded based on conductivity, are the most likely oceans which were determined by evolutionary modeling (Castillo-Rogez et al. 2021). The gray dots are not likely but carried through the analysis for completeness and to show margin. Each of these ocean models is convolved with either of 51 different ionospheric models with various level of height integrated conductivity (HIC), totaling 13,056 different induction responses of Triton. The induced signal from conductive layers, for both ocean and ionosphere, depends on the total conductance, the product of the conductivity and thickness of each layer. A constant overall thickness of Triton's hydrosphere (combined thickness of ice layer and liquid ocean; Figure 8.3 left) of 340 km is assumed for this analysis, taking Io's density ($\sim$3500 kg m^{-3}) as a reference for Triton's core density, based on its likely dehydration following capture (Nimmo & Spencer 2015). As a general rule, thinner oceans with lower conductivity are naturally the most difficult to detect because they induce a weaker signal at the point of measurement (the induced dipolar response decreases by r^{-3}). In the figure, the most difficult ocean to detect within the subset of likely oceans is highlighted by a green dot. This ocean model, referred to as the least-favorable likely ocean, has a thickness of 50 km and conductivity of 9 S m^{-1}. It has the lowest induction response from the subset of likely ocean models and highlighted in this work to illustrate the minimum threshold for detection for different ionosphere HIC.

This parametric space encompasses, with margin, the likely conditions expected in Triton. Due to its origin in the far outer solar system, Triton is expected to have accreted a large fraction of carbon dioxide and monoxide (CO_2 and CO) as well as a few percent of ammonia (NH_3) (e.g., Shock & McKinnon 1993). In solution, the carbon ices tend to form carbonates, with the exact amount being controlled in part by the fraction of accreted ammonia (Castillo-Rogez et al. 2022; Courville et al. 2023). Prior to concentration, the salinity due to carbonates and ammonium (NH_4^+, in equilibrium with NH_3) could represent up to $\sim$70 g l^{-1} (at a reference temperature of 0°C) (e.g., Kobe et al. 1948). Castillo-Rogez et al. (2022) estimated that the corresponding electrical conductivity is about 5.5 S m^{-1}. Depending on the state of freezing of the hydrosphere, the ocean may be significantly more concentrated. Furthermore, carbonate solubility and electrical conductivity both increase with increasing temperature (e.g., Ucok et al. 1980). On the other hand, pressure tends to decrease electrical conductivity (e.g., Schmidt & Manning 2017). Experimental determination of the electrical conductivity of carbonate-ammonium-rich solutions are necessary in order to prepare for a future mission at Triton (and Pluto). The presence of a metallic core in Triton is likely (see Chapter 2). However, the conditions for the onset and long-term activity of a dynamo in the Moon have not been explored. Triton's ionosphere HIC is estimated to be in the range 4–36 kS (Strobel et al. 1990), so the extended range used in this study is to demonstrate

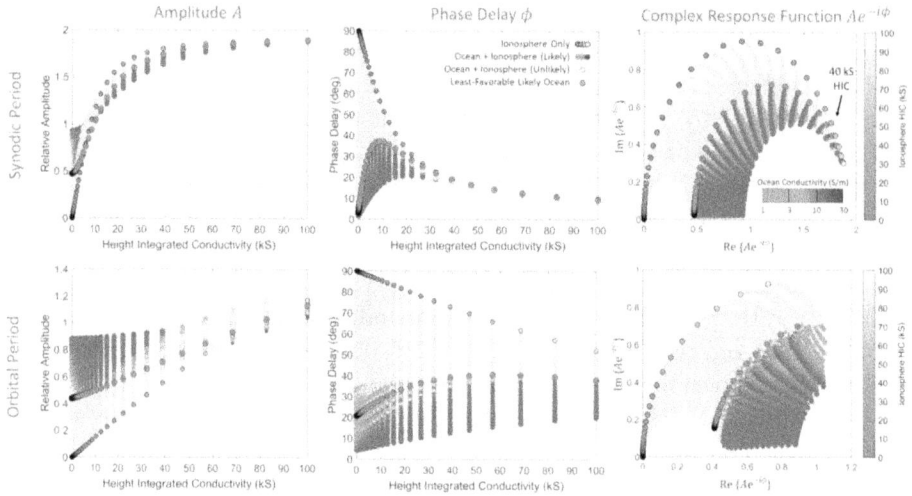

Figure 8.4. Complex response function of Triton interior models. (left column) Model amplitude response and (middle column) phase delay as a function of ionosphere HIC for both the (top) 14 hr synodic and (bottom) the 141 hr orbital periods. Similar to Figure 8.3, the likely ocean (plus ionosphere) models are shaded blue according to their ocean conductivity, the unlikely oceans are colored gray, and the ionosphere-only models are shaded magenta-to-yellow according to their HIC. (right column) Complex response function of the ionosphere-only and ocean-plus-ionosphere models for (top) synodic and (bottom) orbital periods. Adapted with permission from Cochrane et al. (2022). CC BY-NC

detection ability if by chance the HIC has increased since the visit of Voyager 2 in 1989.

Figure 8.4 illustrates the amplitude response and phase delay of all the interior models highlighted in Figure 8.3, plotted as a function of ionosphere HIC for both the 14 hr synodic and the 141 hr orbital periods. The ionosphere-only models (e.g., no ocean) are colored with magenta-to-yellow shading according to their ionosphere HIC to discriminate them between the ocean-plus ionosphere models which are shaded blue (likely ocean) and gray (not likely ocean). Although the separability between the ionosphere-only models and the least-favorable likely ocean models in the amplitude response of the orbital wave is low and almost non-existent for the amplitude response of the synodic wave, the phase delay, especially at the 141 hr period, provides the differences in response needed to facilitate robust classification between the two model classes for the full range of ionosphere HICs used in the study. Also illustrated in Figure 8.4 is the complex response function, plotted in the complex plane for both synodic and orbital waves. Note the clear separation of the ionosphere-only and ionosphere-plus-ocean model classes. The complex response function data for the 14 hr magnetic wave indicate that the primary synodic magnetic wave has limited penetration for an ionosphere intensity greater than ~40,000 S due to the low skin depth associated with a highly conducting ionosphere. This result indicates that detecting a subsurface ocean in the presence of such an intense ionosphere would depend primarily on the 141 hr orbital magnetic wave. The longer period of the 141 hr magnetic wave allows it to penetrate the full range of ionosphere intensities, with clear separation between the least-favorable likely ocean (green circle) and the ionosphere-only models (magenta—yellow).

Using a magnetic induction model initially developed by Zimmer et al. (2000) for analyzing the magnetic induction response from the Jovian moons, the induced magnetic moment $U(t)$, in units of A-m^2, that is generated by conductive interior layers is defined by

$$U(t) = \frac{4\pi}{\mu_0} \frac{R_T^3}{2} M(t), \tag{8.2}$$

where μ_0 is the permeability of free space, $R_T = 1353.4$ km, and the time-varying portion of the moment vector $M(t) = [M_x(t), M_y(t), M_z(t)]$, in units of nT, is defined by a scaled and phase delayed version of the Neptune driving magnetic field,

$$M(t) = \mathrm{Re}\left\{-\sum_k A_k e^{-i\phi_k} B_{N,\,k}(t)\right\}. \tag{8.3}$$

where again, the index k represents the various frequencies in the system. The summation of terms represents the total time-varying component of the induced magnetic moment driven by Neptune's magnetic field $B_{N,\,k}(t)$ where the two primary (synodic and orbital) frequencies dominate.

These equations indicate that the induced magnetic moment from a subsurface ocean is synchronized to Neptune's driving field, but in general, also phase delayed. It also suggests that one must accurately know the magnetic phase of the driving system in addition to the phase delay of the induction signal in order to best interpret the interior from the magnetometer measurements. This constraint complicates single-pass ocean detection because there is uncertainty in Neptune's rotation rate and therefore in its magnetic phase. The magnetic phase is unknown because the rotation of Neptune's core is not discernible from Earth and no spacecraft have visited the Neptune system since Voyager 2 in 1989. To illustrate the various class separation scenarios, the time-varying portion of the induced magnetic moments $M(t)$ are simulated in Figure 8.5 for eight different magnetic phases of Neptune (with respect to the Connerney et al. 1991 magnetic field model of Neptune), each incremented by 45° in phase for a given argument of latitude (AoL) of 50°. (AOL is the angle of Triton in its orbital plane referenced from its orbital ascending node.) Note that for the majority of Neptune magnetic phases, the magnetic separation of the two model classes is clear in the 2D projection shown in Figure 8.5. Even for the classes that appear to overlap in the 2D moment space, (e.g., 135°–225°), there is a small component in the z direction (not shown in the 2D figure) that separates the model classes and makes them discernible.

Although the separation of classes looks promising from a classification perspective for discerning induction responses from the ocean and no ocean classes, the magnetic moment is not what is measured by the magnetometer. The induced magnetic field that is actually measured along the trajectory by the magnetometer is calculated by using the dipole field equation defined by

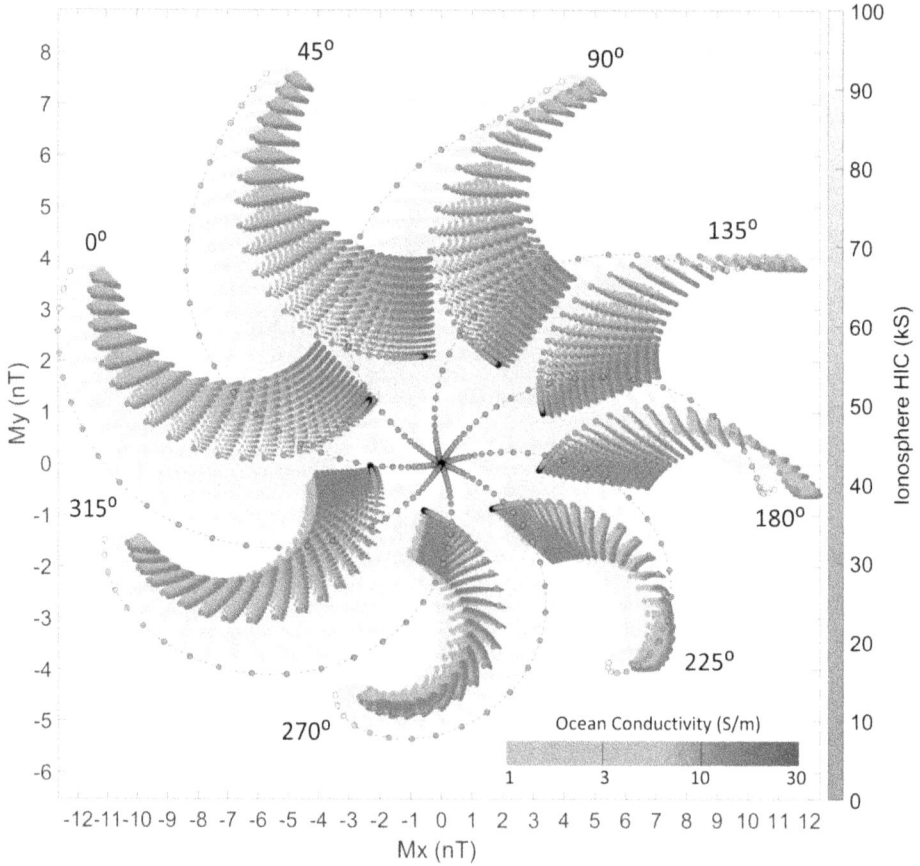

Figure 8.5. Separation of magnetic moment model classes (ocean-plus-ionosphere and ionosphere only) for eight different Neptune magnetic phases (0°, 45°, 90°, 135°, 180°, 225°, 270°, and 315°) for argument of latitude of 50°, on 08-DEC-2038 01:30 ET. Because Neptune's magnetic phase will be unknown at the time of arrival due to the uncertainty in its rotation, the induced magnetic moment of Triton can occupy any space captured by the figure. Note that the coloration of models is consistent with that of Figure 8.4.

$$\boldsymbol{B}_{Ind}(\boldsymbol{r},\,t) = \frac{\mu_0}{4\pi}\frac{3(\boldsymbol{r}(t)\,\cdot\,\boldsymbol{U}(t))\boldsymbol{r}(t)\,-\,r^2(t)\quad\boldsymbol{U}(t)}{r^5(t)}. \tag{8.4}$$

Here, $\boldsymbol{r}(t) = [x(t),\,y(t),\,z(t)]$ is the position of the spacecraft with respect to the center of the Moon. The dipolar nature of the induced field comes from the fact that Neptune's driving field is spatially uniform at Triton. In addition, the magnetometer also measures the superposition of the background field of Neptune, along with other induced magnetic fields from plasma currents in the magnetosphere. Further complicating the extraction of relevant information, the magnetometer also detects magnetic noise and offsets, the spacecraft contaminant magnetic field, and attitude/positioning/timing uncertainties of the spacecraft. Therefore, various methods have been developed that are capable of extracting the relevant information in the midst of all of the field signatures and noise sources.

8.4 Ocean Detection and Characterization Methods

In the last few years, there was a large effort to develop methods to accomplish ocean detection on a single flyby within the moon of Triton under the Trident Discovery mission concept (Frazier et al. 2020). Khurana et al. (2021) developed an inversion technique that determines the amplitude and phase delay of the 14- and 141 hr induced magnetic moments from the data collected on a single flyby. The technique uses the favorable differences and similarities in the anticipated phase delay of the potential ocean and ionosphere models for both synodic and orbital periods (see Figure 8.4) to constrain the induced magnetic moment amplitudes and phase delays of the two primary waves. As illustrated in Figure 8.4, the phase delay of the synodic wave for the ionosphere-only and ocean-plus-ionosphere models with ionosphere HIC > 30 kS are completely identical for all 13,000 models, which indicates that the synodic period cannot be used for discrimination when the ionosphere is highly conducting. However, modeling the response for this range of HIC is much easier as *all* models for *both* class-type converge to the same phase. In this high ionosphere HIC region, one must therefore resort to using the phase of the 141-hr orbital wave for discriminating between the two classes. As illustrated, there is complete separability between the ionosphere-only models and ocean-plus-ionosphere models for any given ionosphere HIC. For models with ionosphere HIC below 30 kS, in addition to small amplitude differences, the differences in the phase delay of the model classes at both periods provide good discrimination ability. However, in order to reliably solve for these parameters, one must also account for the low amplitude harmonics and beats of the two fundamental periods in the system with non-negligible amplitude (e.g., 4 periods with amplitude greater than 0.1 nT as seen in Figure 8.1) as they are a source of structural noise. In order to compensate for this, the amplitudes and phases of a single representative model were first selected from the batch of forward models that best represented the average of the low-amplitude tones and then subtracted from the simulated measurements to partly remove their effects. The inversion was then performed on the resultant dataset, consisting of the 14- and 141-hr components and the residual errors of the low-amplitude tones which yielded more accurate results. The technique was tested against several thousand Triton interior models and found to provide accurate inversions of the dipole moments; it was also shown to work at various Neptune magnetic phases. However, the model assumptions made on these lower order harmonics and beats required to solve for the two main moments cannot perfectly be accounted for on a single flyby. This difficulty results in errors in the inversion methodology, especially when additional noise sources are included, those being instrument noise and offsets, spacecraft attitude, position, timing uncertainties, and structured plasma noise. Multiple flybys are typically needed to acquire additional measurements of the system to better improve the least-square inversion estimates of the parameters at the individual periods.

A second method of ocean detection for the Trident Discovery concept was developed by Cochrane et al. (2022) and is based on principle component analysis (PCA). The method entails forward modeling the combined magnetic induction response from thousands of plausible combinations of ocean and ionosphere configurations, varying the thickness and conductivity of each layer. The method uses thousands of models to identify the principal components of the dataset, thus providing an orthogonal basis that can be used to classify projected measurements. The work demonstrated through a Monte Carlo simulation that the ocean detection on a single-pass is robust against any magnetic phase of Neptune and also in the face of noise sources, including instrument flicker noise and offsets, forward model uncertainty, plasma interaction fields, and spacecraft attitude, position, and timing uncertainty. The simulation indicated that a 1 nT (3 sigma) anticipated error would result when projecting the measurement into PC space when an assumed magneto-meter sample rate of 1 Sample/sec was used. Using this error, it was shown that the range of oceans that are detectable within the simulated parameter space, as illustrated here in Figure 8.6 for convenience. All likely oceans are detectable—even the least-favorable ocean at the least-favorable Neptune magnetic phase of 120°—with at least 1 nT of margin. The authors also showed the ability to characterize three different oceans, for three different Neptune magnetic phases and four different ionosphere HICs, reproduced here in Figure 8.7. Note that some form of characterization on a single flyby is only realizable if the ionosphere HIC is 10 kS or less. However, if more flybys are included, tighter constrains can be placed on the inferred ocean thickness and conductivity, even for large ionospheric HIC.

Other methods have also been recently developed for the Europa Clipper mission that would be applicable to Neptune/Triton. Biersteker et al. (2023) present a method using Bayesian inference to characterize the interiors of icy moons using measurements acquired during multiple flybys and/or while in orbit around a moon. The method estimates the probability distribution for a given set of ocean model parameters O (e.g., ocean thickness, ocean conductivity, and ice thickness) for a given set of magnetometer measurements, M. This is known as the posterior probability distribution, $p(O|M)$. Bayes theorem states that $p(O|M) \propto p(M|O)p(O)$, where $p(M|O)$ is the probability of the measurements for the given set of ocean parameters, known as the likelihood, and $p(O)$ is the probability of these ocean parameters without taking into account any measurements, known as the prior probability. The likelihood depends on how well the model fits the observed data, which can be assessed with goodness-of-fit statistics. The method estimates the posterior probability of many different combinations of ocean parameters using a Markov Chain Monte Carlo (MCMC) sampler. Like the techniques previously discussed, this method provides direct estimates of the ocean parameters. It has two particular advantages of (i) robust quantification of uncertainties, including estima-tion of the degeneracies between parameters, without making assumptions about the form of the posterior probability distribution; and (ii) it naturally incorporates prior constraints from a diversity of other measurements (e.g., bounds on total thickness of the H_2O layer, moment of inertia, and ocean conductivity). A similar MCMC approach has also recently been reported by Petricca et al. (2023).

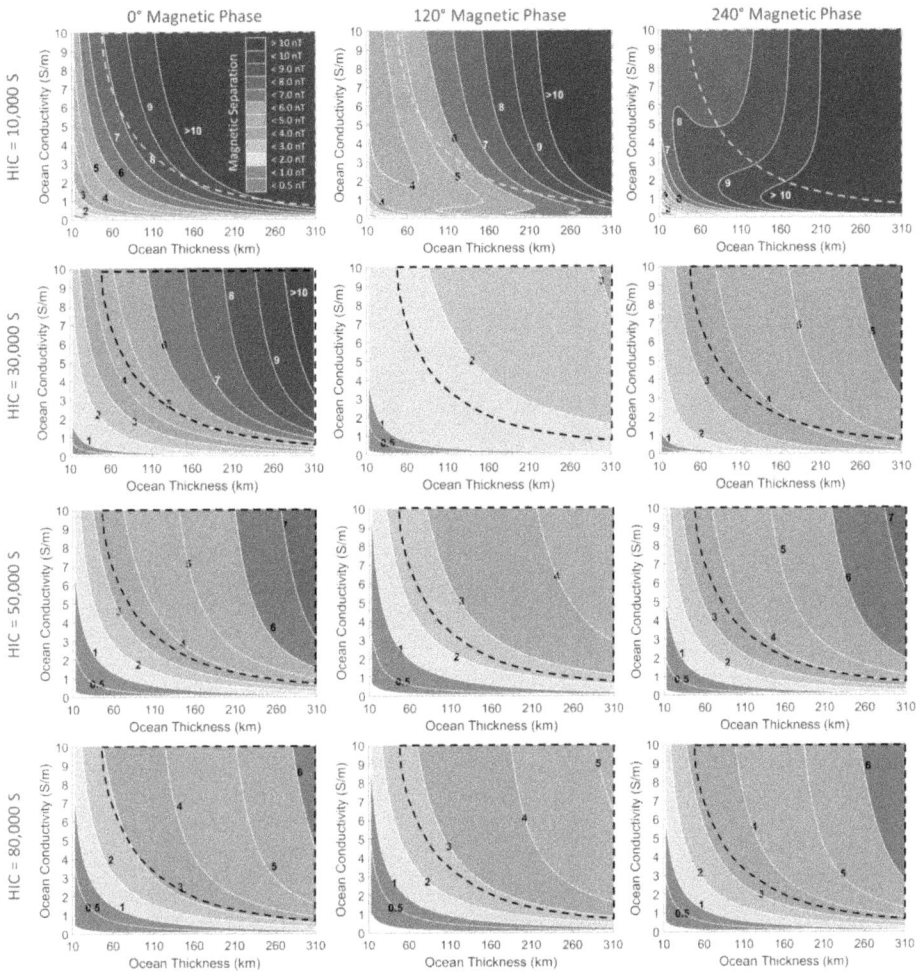

Figure 8.6. Dependence of ocean-to-ionosphere magnetic separation for range of ionosphere HIC and Neptune magnetic phase for a single-flyby. Each panel represents the magnetic separation contours (units of nT) of the ocean-plus-ionosphere models from the ionosphere-only models as a function of ocean thickness and ocean conductivity for (left column) magnetic phase of 0°, (middle column) worst-case magnetic phase of 120°, and (right column) 240°, for ionosphere HIC of (first row) 10, (second row) 30, (third row) 50, and (fourth row) 80 kS. The region outlined by the dashed lines highlights the most likely ocean space. The darkest shades of blue represent the largest magnetic separation and hence best distinguishability. The red contours indicate magnetic separation below 1 nT, where ocean classification or differentiation could not be performed with certainty. Adapted with permission from Cochrane et al. (2022). CC BY-NC

8.5 Trajectory Considerations

There are various flyby characteristics that should be considered when designing a trajectory for a mission with hopes of discovering subsurface water within Triton. For multiple-flyby concepts, the most important aspect of a trajectory is for the flybys to have arrival times with uniform sampling of Neptune longitude and Triton

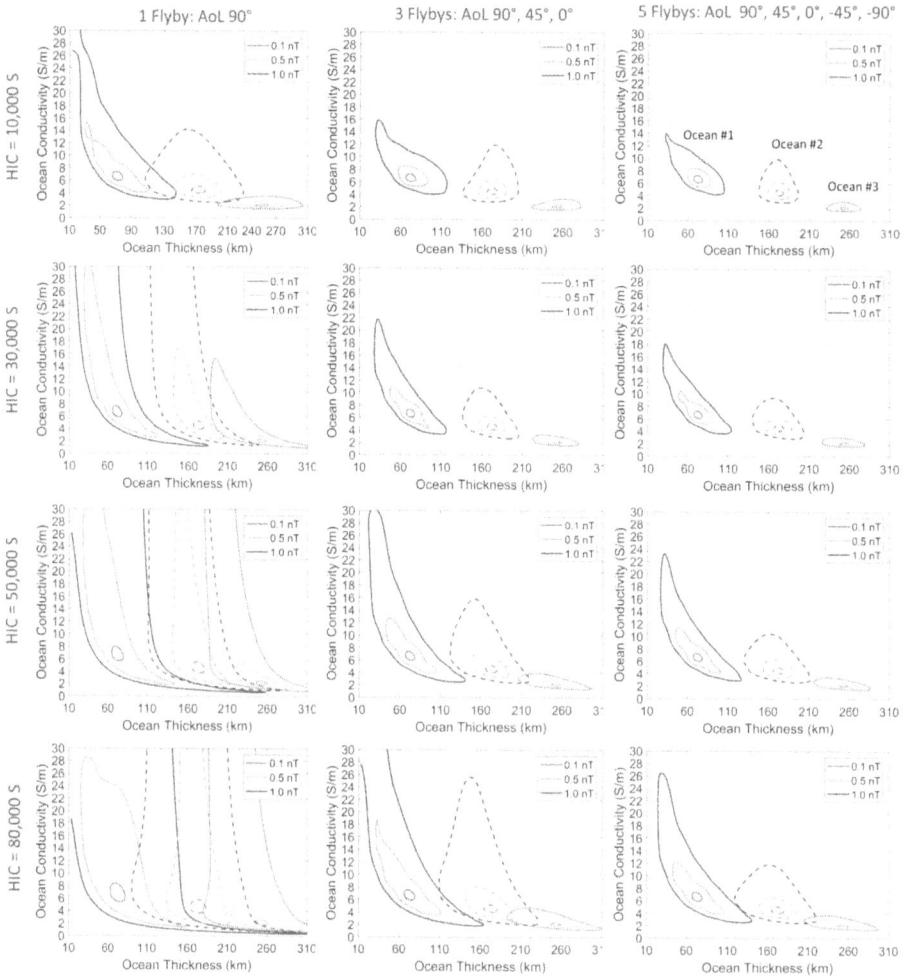

Figure 8.7. Measurements from multiple flybys can be used to improve confidence in the assessment of ocean parameters. Each panel in the figure illustrates characterization contours of three representative oceans for (left column) one flyby at AoL of +90°, (middle column) three flybys at AoL of +90°, +45° and 0° and (right column) five flybys at AoL of +90°, +45°, 0°, −45°, and −90°, where each row corresponds to a different ionosphere HIC scenario. (Note again that AOL is the angle of Triton in its orbital plane referenced from its orbital ascending node.) Ocean #1 ($T = 75$ km, $\sigma = 6.5$ S m^{-1}) is represented by the solid lines, ocean #2 ($T = 175$ km, $\sigma = 4.5$ S m^{-1}) is represented by the dashed lines, and ocean #3 ($T = 255$ km, $\sigma = 2.0$ S m^{-1}) is represented by the dotted lines, where T represents ocean thickness and σ represents ocean conductivity. The three contours shown for each ocean represent the boundaries that encompasses the indistinguishable oceans when three different levels of noise are considered (blue = 1 nT, cyan = 0.5 nT, and red = 0.1 nT). Adapted with permission from Cochrane et al. (2022). CC BY-NC

orbital phase (or argument of latitude) at very low altitudes. Note that a similar approach is being taken for the Europa Clipper mission which will characterize the subsurface ocean within Europa (Kivelson et al. 2023). The low altitude is desired because the induced dipole response falls off at r^{-3} from the conductor source, which is anticipated to be small due to the small driving field, the attenuation effect of the interior response due to the ocean not being a perfect conductor, and because the measurement will likely be made a few 10s or 100s of km above the ice shell, distant from the source. Using the simple relationship for the magnetic field falloff from a dipole source, $B(r) = M/r^3$, and assuming a $B_d = 10$ nT driving field and a conservative scale factor of $A = 0.75$ with ocean present just below the surface, the minimum altitude that should be flown to measure a $B_m = 5$ nT field is computed by $[(AB_d/B_m)^{1/3} - 1]R_T$. This relationship suggests that the flyby altitude at closest approach should be no higher than 200 km. However, flybys less than 300 km would be tolerable, and less than 100 km would be desired (which is also the requirement for the Europa Clipper flybys of Europa as it has a comparable radius, see Kivelson et al. 2023). Acquiring measurements at various Neptune longitude and orbital phases allows for observations of various constructive and destructive interference patterns from both magnetic waves. If by chance the mission comprises of only two flybys of Triton, limited in number to definitively distinguish an induced response from a permanent internal dynamo (similar to the Galileo flybys of Europa, see Kivelson et al. 1999), it should be designed so that the flybys are separated by 180° in synodic phase, which is equivalent to separating the flybys in time by $T_s(2k+1)/2$ hrs, where T_s is the 14 hr synodic period and k is an integer. As illustrated in Figure 8.5, for argument of latitude of 50°, destructive addition of the two dominant waves occurs when the magnetic phase is in the range of 110°–250°. If in the unfortunate event of arriving at Triton when Neptune is in this magnetic phase region, arriving at Triton when Neptune's magnetic phase is 180° away (i.e., 290°–360° and 0°–70°), will yield a more favorable separation of model classes. Similarly, for a three-flyby mission concept, 120° spacing is desired for minimum phase gap coverage, so the second and third passes should be designed to arrive in time after the first flyby by T_s $(3k+1)/3$ hr and $T_s (3k+2)/3$ hr, respectively. For missions with many planned flybys of Triton, the trajectory should instead be designed to have uniform sampling of the 2D space formed by Neptune longitude versus Triton orbital phase, which will maximize variations in the magnetic field patterns observed by the orbiter.

For single-flyby mission concepts, the flyby design is more critical to maximize the chances of detecting an ocean. Unfortunately, because the magnetic phase of Neptune will be unknown upon arrival, due to the uncertainty in rotation rate of the planet, an arrival time that optimizes a constructive interference pattern of the two dominant waves cannot be planned. Therefore, one should attempt to arrive at a time that maximizes the field from the more predictable orbital wave, which occurs when Triton is at an AoL of ±90°. In these locations, the Bx component (IAU Triton frame) of the orbital wave is maximum as illustrated in the left panel of Figure 8.8. Arriving at Triton during the maximum of the orbital wave will allow for optimal separation of classes in case the ionosphere conductivity is stronger than anticipated. Note that the single-flyby mission trajectory developed by the Trident Discovery concept adopted a flyby AoL of +50° because the orbital wave is still sufficiently strong, but also balanced the needs of the other science investigations. One aspect of this specific trajectory AoL that makes it more advantageous than the ±90° case is that it allows the spacecraft to spend more time within Neptune's magnetosphere

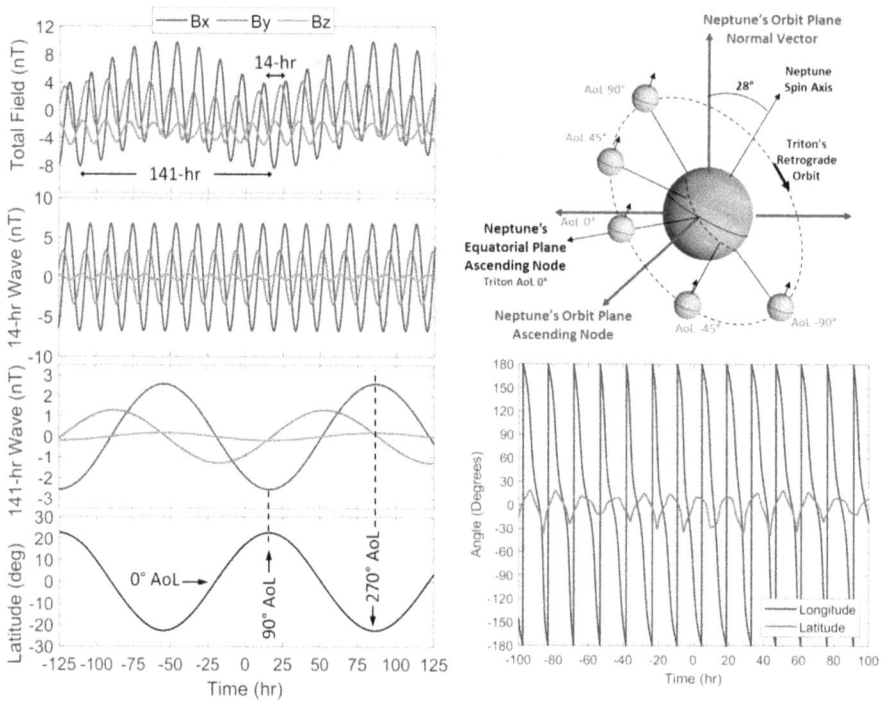

Figure 8.8. Considerations for optimizing flyby trajectories. (Left) Total magnetic field broken down into its 14 and 141 hr components and associated argument of latitude as a function of time. (Upper right) Geometry of the Neptune-Triton system for various argument of latitudes. (Bottom right) Orientation (longitude and latitude) of Neptune's magnetic field in the frame of Triton as a function of time. Note that the induced magnetic moment is always oriented within ±35° in Triton's geographic equator. Adapted with permission from Cochrane et al. (2022). CC BY-NC

(see upper right panel of Figure 8.8), thereby observing multiple rotations of the Planet on its single pass through the magnetosphere. This would allow better characterization of Neptune's driving magnetic field and sense any deviation in its strength and orientation from the observations made by Voyager 2. These measurements are critical, as any potential induced magnetic field from Triton can only be accurately assessed if the driving field is well known. One other parameter that should be considered is the flyby's moon latitude at closest approach. For a single-flyby concept, the spacecraft should be designed to fly as close to the geographic equator as possible at closest approach, as the orientation of Neptune's AC field vector in Triton's frame rotates circularly around Triton to within ±30° in latitude of the geographic equator, see lower right panel of Figure 8.8. The induced magnetic moment generated by the putative ocean would more or less try to cancel this field, and therefore be anti-aligned with this rotating field, suggesting the poles of the induced magnetic moment would also be restricted to within ±30° in latitude of the Moon's geographic equator. Note that this does not guarantee that the spacecraft would fly over a pole of the induced magnetic moment (where the field strength is twice as strong as it would be at the magnetic equator), but would rather increase the chances of it and provide a modest boost in signal strength.

8.6 Instrument Selection

By taking advantage of the natural magnetic variations that are expected in the Neptune magnetosphere, the main instrument that will be key to magnetically sound Triton would be a 3-axis magnetometer. In the analysis performed (Cochrane et al. 2022), it was shown that an assumed 3-axis fluxgate magnetometer with amplitude spectral density of $100 \, pT/\sqrt{Hz}$ at $1 \, Hz$ was sufficient enough to distinguish between the worst-case likely ocean scenario from that of a response solely originating from the ionosphere. However, it is not too uncommon for a fluxgate, and other atomic gas-based magnetometers, to exhibit $<10 \, pT/\sqrt{Hz}$ sensitivity that would guarantee significant margin. Fluxgates have the advantage of decades of heritage in space and regularly achieve noise levels below 100 pT integrated over 0.01–1 Hz (Kivelson et al. 2023). Their main limitation is that their offsets and scale factors drift as a function of time and temperature. Offsets can be tracked by monitoring the magnetic field Alfvénic fluctuations in the solar wind over time (Davis & Smith 1968; Belcher 1973, Acuña 2002; Leinweber et al. 2008) while the gains (and offsets) can be found by two sets of spacecraft rolls about perpendicular axes inside a planetary magnetosphere with sufficient background magnetic field strength (Kepko et al. 1996; Kivelson et al. 2023). Flying an additional optically pumped vapor magnetometer (with scalar and/or vector modes), using metastable ^{4}He or an alkali vapor such as rubidium or cesium, can reduce the amount of spacecraft operations needed to calibrate these sensors as their quasi-absolute nature allows for calibration of the fluxgate vector measurements (Merayo et al. 2000).

As discussed previously, the main contributions to the measurement uncertainty are magnetic fields that arise from yet-to-be determined plasma effects and spacecraft magnetic fields. The latter can be mitigated by using two or more magnetometers along a several-meter long boom in a gradiometer configuration (e.g., Ness et al. 1971; Acuna et al. 2002; Cochrane et al. 2023). Dynamic range should be defined by how close the spacecraft is anticipated to get to Neptune. As Neptune has a surface field of no stronger than approximately 1 Gauss (100,000 nT), an upper limit can be placed on the magnetometer dynamic range. Wide dynamic range can be achieved by developing the magnetometer such that it has multiple gain settings if limited by resolution (e.g., Dougherty et al. 2004; Connerney et al. 2017) or can be achieved with the use of sigma-delta sampling technology, where 20–24-bit resolution can be implemented in firmware (Magnes et al. 2003; Kivelson et al. 2023). The bandwidth, or sampling rate, can be relatively low for planetary field and magnetic induction modeling, as the frequencies are very low (e.g., planetary synodic period and moon orbital components), so that sample rates on order of 1 Sample/sec are sufficient. The measurements of ion-cyclotron waves from electrons and ions will likely drive the required sample rate and bandwidth of the magnetometer towards higher rates. The electron and ion gyrofrequencies are linearly proportional to the magnetic field in which they are immersed. Measurements of the variability of ion-cyclotron waves in the vicinity of the Moon could be used to monitor plume activity, variations in the atmospheric density, and loss rate of ion species.

To accompany the magnetometer, a plasma spectrometer or Faraday Cup to measure the plasma energy, velocity, and concentrations would be advantageous to characterize the induced field contribution from any plasma currents. This is the goal of the Plasma Instrument for Magnetic Sounding investigation for the Europa Clipper mission (Westlake et al. 2023). If a plasma instrument cannot be accommodated, radio science measurements at different bands on ingress and egress of the flyby(s) can be used for measuring the electron concentration in the ionosphere, which could help constrain the ionosphere conductivity, an important input to assess the HIC of the ionosphere. Radio science measurements would also be used to estimate the periodic tidal Love number k_2, as well as providing additional insight into the structure of the interior. Combining gravity and magnetometer measurements thus can be very useful for assessing the interior properties of the Moon (Biersteker et al. 2023; Petricca et al. 2023). As gravitational forces are directly related to distance to the body, low-altitude flybys are also very important with good spatial coverage of the Moon (Mazarico et al. 2023) regarding the gravity science investigation of Europa Clipper. An orbiter mission, such as the Triton Ocean World Survey concept featured in the 2023–2032 Planetary Science and Astrobiology Decadal Survey New Frontiers portfolio, would guarantee extensive spatial and temporal coverage of the geophysical response (tidal, magnetic field) of Triton. An orbiter mission like the Persephone concept (Howett et al. 2021) has also been considered for Pluto but is challenging to implement from a resource standpoint.

8.7 Summary

The decadal survey highlights Triton as a high value candidate ocean world target for a future orbiter mission. A single-flyby mission based on the 2021 Trident proposal (Frazier et al. 2021) may also be considered under the Discovery program. Cochrane et al. (2022) have shown that a deep ocean within Triton, if present, can be detected using magnetic induction on a single flyby assuming a magnetometer with adequate sensitivity ($<100\, pT/\sqrt{Hz}$) is chosen, with spacecraft trajectory characterized with low altitude (<300 km) and low latitude ($<\pm35°$) at closest approach to maximize the signal-to-noise of the measured induction signal. Additionally, the time-of-arrival should be planned such that the argument-of-latitude is close as possible to 90° or 270° to maximize strength of the 141-hr orbital wave. Confirmation that Triton is an ocean world would carry important implications because the Moon is expected to contain a large fraction of organics, per its origin in the Kuiper belt. Although the prospect for either mission to be selected in the future is uncertain, there are other opportunities for ocean worlds in the near-term (Europa Clipper, JUICE) and longer-term (e.g., Enceladus theme in the New Frontiers and Flagship programs.) In particular, the 2023–2032 decadal survey has prioritized a mission to the Uranian system in its Flagship mission portfolio, which is expected to commence in the late-2020s. The latter would extensively investigate the Uranian system, including its five major moons (Miranda, Ariel, Umbriel, Titania, Oberon), which are also considered candidate ocean worlds (Hendrix et al. 2019). The techniques presented in this chapter are relevant to these mission concepts and have the potential to return significant knowledge from these various targets.

References

Acuna, M. H. 2002, RScI, 73, 3717

Agnor, C. B., & Hamilton, D. P. 2006, Natur, 441, 192

Archinal, B. A., Acton, C. H., Hearn, M. F. A., et al. 2018, CeMDA, 130, 22

Arridge, C. S., & Eggington, J. W. B. 2021, Icar, 367, 114562

Bagenal, F., et al. 1997, Pluto and Charon, Space Science Series, ed. S. A. Stern, & D. J. Tholen 523 (Tucson, AZ: Univ of Arizona Press)

Belcher, J. W. 1973, JGR, 78, 6480

Biersteker, J. B., Weiss, B. P., Cochrane, C. J., Harris, C. D. K., Jia, X., Khurana, K. K., Liu, J., Murphy, N., & Raymond, C. A. 2023, PSJ, 4, 62

Belcher, J. W., Bridge, H. S., Bagenal, F., et al. 1989, Sci, 246, 1478

Castillo-Rogez, J., Daswani, M. M., Glein, C., Vance, S., & Cochrane, C. J. 2021, ESSOAr

Castillo-Rogez, J., Daswani, M. M., Glein, C., Vance, S., & Cochrane, C. J. 2022, GeoRL, 49, e97256

Chan, K., Grima, C., Gerekos, C., & Blankenship, D. D. 2023, EGU General Assembly 2023 (Göttingen: Copernicus Publishing) EGU23-10554

Chen, E. M. A., Nimmo, F., & Glatzmaier, G. A. 2014, Icar, 229, 11

Cochrane, C. J., Persinger, R. R., Vance, S. D., et al. 2022, E&SS, 9, e2021EA002034

Cochrane, C. J., Vance, S., Nordheim, T., et al. 2021, JGRE, 126, e2021JE006956

Cochrane, C. J., Murphy, N., Raymond, C. A., et al. 2023, SSRv, 219, 34

Connerney, J. E. P., Benn, M., Bjarno, J. B., et al. 2017, SSRv, 213, 39

Connerney, J., Acuna, M. H., & Ness, N. F. 1991, JGR, 96, 19023

Croft, S., Kargel, J., Kirk, R. L., et al. 1995, Neptune and Triton, (Tucson, AZ: Univ. Arizona Press) 879

Courville, S. W., Castillo-Rogez, J. C., Daswani, M. M., et al. 2023, PSJ, 4, 179

Davis, L., & Smith, E. J. 1968, TAGU, 49, 257

Dougherty, M. K., Khurana, K. K., Neubauer, F. M., et al. 2006, Sci, 311, 1406

Dougherty, M. K., Kellock, S., Southwood, D. J., et al. 2004, SSRv, 114, 331

Frazier, W., Bearden, D., Mitchell, K. L., et al. 2020, 2020 IEEE Aerospace Conf. 1 (Piscataway, NJ: IEEE)

Grimm, R. E. 2009, NRC White Paper, <www8.nationalacademies.org/ssbsurvey/publicview. aspx>

Hansen, C. J., Castillo-Rogez, J., Grundy, W., et al. 2021, PSJ, 2, 137

Hendrix, A. R., Hurford, T. A., Barge, L. M., et al. 2019, AsBio, 19, 1

Howett, C. J., Robbins, S. J., Holler, B. J., et al. 2021, PSJ, 2, 75

Hussmann, H., Sohl, F., & Spohn, T. 2006, Icar, 185, 258

Ivanov, M., & Borozdin, V. 1991, LPI, 22, 623

Kamata, S., Nimmo, F., Sekine, Y., Kuramoto, K., Noguchi, N., Kimura, J., & Tani, A. 2019, NatGeo, 12, 407

Keane, J. T., Matsuyama, I., Kamata, S., & Steckloff, J. K. 2016, Natur, 540, 90

Kepko, E. L., Khurana, K. K., Kivelson, M. G., Elphic, R. C., & Russell, C. T. 1996, ITM, 32, 2

Kimura, J., & Kamata, S. 2020, PSS, 181, 104828

Kivelson, M. G., Khurana, K. K., Stevenson, D. J., et al. 1999, JGR, 104, 4609

Kivelson, M. G., et al. 2023, SSRv, 219, 48

Kivelson, M. G., Khurana, K. K., & Volwerk, M. 2009, Europa, ed. R. T. Pappalardo, W. B. McKinnon, & K. K. Khurana (Tucson, AZ: Univ. Arizona Press) 545

Khurana, K. K., Cochrane, C. J., Mitchell, K., et al. 2021, Magnetospheres of the Outer Planets Virtual Conf. *(July 12, 2021)*

Khurana, K. K., Jia, X., Kivelson, M. G., et al. 2011, Sci, 332, 1186

Kirk, R. L., Soderblom, L. A., Brown, R., Kieffer, S., & Kargel, J. 1995, Neptune and Triton, (Tucson, AZ: Univ. Arizona Press) 949

Kobe, K. A., & Sheehy, T. M. 1948, Ind. Eng. Chem., 40, 99

Leinweber, H. K., Russell, C. T., Torkar, K., Zhang, T. L., & Angelopoulos, V. 2008, MeScT, 19, 055104

Liuzzo, L., Paty, C., Cochrane, C., et al. 2021, JGRA, 126, e2021JA029740

Magnes, W., Pierce, D., Valavanoglou, A., et al. 2003, MeScT, 14, 1003

Mazarico, E., Buccino, D., Castillo-Rogez, J., et al. 2023, SSRv, 219, 30

Merayo, J. M. G., et al. 2000, MeScT, 11, 120

Ness, N. F., Behannon, K. W., Lepping, R. P., & Schatten, K. H. 1971, JGR, 76, 3564

Nimmo, F., & Pappalardo, R. T. 2016, JGRE, 121, 1378

Nimmo, F., & Spencer, J. R. 2015, Icar, 246, 2

Peters, S. T., Schroeder, D. M., Haynes, M. S., Castelletti, D., & Romero-Wolf, A. 2021, ITGRS, 59, 9144

Petricca, F., Genova, A., Castillo-Rogez, J. C., et al. 2023, GeoRL, 50, e2023GL104016

Picardi, G., Plaut, J. J., Biccari, D., et al. 2005, Sci, 310, 1925

Prockter, L. M., Nimmo, F., & Pappalardo, R. T. 2005, GeoRL, 32, L14202

Richardson, J. D., & McNutt, R. L. 1990, GeoRL, 17, 1689

Richardson, J. D. 1993, GeoRL, 20, 1467

Romero-Wolf, A., Steinbruegge, G., Castillo-Rogez, J., et al. 2023, Feasibility of Passive Sounding of Uranian Moons using Uranian Kilometric Radiation, arXiv:2305.05382

Romero-Wolf, A., Vance, S., Maiwald, F., et al. 2015, Icar, 248, 463

Ruiz, J. 2003, Icar, 166, 436

Saur, J., Neubauer, F. M., & Glassmeier, K. H. 2010, SSRv, 152, 391

Schenk, P. M., & Zahnle, K. 2007, Icar, 192, 135

Schenk, P. M., Beddingfield, C. B., Bertrand, T., et al. 2021, RemS, 13, 3476

Schilling, N., Neubauer, F. M., & Saur, J. 2007, Icar, 192, 2007, 41

Schmidt, C., & Manning, C. 2017, GChPL, 3, 66

Schroeder, D. M., Bingham, R. G., Blankenship, D. D., et al. 2020, AnGla, 61, 1

Seufert, M., Saur, J., & Neubauer, F. M. 2011, Icar, 214, 477

Shock, E. L., & McKinnon, W. B. 1993, Icar, 106, 464

Singer, K. N., Greenstreet, S., Schenk, P. M., Robbins, S. J., & Bray, V. J. 2021, The Pluto System eds. S. A. Stern, et al. (Tucson, AZ: Univ. Arizona Press) 121

Singer, K. N., White, O. L., Schmitt, B., et al. 2022, NatCO, 13, 1542

Sittler, E., & Hartle, R. 1996, JGRA, 101, 10863

Strobel, D. F., Cheng, A. F., Summers, M. E., & Strickland, D. J. 1990, GeoRL, 917, 1661

Styczinski, M. J., Vance, S. D., Harnett, E. M., & Cochrane, C. J. 2022, Icar, 376, 114840

Tyler, G., Sweetnam, D., Anderson, J., et al. 1989, Sci, 246, 1466

Ucok, H., Ershaghi, I., & Olhoeft, G. 1980, JPT, 32, 717

Vance, S. D., Styczinski, M. J., Bills, B. G., et al. 2021, JGRE, 126, e2020JE006418

Vance, S., Styczinski, M., Castillo-Rogez, J., et al. 2021, in AGU Fall Meeting Abstracts 2021 (Washington, DC: AGU) P31A

Vance, S. D., Kedar, S., Panning, M. P., et al. 2018, AsBio, 18, 37

Weiss, B. P., Biersteker, J. B., Colicci, V., et al. 2021, GeoRL, 48, e2021GL094758

Westlake, J., et al. 2023, SSRv, 219, 62

Winkenstern, J., & Saur, J. 2023, JGRE, 128, e2023JE007992

Zhang, M., Richardson, J. D., & Sittler, E. C. 1991, JGR, 96, 19085

Zimmer, C., Khurana, K. K., & Kivelson, M. G. 2000, Icar, 147, 329

Chapter 9

Future Measurement Needs: Surface Processes

Alessandra Migliorini, Bryan J Holler, Will M Grundy, Tom S Stallard, Federico Tosi and Leslie A Young

The exploration of Triton began over three decades ago, with the Voyager 2 flyby in August 1989. This unique flyby provided a first glimpse of a very unusual object, with a complex and young surface. As a captured Kuiper Belt Object (KBO), Triton shares much in common with Pluto and possibly other dwarf planet class KBOs. Only recently, Pluto was the objective of NASA's New Horizons spacecraft that flew by the Pluto-Charon system in July 2015. However, for the majority of the last 40 years, barring the few days leading up to the flybys of Triton and Pluto, the only source of photometric and spectroscopic data for these objects was supplied by Earth-based telescopes, which allowed us to intermittently monitor the processes acting on their surfaces and in their atmospheres. Despite the vast dataset obtained with telescopic measurements, several areas are still to be investigated for Triton and Pluto, such as the latitudinal distribution of volatile and non-volatile ices on Triton, the inventory of minor species, including CO, HCN, and ethane and other CH_4 by-products on both surfaces, the atmospheric evolution of Pluto, and finally their status as ocean worlds. In this work we identify some key areas where the next generation of ground- and space-based telescope facilities could greatly contribute to the science related to Pluto, Triton, and related KBOs. A review of *in situ* exploration with mission architectures involving orbiters and/or landers is also outlined. However, even new flyby missions would be incredibly valuable in order to study regions that were unilluminated and/or mapped only at low spatial resolution during previous flybys. The next round of *in situ* exploration would benefit considerably from the application of new instrumentation and from exploiting the passage of time since the initial flybys with Voyager 2 and New Horizons to better understand the seasonal evolution of both Triton and Pluto.

9.1 Introduction

As a captured Kuiper Belt Object (KBO), Triton is the ideal object for comparative planetology studies including Pluto and other dwarf planet class KBOs. Triton and

doi:10.1088/2514-3433/ad5278ch9

Pluto are very similar in many aspects. Their surfaces have a similar composition, which includes the volatile ices N_2, CH_4, CO, and non-volatile ices such as H_2O, and complex hydrocarbons. They also have tenuous atmospheres composed primarily of N_2, which appears to be the dominant ice species on both Triton and Pluto.

Both Triton and Pluto have very peculiar orbits, inclined by 26° with respect to Neptune's orbit in case of Triton and 17° to the ecliptic in the case of Pluto, which gives rise to unique seasons. Triton's seasonal cycle is complex due to Neptune's obliquity, Triton's orbital inclination around Neptune, and the rapid precession of Triton's orbital node, with alternating epochs of extreme and mild solstices and a beat pattern between the nodal precession and Neptune's 165-year orbit (Trafton 1984). For Pluto, the combination of an axial tilt of 119.5°, a high eccentricity of 0.25, and the amplitude of obliquity oscillation leads to the so-called "extreme seasons" over Pluto's 248 Earth-year orbit (Binzel et al. 2017; Stern et al. 2017), during which the subsolar latitude falls nearest to the poles, as a consequence of the beat frequency between Pluto's obliquity period (3 Myr) and the longitude of perihelion precession period (3.7 Myr; Dobrovolskis et al. 1997; Earle & Binzel 2015; Earle et al. 2016). We are still in the early stages of understanding the observable effects these seasonal changes have on the composition and distribution of ices across the surfaces of Triton and Pluto. The present day provides unique opportunities to enhance this understanding, with Triton rapidly moving towards equinox in 2046 after the most extreme summer solstice in nearly 700 years and Pluto approaching northern hemisphere summer solstice in 2029.

In situ exploration of these two peculiar objects has been limited. Triton was visited by the Voyager 2 mission in August 1989, and the entirety of what we know about this object was acquired during this single flyby and using Earth-based telescopes. Pluto was recently investigated with the NASA New Horizons mission, which flew through the Pluto-Charon system in July 2015. Apart from these events, no other spacecraft has visited the icy bodies of the outer solar system. To date, *in situ* exploration of the outer solar system has been possible only with single flybys, which can provide only a short glimpse of ongoing processes and are limited in both space and time. Earth-based observations, made with ground-based telescopes, the Hubble Space Telescope (HST), and the James Webb Space Telescope (JWST), provide the necessary time-baseline to contextualize the high-resolution imagery obtained during short spacecraft flybys.

In this chapter we review the processes occurring on the surfaces of Triton and Pluto, with the aim of outlining where our knowledge is lacking with respect to these two related bodies. We also propose future measurements from ground and space that could help fill these gaps.

9.1.1 Triton

Triton, Neptune's largest moon, orbits around Neptune in a retrograde sense in 5.877 days on a perfectly circular ($e = 0$), exceptionally inclined ($\sim$157°) orbit with respect to Neptune's equatorial plane. Given this retrograde orbit, it is suspected to be a captured Kuiper Belt Object (McCord 1966; McKinnon 1984;

Figure 9.1. Triton's surface color map as obtained by Voyager 2 in 1989. Credit: NASA/JPL-Caltech/Lunar & Planetary Institute.

Agnor & Hamilton 2006). Due to intense tidal heating following its capture, the icy material that composes Triton may have melted to form a subsurface ocean (Hussmann et al. 2006; Gaeman et al. 2012). To date, Triton has only been visited by the Voyager 2 spacecraft in August 1989. The images returned from Voyager 2 cover only 40% of the total surface of the satellite, concentrated in the illuminated southern hemisphere, which was at that time approaching summer solstice. Figure 9.1 shows a color map of Triton's surface obtained with Voyager 2.

Voyager 2 showed that Triton's southern hemisphere is diversified in terms of albedo, colors, texture, and geologic features (McEwen 1990). The leading side is dominated by *Volcanic Plains* (Croft et al. 1995), while the trailing side is characterized by so-called *Cantaloupe Terrain* (Schenk & Jackson 1993), as a result of either geological or atmospheric-driven processes.

The observed structures on Triton's surface are remarkably unusual in the solar system and provide evidence of ongoing activity (e.g., Kargel & Strom 1990, Kargel 1994, Prockter et al. 2005). As an example, the smooth structures located at low latitudes bounded by about 200 m high chains of mountains may be the result of low-viscosity cryovolcanism, possibly from an ocean, or through glacial processes or volatile sublimation. Exogenic processes, including atmosphere-surface interactions or solar-driven ones, as well as endogenic ones, including transport of material through the ice shell and/or atmospheric precipitation of organics, have been investigated to identify an explanation for the observed features. See (Chapter 4) for a complete description of the geologic properties of Triton.

Two plumes of gas and dust were observed with Voyager 2 emanating from Triton's surface close to the subsolar latitude in the Southern hemisphere at the time of the flyby, extending up to 8 km in altitude and over 100 km long and then dragged by atmospheric winds (Soderblom et al. 1990). More than 120 streaks, called fans, were also imaged in the Southern hemisphere, with similar properties to the plumes

(Hansen et al. 1990). Possible theories to explain the observed plumes and fans involve exogenic activity, like insolation-driven heating of the N_2 polar cap which results in emissions similar to geysers (Soderblom et al. 1990), or an endogenic origin, which invokes internal heat as the driver of the observed erupting material (Laufer et al. 2013). A recent study reanalyzed the major eruption hypotheses in light of additional data from Mars, Enceladus, and Pluto to explain the plume distribution over the surface of Triton, and all hypotheses remain plausible (Hofgartner et al. 2022). The authors demonstrate that there is no evidence to consider the solar-insolation process as the favored one, and that all extant hypotheses have to be investigated further to better address the origin of the observed plumes. They propose that observations with a single-flyby mission, like the Trident concept discussed in the following section, would greatly help in understanding the nature of plumes and fans. In addition, future extremely large telescopes (ELTs) would be sufficient to obtain a spatial resolution on the order of 100 km or greater and to shed light on the composition and thickness of the southern hemisphere terrains, where plumes and fans have been detected. More details on plumes and cryovolcanism on Triton are reported in (Chapter 3).

Voyager 2 was not able to investigate Triton's surface composition because it did not carry a near-infrared spectrometer onboard. Ground-based spectral measurements have therefore been the only source of compositional information over the last four decades, providing globally-averaged information in the visible and near-infrared spectral ranges. From these data, we know that the surface of Triton displays a variety of ices: N_2, CO, CO_2, CH_4, and H_2O (Figure 9.2; Owen et al. 1993; Quirico et al. 1999; Cruikshank et al. 1993, 2000). CO_2 and H_2O, which are non-volatile ices at typical dayside temperatures of Triton (Fray & Schmitt 2009), cover $\sim$45% of the surface ($\sim$10%–20% is CO_2 ice) and have no apparent variation with longitude (Grundy et al. 2010; Holler et al. 2016). They may be the primary constituents of the underlying bedrock exposed due to seasonal transportation of more volatile materials. N_2 and CO are the most volatile species on Triton (CH_4 is

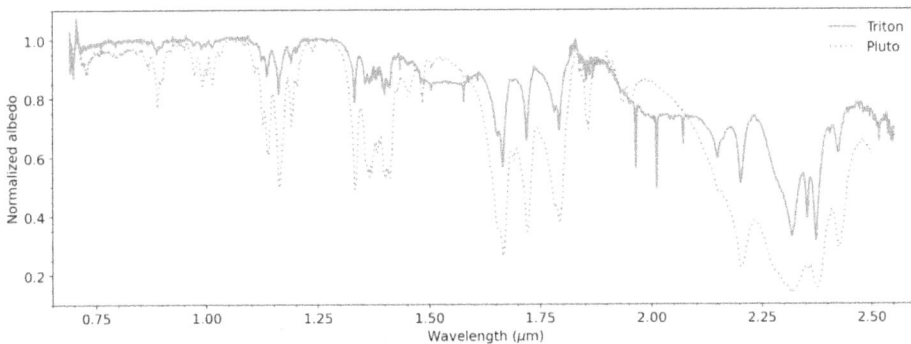

Figure 9.2. Comparison of the "grand average" near-infrared spectra ($\sim$0.7–2.5 μm) of Triton (solid red line) and Pluto (dashed blue line) from IRTF/SpeX (e.g., Grundy et al. 2010; Holler et al. 2016; Broadfoot et al. 1989; Holler et al. 2014). Both spectra are normalized near 1.05 μm. The Triton spectra included in this average were obtained from 2001 to 2022 and the Pluto spectra were obtained from 2002 to 2022.

less so) and dominate the composition of Triton's tenuous atmosphere (Broadfoot et al. 1989; Tyler et al. 1989; Lellouch et al. 2010) and the seasonal volatile transport across the surface. Spectral modeling suggests that, at the surface, CH_4 is mostly diluted in N_2 ice but a small fraction of CH_4 may still be present as discrete patches of pure ice, possibly correlating with specific geologic terrains (Quirico et al. 1999; Merlin et al. 2018). Similarly, most of the CO is diluted in N_2 ice rather than existing as patches of pure CO ice (Merlin et al. 2018; Tegler et al. 2019).

The sublimation and recondensation of volatile species can lead to geographical and temporal variations of the surface composition, which in turn is revealed by both albedo variations and in the formation of compounds via atmospheric photochemistry (Krasnopolsky & Cruikshank 1999; Krasnopolsky 2020) or by the irradiation of a three-component layer ("alloy") of N_2:CH_4:CO ices (Moore & Hudson 2003). Among these products are hydrocarbons heavier than methane, of which ethane (C_2H_6), ethylene (C_2H_4), and perhaps acetylene (C_2H_2) appear to be the most chemically obvious. However, although some spectral signatures suggest the presence of ethane dissolved in nitrogen (Holler et al. 2016; DeMeo et al. 2010), the simultaneous presence of two CO isotopes (^{12}CO and ^{13}CO) complicates a confident spectral assignment and quantification of ethane abundance (Merlin et al. 2018). Ethane was also suggested to be present on Pluto's surface (Nakamura et al. 2000) to explain weak absorption bands at 2.28, 2.32 and 2.40 μm, although the proposed model was not able to properly reproduce the band depth at 2.405 μm. The analysis of the longitudinal distribution of the absorption band at 2.405 μm on Triton showed that it is consistent with ethane (Holler et al. 2016).

The variations in albedo, texture and structure on Triton's surface revealed in the Voyager images, and by the Hubble Space Telescope (McEwen 1990; Hicks & Buratti 2004; Bauer et al. 2010; Hicks et al. 2022), may be related to a non-uniform distribution of N_2, the dominant surface component. A global spectral study of the surface of Triton by Grundy et al. (2010) and Grundy & Young (2004) demonstrated that the diagnostic 2.15-μm N_2 ice spectral feature varies in intensity with longitude on Triton, being deeper in the sub-Neptune hemisphere. The same rotational variability is observed for CO, while the longitudinal variation of the CH_4 bands does not follow this behavior (Grundy et al. 2010). These results were recently confirmed and expanded considering 63 nights of spectroscopic observations in the period 2002–2014 (Holler et al. 2016). They also concluded that the southern hemisphere is currently denuded of volatile ices, while the high albedo measured during the Voyager 2 flyby suggested it was covered by volatile ices in 1989. A similar result was derived from the albedo maps, obtained over a period of 84 years, which combine ground-based observations with the color maps of Voyager 2 (Hicks et al. 2022). Triton's spectral slope varies on timescales of years at visible wave-lengths (Buratti et al. 1999), suggesting that organic compounds, possibly tholins, are present on the surface just as on Titan (McCord et al. 2006) and that their distribution changes over time. This may be due to time-varying atmospheric radiolytic production or episodic excavation from underground (e.g., Sulcanese et al. 2023).

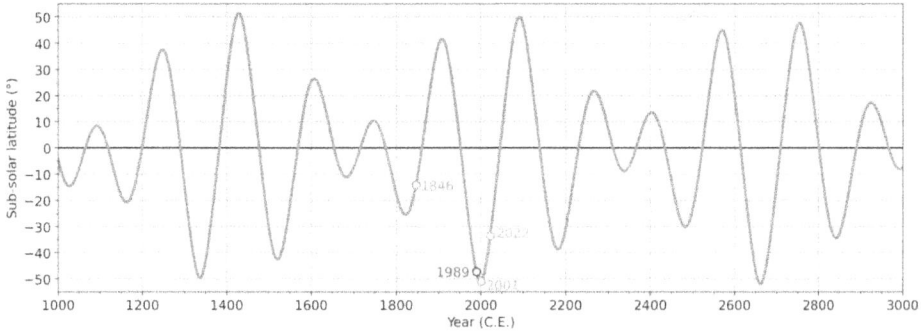

Figure 9.3. Evolution of the subsolar latitude on Triton with time, covering 1000–3000 C.E. Equation from Trafton (1984). The complex seasonal evolution is a consequence of Triton's orbital parameters and evolution, combined with Neptune's 165 Earth-year heliocentric orbit. The green circle indicates the year of Triton's discovery (1846), the purple circle marks the Voyager 2 flyby in 1989, and the red circles and curve mark the period of IRTF/SpeX observations (2001–2022).

Alternatively, the observed rotational variability could instead be described by a persistent southern polar cap of N_2 ice, as predicted by long-term global circulation models (GCMs; Bertrand et al. 2020). In this scenario, the ubiquitous non-volatile ices (H_2O and CO_2) occupy a low-latitude collar, with the observed increases in N_2 and CH_4 ice absorption strength (Holler et al. 2016) ascribed to volatile-rich regions of the northern hemisphere experiencing solar insolation for the first time in decades. The lack of spatially resolved spectroscopy of Triton, either from *in situ* or Earth-based observatories, significantly complicates the evaluation of Triton's surface ice distribution.

Figure 9.3 shows the complex seasonal evolution in the case of Triton (from JPL Horizons), with a variation over a 688-year period, considering its spin pole precession in the revolution around Neptune. The purple dot indicates the epoch of the Voyager visit to the Neptune-Triton system, in 1989, when Triton was nearing the end of southern hemisphere spring. Southern summer solstice occurred in 2000, and from that date to 2050 the southern hemisphere will still be illuminated, providing the chance of comparisons with the Voyager images of the surface. In the figure, we also show the period of IRTF/SpeX observations, for which a good temporal coverage is obtained.

In 2050 the subsolar latitude will cross the equator, and the northern hemisphere will experience direct solar illumination. For observations after this date, it will be more challenging to compare the same regions as those in the data and images obtained with Voyager, but would allow exploring structures and composition in the Northern hemisphere.

Furthermore, the observation of Triton's leading hemisphere in the 1–5 μm range with VLT/NACO (Protopapa et al. 2007) allowed identification of two unknown absorption features, close to 4.0 and 4.6 μm, which deserve further investigation. The feature might be due to globally-distributed non-volatile SO_2 or CO_2 over Triton's surface, possibly evidence of aeolian transport, although not yet clarified. This volatile transport over Triton's surface was discovered to undergo a seasonal variation by comparing UV albedo maps obtained by the Hubble Space Telescope (Bauer et al. 2010). A similar volatile transport and possibly related seasonality was suggested also for Pluto (Buie et al. 2010), proving the validity of investigating these two bodies as archetypes of the dwarf planet class of Kuiper Belt Objects (KBOs).

9.1.2 Pluto

The Pluto system consists of two planet-sized bodies Pluto and Charon, accompanied by four much smaller satellites: Styx, Nix, Kerberos, and Hydra. All six orbit in co-planar, near-circular orbits around a common barycenter that is external to the surface of Pluto, although Pluto accounts for 89% of the mass in the system. This binary planet configuration is unlike the more familiar arrangement of one dominant planet plus a retinue of much smaller satellites. Physical and orbital parameters are listed in Table 9.1.

The Pluto system orbits the Sun on an inclined, eccentric orbit with a period of 248 Earth years. Although the orbit crosses that of Neptune, it is protected from being gravitationally scattered by Neptune by means of a 2:3 mean motion resonance with Neptune (Milani & Nobili). For every two heliocentric orbits of Pluto, Neptune completes three, and whenever Pluto is near its perihelion and thus close to Neptune's orbit, Neptune is elsewhere in its orbit so that they never experience close encounters.

Pluto and Charon are tidally locked, with both spin axes and the mutual orbital axis aligned. Each spins on its axis with the same 6.4 Earth-day period as their mutual orbital period. The centers of the hemispheres of Pluto and Charon that are permanently oriented toward each other define the zero longitudes on both bodies.

Pluto probably formed as one of numerous similar objects in a disk of planetesimals between about 20 and 30 AU from the Sun, just outside of the giant-planet forming region. Eventually, Neptune migrated outward through that disk, leading to the loss of most of the outer planetesimal population, though some still survive as members of the present-day Kuiper belt (e.g., Malhotra 1995; Levison et al. 2008). Pluto's satellite system is thought to have formed in the aftermath of a giant collision

Table 9.1. Physical and Orbital Parameters of the Pluto-Charon System

Object	Dimensions (km)	Orbital Period (Days)	Rotation Period (Days)	Mean Distance from Barycenter (km)	Density (kg m^{-3})
Pluto	2377	6.387	6.387	2130	1854
Charon	1212	6.387	6.387	17,440	1702
Styx	$16 \times 9 \times 8$	20.16	3.24	42,700	–
Nix	$48 \times 33 \times 30$	24.85	1.83	48,700	–
Kerberos	$19 \times 10 \times 9$	32.17	5.3	57,800	–
Hydra	$50 \times 36 \times 32$	38.20	0.4295	64,700	–

Pluto and Charon dimensions and densities from Nimmo et al. (2017). Minor satellite dimensions, orbital periods, and spin periods from Weaver et al. (2016).

between proto-Pluto and another body (e.g., Canup 2011), most likely during the era when Neptune's migration destabilized orbits in that region.

Knowledge of Pluto's surface composition and atmosphere accumulated as telescope facilities and instrumentation advanced, with infrared spectral observations revealing the presence of CH_4, N_2, and CO ices on Pluto, along with water ice on Charon (e.g., Owen et al. 1993, Cruikshank et al. 1976, Buie 1987). In Figure 9.2 we report an average spectrum of Pluto. A stellar occultation provided the first direct detection of Pluto's atmosphere (Hubbard et al. 1988), as well as a way to monitor its seasonal evolution in subsequent years (Elliot et al. 1989, 2007; Meza et al. 2019). The presence of volatile ices in equilibrium with a thin atmosphere suggested Pluto and Triton should be thought of as planetary cousins (eventually to be joined by the likes of Eris and Makemake), while Charon's water ice-dominated surface invited comparisons with comparably-sized uranian and saturnian icy moons.

Interest in spacecraft exploration of the Pluto system was boosted in 1989 when Triton's complex, youthful surface was revealed by the Voyager 2 flyby, and continued to grow as NASA explored a variety of potential architectures for such a mission through the 1990s. This effort finally culminated with the launch of NASA's New Horizons probe in 2006 (e.g., Neufeld 2014, Stern & Grinspoon, 2018). New Horizons flew through the Pluto system in July 2015, yielding a huge leap in knowledge. The probe obtained its highest resolution views of Pluto's anti-Charon hemisphere, the center of which is dominated by *Sputnik Planitia*, a thousand-kilometer glacier of N_2, CO, and CH_4 ices filling an ancient impact basin (Stern et al. 2015; Grundy et al. 2016; Moore et al. 2016). The global map of Pluto obtained with New Horizons is shown in Figure 9.4.

Pluto's surface geology is described in much greater detail in a companion chapter (Schenk et al. this issue). Unlike Triton, Pluto has a mix of ancient, heavily

Figure 9.4. Global color mosaic map of Pluto from New Horizons. Credit: NASA/Johns Hopkins University Applied Physics Laboratory/Southwest Research Institute.

cratered regions alongside youthful terrains that are clearly still being shaped by ongoing processes. The presence of gravitationally bound volatiles in vapor pressure equilibrium with Pluto's atmosphere is key to enabling much of the present-day activity, as described in greater detail in another chapter (Chapter 5). The key property of these volatiles is that by undergoing phase changes, they enable energy from sunlight or decay of radionuclides to transport material, transforming heat energy into potential energy that can transform the landscape. In addition to the convective overturn in *Sputnik Planitia* (McKinnon et al. 2016; Trowbridge et al. 2016) and the valley glaciers flowing down into it from the east (Howard et al. 2017a; Umurhan et al. 2017), volatiles are implicated in the formation of Pluto's penitente-like bladed terrains (Moore et al. 2017), and complex patterns of pits (Howard et al. 2017b). They likely also play a role in erosion at spatial scales below those resolved in New Horizons images, and could play an important role below Pluto's surface, too. In addition to transporting volatiles in response to changing thermal energy inputs, Pluto's atmosphere directly modifies the surface through formation of dunes (Telfer et al. 2018). It also acts as a complex chemical factory, with photochemistry in the upper atmosphere producing haze particles composed of a mix of heavier hydrocarbons and nitriles and more complex tholin-like macro-molecules (e.g., Cheng et al. 2017; Lavvas et al. 2021). The haze particles settle to the surface where they likely undergo additional processing over time, and are incorporated into Pluto's distinct surface environments in various ways (Grundy et al. 2018), with a concentration of this processed material in the *Cthulhu Macula* region (Holler et al. 2014; Protopapa et al. 2017).

Global circulation models predict that Pluto's northern hemisphere will be completely denuded of volatile ices by 2030 (i.e., at the time of northern hemisphere summer solstice, see Figure 9.5; Bertrand & Forget 2016). Over the past two decades, a monotonic decrease in the strength of the 2.15 μm N_2 feature has been observed, coinciding with a similar decrease in CO band strength and a redshifting of the CH_4 band centers (Grundy et al. 2014; Young et al. 2021). The latter indicates

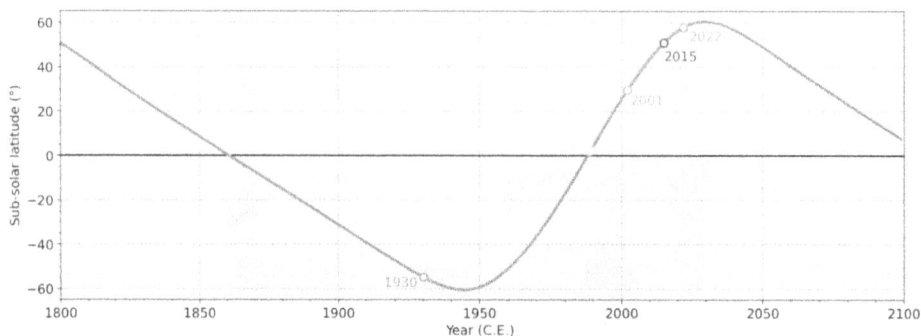

Figure 9.5. Evolution of Pluto's subsolar latitude from the period 1800–2100 C.E., which covers over a full Pluto year (248 Earth years). The green circle marks the year of Pluto's discovery (1930), the yellow circle marks the perihelion date (1989), the purple circle indicates the New Horizons flyby in 2015, and the red circles and curve cover the period of IRTF/SpeX observations (2002–2022). The background is color-coded by northern hemisphere season.

an increase in CH_4 concentration in N_2:CH_4 and CH_4:N_2 mixtures (e.g., Protopapa et al. 2015). Meza et al. (2019) used stellar occultations over a long time-baseline to determine that Pluto's atmospheric pressure had yet to level off in 2016 and would in fact continue to increase for an additional few years. Combined, all of these pieces of evidence indicate the preferential sublimation of more volatile N_2 from Pluto's northern latitudes (Holler et al. 2022). The onset of wholesale CH_4 sublimation from the same region has yet to be observed, but this process could occur on very short timescales of 1–2 Earth years, similar to previously observed color, light curve, and spectral changes (Buie et al. 2010; Holler et al. 2022). GCM predictions imply that spectral observations in the next few years will be crucial for understanding Pluto's total volatile inventory, the important evolutionary processes in each season, and the fate of its atmosphere as it moves further and further from the Sun.

9.2 Open Questions

Overall, the summaries above demonstrate that our knowledge of the surfaces of Triton and Pluto, including coverage and composition, is largely incomplete and that, apart from the NASA Voyager 2 and New Horizons missions, most of their investigation is left to telescopic observations. While telescopic observations allow the study of the disk-averaged surface compositions of Pluto and Triton, the spatial resolution is not sufficient to infer the detailed ice distribution across the surface and the composition of specific geologic structures. Moreover, considering the dynamical history of the two bodies and large seasonal variability, their surface compositions are still an open question and deserve further investigation both from space- and ground-based facilities.

Specifically, we identify the following key gaps in our understanding of the surface compositions of Triton and Pluto and the evolutionary processes currently shaping their surfaces and atmospheres:

- *Triton's volatile ice distribution*: The largest open question pertaining to Triton is the latitudinal distribution of its volatile and non-volatile ices. As touched on earlier, long time-baseline, ground-based photometric and spectroscopic observations suggest that Triton's south polar region is dominated by non-volatile H_2O and CO_2 and a longitudinally variable "collar" of volatile N_2, CO, and CH_4 at lower latitudes (e.g., Grundy et al. 2010; Holler et al. 2016; Hicks et al. 2022). On the other hand, GCMs predict a long-term reservoir of volatile N_2 across the south polar terrain and a longitudinally uniform collar of non-volatile H_2O and CO_2 at lower latitudes (Bertrand et al. 2022). The true ice distribution has significant implications for the origins of Triton's plumes, the seasonal evolution of the atmosphere, and the accuracy of the initial conditions used in GCMs (e.g., total volatile abundance, thermal inertia, etc.). This discrepancy is challenging to address without the proper combination of longitudinal coverage, wavelength range, plate scale, and point spread function (PSF) stability. The NIRCam imager on JWST (short-wavelength plate scale of $\sim$31 mas) and 10-meter-class ground-based telescopes equipped with adaptive optics (AO) are capable of providing a few resolution elements across Triton's disk ($\sim$0.13$''$), enough to distinguish the northern and southern hemispheres. JWST's NIRCam imager provides a wide range of filters of various bandwidths, which could be selected to target portions of Triton's spectrum where the various volatile and non-volatile ices absorb (as well as continuum regions for comparison). Deconvolution of Triton's disk using PSF reference stars, combined with observations at multiple sub-observer longitudes, would increase the spatial sampling, possibly providing regional-scale compositional maps. A similar deconvolution process could be performed on each slice in the spectral data cubes produced by the Keck Observatory's OSIRIS integral field unit (Larkin et al. 2006), for example, which is capable of providing hemispherical-scale spectroscopy of Triton's surface (plate scales as small as 20 mas, but with larger PSF FWHMs due to the atmosphere, even with adaptive optics).

- *Pluto's atmospheric evolution*: As shown in Figure 9.5, Pluto is currently approaching northern hemisphere summer solstice after a relatively short spring (1988–2029). Everything we know about Pluto since its discovery in 1930 has come during a period of decreasing heliocentric distance (perihelion was in 1989), increasing subsolar latitude, or both. Pluto's atmosphere was first detected in 1988 via a stellar occultation (Elliot et al. 1989) and has been subject to predictions of collapse for decades, an event that is yet to occur (e.g., Meza et al. 2019; Young et al. 2021). Post-New Horizons GCMs predicted the complete loss of volatile ices from Pluto's northern latitudes (Bertrand & Forget 2016), but the question remains of whether this ice will simply be transported to lower latitudes (detectable through ground-based monitoring campaigns, e.g, Holler et al. 2022). or enter the atmosphere (detectable through stellar occultations). Increased CH_4 in the atmosphere could lead to increased haze production and deposition, which could be detected with disk-averaged spectroscopy (Grundy et al. 2018). With Pluto's heliocentric distance continuing to increase and a gradual decrease in the subsolar latitude starting in 2029, GCMs and other atmospheric models will be put to the test in ways that were not previously possible. Continued efforts to carry out occultation campaigns will provide valuable information on Pluto's atmospheric pressure over time, a necessary input for GCMs to model Pluto's future atmospheric evolution. Depending on the brightness of the occulted star and the exact shadow path, such campaigns could make use of professional telescopic facilities, fixed-position university and amateur telescopes, or mobile telescope deployments specifically to capture an event in an area of the world with lower telescope density.

- *Status of Triton and Pluto as ocean worlds*: A primary focus in astrobiology is the identification and study of ocean worlds (Hendrix et al. 2019). Europa is the most obvious candidate, with Ceres, Ganymede, Callisto, Titan, Enceladus, Dione, Triton, and Pluto also making the list. Triton's plumes are potentially endogenic (Hofgartner et al. 2022), with further *in situ* study of their origins and composition revealing properties of a subsurface ocean. Pluto presents geologic hints of past cryovolcanism (e.g., Nimmo et al. 2021;

Singer et al. 2022), but no active eruptions or resurfacing events were observed during the New Horizons flyby (Hofgartner et al. 2018). While surface composition can provide indications of interior composition and processes, it is not a sufficient indicator of a subsurface ocean by itself; only *in situ* spacecraft can provide the high spatial resolution imaging and magnetic field data required to confirm an ocean's presence.

- *Understanding the full inventory of minor species on Triton and Pluto*: Ground-based spectroscopy is limited by telluric absorption and lower target flux to wavelengths <2.5 μm. Absorption features of many organic species and haze products (e.g., nitriles like HCN, see Table 9.3) are found at these longer wavelengths (e.g., Moore et al. 2010; Hudson et al. 2014a, 2014b), preventing a full accounting of the surface compositions of Triton and Pluto and the irradiation chemistry occurring in their atmospheres and on their surfaces. Specifically, questions remain about the presence of ethane (C_2H_6), one of the most easily produced irradiation by-products of CH_4, on the surface of Triton. The 2.405 μm feature is accessible from ground-based facilities and is attributed to ethane on Pluto due to its longitudinal variability and increased concentration in the dark *Cthulhu Macula* region (Holler et al. 2014; DeMeo et al. 2010). This same feature is present in Triton's spectrum but could instead be due to ^{13}CO (Holler et al. 2016). If ethane, and other hydrocarbons and nitriles, are truly present on Triton in detectable amounts, this would have important implications for resurfacing rates. Detections of these species require space-based facilities to probe wavelengths beyond 2.5 μm, and the sensitivity to observe outer solar system targets where reflected solar flux is diminished.

In addition to this, one must also recall that Triton and Pluto have each been the subject of single, multi-hour flybys by robotic spacecraft and that currently no space missions to the outer solar system are under consideration by the major space agencies. New measurements, both with telescopes and dedicated space missions, would be potentially transformative for models about atmospheric circulation and surface variability. In this respect, the continuous time coverage with observations of the surface composition of Triton and Pluto is crucial to understanding the complex seasonality experienced by the icy surface of these bodies, and to predict volatile transport and plume activity.

9.3 Measurements in the Next Decade

In this section, we summarize the measurements that can be performed with ground-based and space-based telescopes, focusing on the capabilities of the new facilities, such as the Extremely Large Telescopes (ELTs), that will begin operations in the next few years. Despite the limitations in spatial resolution achievable from the Earth, telescopic observations remain a great resource for long-term monitoring of the surface composition and icy variability of both Triton and Pluto, as demonstrated by the long-term ground-based campaigns with IRTF/SpeX over the past 20+ years (e.g., Grundy et al. 2010; Holler et al. 2016; Broadfoot et al. 1989; Holler et al. 2014; Grundy et al. 2014, 2013).

9.3.1 Ground-based Facilities

Even with the best adaptive optics (AO) correction, current ground-based telescopic observations struggle to resolve the surface of either Triton or Pluto. The highest resolution Neptune images, taken by the Keck telescope, have a spatial resolution of 360 km per pixel, or a diameter of ~135 pixels. This would allow Triton to be resolved with only ~7 pixels, too small to properly monitor either changes in the surface morphology since Voyager 2, or provide insight into the revealed surface structures with changing seasons. However, while Voyager 2 provided detailed images of Triton's surface, it could not reveal the visible or near-infrared spectra of different geological units. Spectroscopic studies from Earth provide a vastly improved spectral resolution than is possible from space. Typical infrared instrumentation in the outer solar system has a spectral resolving power of ~400, in stark contrast with, e.g., VLT-CRIRES, which can observe with a resolving power of ~100,000. However, unfortunately the limitations of adaptive optics often restrict the capabilities of ground-based instruments, depending upon the level of correction needed. CRIRES, for example, is able to attain a spatial resolution in the near-infrared of only ~0.1″ when closing the loop on extended objects the size of Triton and Pluto. As relatively small objects, Triton with an angular diameter of ~0.13″ and Pluto at ~0.095″, it is challenging to spatially resolve surface structures at high spectral resolving powers.

More recent AO-dedicated instrumentation provides unprecedented clarity of extended astronomical objects, often able to spatially resolve features close to the diffraction limit of a telescope. However, these instruments are often much more limited in their spectral resolving power. For instance, VLT's SPHERE provides spectra between $0.95-2.32 \mu m$ at a resolving power of only R~50, and between the more limited range of $0.95-1.65 \mu m$ at R~350. The nominal resolution for SPHERE is 7.4 mas^2 pixel^{-1}, which would provide ~18-pixel resolution for Triton at opposition. However, more detailed analysis of the true spatial resolution of extended solar system objects based upon past observations of Ganymede and Europa (King & Fletcher 2022) show that in the near infrared, SPHERE's uniquely resolved spatial elements are around 5 pixels across, placing Triton and Pluto at the very limit of resolvable targets, approximately 3 and 2 resolvable units across each. This would be enough to clearly identify the unique spectral characteristics of Triton's northern and southern hemispheric units, providing the first clear distinction of these geological units, but is likely to prove to be a challenging observation. Using visible light, rather than infrared spectra, where the smaller diffraction limit would allow accuracy closer to (7 mas)^2 pixel^{-1}, this spatial resolution should improve significantly, but it is not clear whether this would provide more spectral information than is available from the color images taken by Voyager 2.

As such, the advent of the next generation of giant ground-based telescopes would greatly increase our knowledge of Triton and Pluto. In particular, the imaging and spectroscopic capabilities expected for the 30–100-m class telescopes in the coming years would significantly improve the diffraction limit across all wavelengths, allowing an insight into the surface changes on Triton and the atmospheric evolution of Pluto, by acquiring a series of measurements. The repeatability of these observations would also be an important complement to a dedicated space mission to either of these bodies, due to the expected spatial and spectral resolution of the next generation ground-based instruments and the time series of the measurements.

With a spatial resolution of 45 km pixel^{-1} at Neptune and Triton, the latter could be observed over about 100 pixels. Such resolution would be very useful for monitoring changes in the surface composition of Triton, as well as gas geysers activity and atmospheric variability, although it would not be sufficient to identify plumes and fans. Similarly, the spatial resolution at the distance of Pluto would be on the order of 60 km pixel^{-1} and its disk would be imaged with approximately 90 pixels using a 100 m telescope like the European Extremely Large Telescope (E-ELT). Although worse than the spatial resolution obtained with the NASA New Horizons spacecraft, which was was on the order of a few hundred meters per pixel with MVIC and LORRI at closest approach, the complementarity of these measurements with E-ELT would allow one to monitor surface changes over several years and enable direct comparison with the New Horizons dataset. In Table 9.2, we summarize the expected surface resolution at Triton with the biggest ground-based facilities, currently available or under implementation.

Although more effective in the investigation of atmosphere and cloud properties, stellar occultation measurements, including the surface pressure and temperature variations, can be useful to study seasonal changes of volatile species in the case of Pluto (Meza et al 2019; Sicardy et al. 2003) and Triton (Oliveira et al. 2022). The expected occultation candidates for Pluto and KBOs have increased in the last decades thanks to the faint star survey at the European Southern Observatory (Assafin et al. 2010) and the Gaia Data Release 2 in April 2018 (Prusti et al. 2016; Brown et al. 2018), while the number of possible events in the case of Triton is quite limited due to systematic errors in Neptune's orbit prediction, which makes the position of the moon quite uncertain, and because of the high contrast between Neptune and Triton.

Table 9.2. Expected Spatial Resolution at Triton and Pluto with Current and Future Ground-based Facilities

Facility	Diameter (m)	Triton Surface resolution (km)	Triton Pixels across typical disk	Pluto Surface resolution (km)	Pluto Pixels across typical disk
Keck	10	375 (in J band)	~7	~510 (in J band)	~5
GTC	24.5	215 (at 1 μm)	~12	~290 (at 1 μm)	~9
TMG	30	175 (at 1 μm)	~15	250 (in H band)	~9.5
E-ELT	100	40	~100	60	~90

At Pluto, stellar occultations and modeling allowed for a link between the observed surface pressure increase and the heating of the N_2 ice in Sputnik Planitia and the other areas exposed to the Sun (Meza et al. 2019).

From the available measurements of Triton reported so far, it was possible to study the seasonal variation in pressure over a period of about 30 years, which is linked to distribution on the surface of N_2 ice and its mixture with CH_4 (Oliveira et al. 2022).

Being able to probe the atmosphere up to about 190 km above Triton's surface in the best cases, stellar occultation measurements could allow the serendipitous detection of plumes if present.

9.3.2 Space-based Facilities

The high spatial resolution achievable with space-based facilities was demonstrated by Bauer et al. (2010) using the Hubble Space Telescope Advanced Camera for Surveys (HST/ACS) High-Resolution Channel (HRC). This mode, which is now unfortunately defunct, had a plate scale of $0.025''$ pixel^{-1}, equating to $\sim$5 pixels across Triton's diameter and $\sim$21 pixels covering the entire disk. Observations at a handful of sub-observer longitudes helped to further increase the spatial sampling across Triton's surface, with results pointing to an increase in visible albedo at equatorial latitudes on the Neptune-facing hemisphere and a decrease in albedo on the anti-Neptune hemisphere, compared to the Voyager 2 maps. Future attempts to obtain HST/ACS data for comparison to results from Voyager 2 and Bauer et al. (2010) have been frustrated by the unavailability of the HRC, with the next best option being the Wide Field Camera 3 (WFC3) UVIS channel, which provides a plate scale of $0.04''$ pixel^{-1}. A search for regional- and hemispherical-scale changes in visible albedo is still possible with $\sim$3 pixels across Triton's diameter, but such observations have yet to be attempted. Pluto is essentially unresolved at the WFC3/UVIS plate scale.

Additional observations of Triton's visible albedo distribution are useful for comparison to previous observations going back to 1989, but are still unable to provide compositional information. The James Webb Space Telescope (JWST) and the NIRCam instrument will open new avenues to studying regional-scale changes on both Triton and Pluto. The short-wavelength channel of NIRCam has a plate scale of $0.031''$ pixel^{-1} ($\sim$4 pixels across Triton's diameter and $\sim$3 across Pluto) and covers a wavelength range from 0.6–2.3 μm. Appropriate filter selection could reveal regional-scale compositional distributions of key ices such as N_2, CH_4, H_2O, and CO_2; multiple such observations over the lifetime of JWST (at least 10 years) would produce valuable temporal information as the subsolar latitude changes rapidly on Triton and Pluto passes through northern hemisphere summer solstice. There were no Cycle 1 programs to observe Triton with NIRCam, but 3 NIRCam imaging observations at different sub-observer longitudes were included in program #1658[1]; these data were subject to a proprietary period and were not published at the time of writing.

---

[1] https://www.stsci.edu/jwst/science-execution/program-information?id=1658

JWST's NIRSpec instrument and its Integral Field Unit (IFU) provide near-infrared imaging spectroscopy from 0.6–5.3 μm on a space-based platform. However, the spatial pixels ("spaxels") of NIRSpec are 0.1″ on a side, which are unable to provide meaningful resolution across Triton, and Pluto is smaller than a single spaxel. Thus, high spatial resolution spectroscopy is best left to the ELTs in the coming years. The NIRSpec IFU can, however, address other open questions regarding Triton and Pluto, specifically those requiring a broader spectral grasp, such as taking a more complete inventory of the minor species present on both bodies. High resolving-power (R~2700) NIRSpec IFU observations of both Triton (program #1272, PI: D. Hines) and Pluto (program #1171, PI: J. Stansberry), covering the full near-infrared spectral range, were included in JWST Cycle 1. These data were also subject to a proprietary period and were not published at the time of writing.

9.4 In Situ Exploration

Future *in situ* studies using spacecraft would advance understanding of both bodies considerably. Following the initial exploration of both Triton and Pluto by Voyager 2 and New Horizons, respectively, more ambitious mission architectures involving orbiters and/or landed elements would be compelling next steps. However, even new flyby missions would be incredibly valuable, with flyby missions of Triton recently proposed and a feasibility study of a Pluto orbiter completed under the NASA Planetary Decadal Mission Concept Studies program (Howett et al. 2021). In addition to studying regions that were unlit and/or mapped only at low spatial resolution, there could be considerable benefit from the application of new instrumentation and from exploiting the passage of time since the initial flybys to elucidate ongoing seasonal evolution.

9.4.1 Spectroscopic Observation Needs

The composition of Triton's surface and thin atmosphere provides constraints on the spectral range and spectral sampling of a spectrometer carried aboard a future mission to the Neptune-Triton system. Such an instrument should span a wide enough spectral range to reveal those signatures most diagnostic of the compounds mentioned so far, and of further potential contaminants. This kind of instrument is highly desirable for future *in situ* exploration of Triton, due to the lack of any spectroscopic capability of the Voyager 2 mission, but also in the case of Pluto because the spectral coverage of the *Linear Etalon Imaging Spectral Array* (LEISA) onboard New Horizons was limited to 2.5 μm.

Table 9.3 summarizes the spectral position of the main signatures known or expected on the surface of Triton, based on a vast laboratory literature. Considering the abovementioned similarities, the same applies to Pluto. The values are expressed both in wavelength and in wavenumber and refer to the pure compound.

From this survey, it turns out that the whole spectral range 0.7–5.1 μm is extremely diagnostic to spectroscopically identify and map most chemical species known or expected to exist on the surface. In particular, the range 0.7–2.0 μm is very

Table 9.3. Relevant Near-infrared Ice Absorption Features for Triton and Pluto

Compound	Mode	Signature Position	
		μm	cm^{-1}
H_2O (water)	(v_1, v_3) Stretch O−H	3.045	3277
	(v_2) Bend H−O−H	6.024	1660
	Combination $2v_1 + v_3$	1.04	9615
	Combination $v_1 + v_2 + v_3$	1.25	8000
	Combination $v_1 + v_3$	1.50	6667
	Crystalline water ice signature	1.65	6056
	Combination $v_2 + v_3$	2.02	4950
CO_2 (carbon dioxide)	(v_3) Asymmetric stretch C=O	4.268	2343
	(v_1) Symmetric stretch C=O	7.46	1340
	Combination $v_1 + v_3$	2.697	3708
	Combination $2v_2 + v_3$	2.778	3600
CO (carbon monoxide)	(v_1) Stretch ^{12}CO	4.675	2139
	(v_1) Stretch ^{13}CO	4.780	2092
	Overtone $2v_1$	2.35	4260
N_2 (nitrogen)	2−0 collision-induced band	2.15	4651
CH_4 (methane)	About 40 spectral features between 0.7 and 4.6 μm, with variable strength	0.7–4.6	2193–13,706
C−H bond	Stretch	3.4	2941
	Overtone	1.73	5780
−C≡N functional group	Stretch	4.35 < λ < 4.90 (4.57 for HCN)	~2188
	Overtone	2.2	4545
C_2H_6 (ethane)	Several spectral features between 1.6 and 3.5 μm, with variable strength	1.689	5921
		2.015	4963
		2.274	4398
		2.314	4322
		2.405	4160
		2.460	4065
		3.365	2971
		3.472	2880
C_2H_4 (ethylene)	Several spectral features between 2.0 and 3.5 μm, with variable strength	2.125	4705
		2.225	4495
		2.388	4188
		3.237	3089
		3.362	2974
	(v_3) Asymmetric C−H stretch	3.042	3287

(Continued)

Table 9.3. (*Continued*)

Compound	Mode	Signature Position	
		μm	cm^{-1}
C_2H_2	(v_3) Asymmetric stretch $^{13}C-H$	3.105	3220
(acetylene)	(v_2) C≡C stretch	5.066	1974
	Combination $v_2+v_4+v_5$	3.005	3328
	Combination v_3+v_4	2.589	3862
	Combination v_1+v_5	2.453	4077

Where possible, an assignment is provided and the wavelength and wavenumber of the center of the feature are specified.

diagnostic for identifying water and methane ices. The range from 2.0 to 4.0 μm is complementary to the first one for mapping water and methane ices, for discriminating between crystalline and amorphous water ice, and it is key to identify CO_2 ice, N_2 ice, and organic compounds such as ethane, acetylene, and others. Finally, the spectral range from 4.0 to 5.1 μm is essential for mapping the main signatures of CO_2, CO, nitrogen-rich organics such as nitriles and tholins.

It should be noted that in addition to those known from the literature and mentioned above, and similar to other icy satellites of the solar system where geological activity occurs or has recently occurred, for Triton one cannot exclude the existence of additional non-ice compounds at the local scale such as hydrated mineral salts, for which the same spectral range 0.7−5.1 μm is also extremely diagnostic. Since it is difficult to explore with ground-based facilities, the wavelength range beyond 4.0 μm deserves thorough investigation, considering also that two unknown absorption features, close to 4.0 and 4.6 μm, were identified on Triton's surface (Protopapa et al. 2007), possibly due to globally-distributed non-volatile SO_2 or CO_2 and evidence of possible aeolian transport. The NIMS imaging spectrometer aboard the NASA Galileo mission (Carlson et al. 1992), whose data were the first to suggest the presence of hydrated mineral salts on Europa and Ganymede, was also sensitive in the 0.7−5.2 μm range. If one were forced to give up a portion of this range for any reason, the shorter wavelength range between 0.7 and 1.0 μm could be sacrificed only if recovered by multispectral data obtained by a framing camera equipped with color filters covering this range.

The average spectral sampling step of a spectrometer is given by the spectral range divided by the number of spectral channels in which the signal is sampled (the larger the interval, the broader each sampling step is). For example, by assuming a spectral interval from 0.7 to 5.1 μm and a detector with 500 effective spectral channels (possibly obtained by binning larger formats), this translates into an average sampling step size between 8 and 9 nm. This sampling step size, and its associated spectral resolving power, must be compared with the need to resolve weak-but-diagnostic spectral signatures. In this regard, an upper limit to the sampling step size for a spectrometer to be carried onboard a spacecraft meant to perform a close and in-depth exploration of Triton and Pluto is 10 nm, meaning that larger values would not allow a significant advance compared to Earth-based telescopic observations (although the proximity to Triton and Pluto would have the clear advantage of a much better spatial resolution than the one obtainable with any Earth-based facility).

9.4.2 Flyby Missions

The *Trident* mission concept, proposed to the NASA Discovery program in 2019, aimed to investigate Neptune and Triton in detail during a 10-day flyby through the system (Frazier et al. 2020). A strength of the mission was its ability to take advantage of a very fortuitous planetary alignment that would bring the spacecraft to Neptune within 14 years. By arriving at Neptune and Triton in 2039, it would have allowed investigation of the system roughly one full season after that observed by the Voyager 2 spacecraft in 1989 (Figure 9.3) and enabled seasonal comparisons and exploration of a previously unilluminated region of Triton. The objectives of this mission covered the major themes of the NASA OPAG panel (Hendrix et al. 2019) for Triton and Neptune, including the exploration of evolutionary pathways towards habitable worlds, understanding the processes on active worlds, and the broad goal to "explore vast, unseen lands". The payload included a dual-sensor magnetometer (MAG), an IR Spectrometer and Narrow-Angle Camera (IRS/ NAC), a Wide-Angle Camera (WAC), a Plasma Spectrometer (PS), and a Radio Science (RS) suite of instruments. The innovative design of the mission and the careful determination of the trajectory allowed *Trident* to fit within a Discovery budget, providing measurements of the surface features, detecting the presence of a subsurface ocean, and enhancing our understanding of the ionospheric interaction with Neptune's surrounding space environment. Unfortunately, the *Trident* concept was not selected for implementation, but it did lay the groundwork for future low-cost, high-impact missions to Triton and the outer solar system.

An additional strength of *Trident* was to spectrally explore Triton's surface *in situ* for the first time using a near-infrared spectrometer. The regional mapping provided by such an instrument during a flyby requires small phase angle values (<20°) to observe most of the dayside (percentage depending on the geometry of the flyby and on the allotted data volume) and to reduce shadows. Spatial resolution is the real key factor for achieving breakthrough discoveries at Triton, which are not possible with Earth-based telescopic observatioons. Observing the same regions over a wide range of phase angles, with one or multiple flybys, is of great interest to evaluate the photometric properties of the surface material. Moreover, high phase angle observations (>160°) are crucial for the investigation of the thin atmosphere on the dayside limb of the satellite.

Reflectance spectroscopy at Triton and Pluto is challenged by the large helio-centric distance inducing low solar flux, combined with some regions of the satellite presenting an inherently smaller albedo. Adequate exposure times (to be predicted on the basis of a consolidated radiometric performance model), combined with small phase angles and therefore with a higher signal level, would guarantee the signal-to-noise ratio (SNR) needed to achieve the scientific objective of revealing and mapping the surface composition of Triton. Compared to the icy satellites of Jupiter and Saturn, one advantage is that most of Triton's surface is not coated with water ice, which has a sharp drop in reflectance beyond 2.7 μm and poses a serious challenge for the identification of organic compounds in the spectral region between 3.0 and 3.7 μm.

Many of the arguments for a Triton flyby mission like *Trident* can be made for a Pluto flyby, too. Although active eruptions or plumes were not detected by New Horizons at Pluto (Hofgartner et al. 2018), there was abundant evidence for geologically recent eruptive activity (e.g., Singer et al. 2022; Cruikshank et al. 2021). A new probe could search for evidence of eruptions occurring in previously poorly resolved regions and could also search for evidence of changes around the sites of recent activity identified in the New Horizons data.

9.4.3 Orbiters

Anything that can be accomplished with a flyby can also be done with an orbiter, and an orbiter can generally do it better, since such a platform offers more time to execute diverse observations and there is far more scope for optimizing the geometry to satisfy the competing needs of different types of observations. Orbiters enable systematic global coverage, rather than the awkward dichotomy between well-studied encounter hemisphere and comparatively neglected non-encounter hemisphere that is intrinsic to a flyby. Repeated observations from an orbiter can reveal short-term variability of surface features and atmospheric phenomena, shedding light on the processes involved. Orbiters also facilitate observations of the surface at multiple illumination and observation orientations which enables better reconstruction of topography as well as revealing subtle spectral and photometric effects associated with compositional or structural strata in the optically active surface.

In addition to all the usual remote sensing instruments, instruments that require more proximity are especially useful in orbiter missions. These include mass spectrometers for directly sampling the atmosphere. Dips into the uppermost atmosphere enable direct measurement of atmospheric composition including photolytic and radiolytic products (e.g., hazes), can measure isotopic ratios, and could potentially detect chemical inputs from in-falling Kuiper belt dust (and circum-neptunian dust, in the case of Triton). Radar systems can probe the subsurface to detect layering related to prior climate epochs and could reveal internal plumbing associated with eruptive processes or subsurface circulation of volatile fluids. A laser altimeter can provide very accurate topography and potentially detect changes in topography due to sublimation or deposition, or subsurface activity or tidal flexing in the case of Triton. It also enables exploration of the night side, including unilluminated polar winter regions, and can be used to monitor clouds and haze layers in the atmosphere. Multiple radio occultations can probe atmospheric temperature and pressure structure over distinct regions and can also assess its variability from one day to the next, and could perhaps detect seasonal trends, depending on the lifetime of the mission. Solar occultations using an ultraviolet spectrometer can do the same for atmospheric compositional structure. Stellar occultations offer the opportunity to investigate different local times of day, albeit with much fainter sources (solar and radio occultations can only access sunrise and sunset). Gravity measurements can reveal the moments of inertia and internal structure, and magnetometer measurements can investigate the interior oceans thought to be present inside both bodies.

Flagship missions to explore the whole Neptune-Triton system have been proposed to multiple space agencies, although it would require a longer time frame to be implemented with respect to a Discovery mission concept. One example is the NASA *Neptune Odyssey* mission to the Neptune-Triton system, aiming to explore the planet, its ring system, Triton and the small satellites, and the space environment surrounding the system (Rymer et al. 2021).

Another example of a mission concept to the Neptune-Triton system is the one presented to ESA in the framework of the ESA *Cosmic Vision* Programme 2015–2025, described in Masters et al. (2014). It presents a 2-year tour around the system as a possible avenue for in-depth investigations of Neptune and Triton to provide answers to the key system-related questions. The described trajectory includes a series of flybys with Triton at different altitudes to investigate its atmospheric properties, the surface composition and variegation, and Triton's interaction with Neptune's magnetosphere.

The natural progression in the exploration of the Pluto-Charon system, after the success of New Horizons, would be possible with an orbiter, like the mission concept reported in, e.g., Howett et al. (2021) and Buie et al. (2021). The authors envision an orbiter around Pluto able to globally explore its surface, focusing in particular on the southern hemisphere regions, which were not covered with New Horizons, and with a spatial resolution sufficient to study the surface composition and possible differences with respect to the northern regions, and to eventually identify areas of volatile accumulations. The privileged position of an orbiter would also allow measurements with a spatial resolution of a few meters per pixel, with a great improvement in the investigation of the surface, atmospheric composition, and precise crater counting.

Observations during a time frame of 1–2 years, as could be achieved from an orbiter mission, is essential to study longer term regional variability at a proper spatial scale. The instrument suite identified in these science case studies include remote-sensing instruments, like Narrow-Angle and Wide-Angle cameras, spectrometers in the UV-Vis-IR ranges, a mass spectrometer, and a radio science package (Buie et al. 2021) also lists a series of challenges that an orbiter would experience at the heliocentric distances of the Pluto-Charon system. In particular, the large distance from the Sun greatly impacts the long journey required to reach the Pluto system, the downlink to Earth and the required power which is limited at Pluto's distance. Nonetheless, an orbiter would be the natural successor to the New Horizons mission, with its ability to answer questions raised by the 2015 flyby by providing more complete surface coverage, observations of new surface regions, and observing surface and atmospheric phenomena over longer timescales.

9.4.4 Landers

Landers can provide a compelling close-up view of the surface and the lower atmosphere, and can provide a wealth of information on the processes active at that interface. Their proximity enables them to study structures at much smaller spatial scales than can easily be resolved from orbit, which can be key to understanding the long-term evolution of surface terrains. Landers enable the use of a great variety of laboratory-type instrumentation that requires direct access to the material under study. These include microscopy to reveal small scale structure of surface materials, gas chromatography (GC) to measure chemistry of trace materials, X-ray diffraction (XRD) to derive the crystal structure of solid materials, and alpha-particle X-ray spectrometry (APXS) and gamma ray spectrometry (GRS) to obtain elemental abundances. Raman, infrared, and laser-induced breakdown (LIBS) spectrometry can measure surface compositions, even at some distance from the lander. Meteorological instruments can monitor atmospheric phenomena, including wind speed and direction, pressure, and temperature, as well as the presence of suspended particles. Mass spectrometers (MS) can be used to measure local atmospheric composition including isotopic ratios, and can also be used to study solids that are vaporized by warming or laser ablation. Landers can directly investigate the subsurface below them via drills, scoops, and ground-penetrating radar. Active seismic measurements where the lander generates a seismic pulse and listens for the echoes are another way to do this, and passive seismic investigations offer a powerful way to learn about the deeper interior and to assess the frequency and energy of internal geological activity.

An especially exciting class of lander is mobile rovers that are able to explore more widely than a fixed lander. The remarkable diversity of geological provinces that exist on both Pluto and Triton makes this capability especially desirable. Based on a NASA Innovative Advanced Concepts (NIAC) study, Masters et al. (2014) (and see also Izenberg et al. 2021) a hopper design was proposed that could exploit the N_2 ice that is widely distributed on both Triton and Pluto as a propellant, using heat from a radioisotope power supply to produce compressed gas from the ice that could be used as an exhaust gas to propel the vehicle on flights of up to 20 km per hop. A fixed-lander configuration could reap some of the same benefits of mobility if more than one lander could be deployed (e.g., Balint 2005). Multi-lander missions are especially advantageous for seismic studies.

A lander is a much more advanced mission concept compared to flybys and orbiters, and is the obvious next step after an orbiter. While flybys present the most promising route to studying Triton and Pluto in the near future, due to their shorter transit times, orbiters require longer transit times to make orbital insertion feasible within the spacecraft mass constraints. There is also the additional complication of balancing the amount of mass used for fuel and the amount dedicated to the instrument suite, which can significantly extend transit times with existing rockets (e.g., Howett et al. 2021). Landers face the same challenge as orbiters in this regard, with the added challenge of successfully landing on the surface of an object with an atmospheric pressure on the order of 10 μbar, nearly 1000 times lower than Mars' atmospheric pressure. The scientific return from a Triton or Pluto lander would be extraordinary, but there remain significant engineering challenges to overcome before such a mission would be feasible.

9.5 Conclusions

The exploration of Triton began several decades ago with the Voyager 2 flyby in 1989, and revealed a moon with a young, complex surface. Its similarity to Pluto and the KBOs put further attention on the exploration of Pluto, which was ultimately visited in 2015 by NASA's New Horizons spacecraft.

The study of these bodies is continued with telescopic observations that provide the opportunity to continuously monitor their surface composition and infer chemical and dynamical processes acting on their surfaces. Climate models of Triton and Pluto (Holler et al. 2022), based on inputs from ground-based observations (Grundy et al. 2010; Holler et al. 2016; Hicks et al. 2022), allow predictions of the seasonal evolution of volatile species across their surfaces. However, one strong limitation of these measurements is the spatial resolution and observing geometries, which are insufficient to resolve fine structures on the surfaces. It is expected that the telescopic observations of these very distant objects will experience significant improvement with the advent of 30 m class telescopes later in the 2020s and early 2030s, and the expanded use of AO systems.

Being identified as the highest priority candidate ocean world by the NASA OPAG panel, Triton, and the whole Neptune system, has been the subject of several space mission concepts, proposed to multiple international space agencies for future exploration. Similarly, the Pluto exploration by New Horizons has opened several new questions that deserve further investigation both from the ground and from space. Future space exploration, either through a flyby, an orbiter, or a lander mission, would address some of the open questions about Triton and Pluto outlined in this work.

References

Agnor, C. B., & Hamilton, D. P. 2006, Natur, 411, 192

Assafin, M., Camargo, J. I. B., Vieria Martins, R., et al. 2010, A&A, 515, A32

Balint, 2005, Jupiter Icy Moons Orbiter (JIMO) white paper, NASA https://trs.jpl.nasa.gov/handle/2014/39526

Bauer, J. M., Buratti, B. J., Li , J.-Y., et al. 2010, ApJL, 723, L49

Bertrand, T., Lellouch, E., Holler, B. J., et al. 2022, Icar, 373, id. 114764

Bertrand, T., & Forget, F. 2016, Natur, 540, 86

Bertrand, T., Lellouch, E., Holler, B. J., et al. 2020, Icar, 373, 114764

Binzel, R. P., Earle, A. M., Buie, M. W., et al. 2017, Icar, 287, 30

Broadfoot, A. L., Atreya, S. K., Bertaux, J. L., et al. 1989, Sci, 246, 1459

Brown, A. G. A., Vallenari, A., Prusti, T., et al. 2018, A&A, 616, A1

Buie, M. W., Cruikshank, D. P., Lebofsky, L. A., et al. 1987, Natur, 329, 522

Buie, M. W., Hofgartner, J. D., Bray, V. J., & Lellouch, E. 2021, The Pluto System After New Horizons, ed. S. A. Stern, J. M. Moore, W. M. Grundy, L. A. Young, & R. P. Binzel (Tucson, AZ: Univ. Arizona) 569

Buie, M. W., Grundy, W. M., Young, E. F., Young, L. A., & Stern, S. A. 2010, AJ, 139, 1117

Buratti, B. J., Hicks, M. D., & Newburn, R. L. J. 1999, Natur, 397, 219

Canup, R. M. 2011, AJ, 141, 35

Carlson, R. W., Weissman, P. R., Smythe, W. D., & Mahoney, J. C. 1992, SSRv, 60, 457

Cheng, A. F., Summers, M. E., Gladstone, G. R., et al. 2017, Icar, 290, 112

Croft, S. K., Kargel, J. S., Kirk, R. L., et al. 1995, Neptune and Triton (Tuscon, AZ: Univ. Arizona Press) 879

Cruikshank, D. P., Pilcher, C. B., Morrison, D., et al. 1976, Sci, 194, 835

Cruikshank, D. P., Roush, T. L., Owen, T. C., et al. 1993, Sci, 261, 742

Cruikshank, D. P., Schmitt, B., Roush, T. L., et al. 2000, Icar, 147, 309

Cruikshank, D. P., Dalle Ore, C. M., Scipioni, F., et al. 2021, Icar, 356, 113786

DeMeo, F. E., Dumas, C., de Bergh, C., et al. 2010, Icar, 208, 2010

Dobrovolskis, A. R., Peale, S. J., Harris, A. W., et al. 1997, Dynamics of the Pluto-Charon Binary (Tucson, AZ: Univ. Arizona Press) 159

Earle, A. M., Binzel, R. P., Young, L. A., et al. 2017, Icar, 287, 37

Earle, A. M., & Binzel, R. P. 2015, Icar, 250, 405

Elliot, J. L., Dunham, E. W., Bosh, A. S., et al. 1989, Icar, 77, 148

Elliot, J. L., Person, M. J., Gulbis, A. A. S., et al. 2007, AJ, 134, 1

Fray, N., & Schmitt, B. 2009, P&SS, 57, 2053

Frazier, W., Bearden, D., Mitchell, K. L., et al. 2020, 2020 IEEE Aerospace Conf. (Piscataway, NJ: IEEE) 1

Gaeman, J., Hier-Majumder, S., & Roberts, J. H. 2012, Icar, 220, 339

Grundy, W. M., Young, L. A., Stansberry, J. A., et al. 2010, Icar, 205, 594

Grundy, W. M., Binzel, R. P., Buratti, B. J., et al. 2016, Sci, 351, id.aad9189

Grundy, W. M., Bertrand, T., Binzel, R. P., et al. 2018, Icar, 314, 232

Grundy, W. M., Olkin, C. B., Young, L. A., & Holler, B. J. 2014, Icar, 235, 220

Grundy, W. M., Olkin, C. B., Young, L. A., Buie, M. W., & Young, E. F. 2013, Icar, 233, 710

Grundy, W. M., & Young, L. A. 2004, Icar, 172, 455

Hansen, C. J., McEwen, A. S., Ingersoll, A. P., & Terrile, R. J. 1990, Sci, 250, 421

Hendrix, A. R., Hurford, T. A., Barge, L. M., et al. 2019, AsBio, 19, 1

Hicks, M. D., & Buratti, B. J. 2004, Icar, 171, 210

Hicks, M. D., Buratti, B. J., & Dombroski, D. 2022, PSJ, 3, 84

Hofgartner, J. D., Buratti, B. J., Devins, S. L., et al. 2018, Icar, 302, 273

Hofgartner, J. D., Birch, S. P. D., Castillo, J., et al. 2022, Icar, 375, 114835

Holler, B. J., Young, L. A., Grundy, W. M., & Olkin, C. B. 2016, Icar, 267, 255

Holler, B. J., Young, L. A., Grundy, W. M., Olkin, C. B., & Cook, J. C. 2014, Icar, 243, 104

Holler, B. J., Yanez, M. D., Protopapa, S., et al. 2022, Icar, 373, id. 114729

Howard, A. D., Moore, J. M., Umurhan, O. M., et al. 2017a, Icar, 287, 287

Howard, A. D., Moore, J. M., White, O. L., et al. 2017b, Icar, 293, 218

Howett, C. J. A., Robbins, S. J., Holler, B. J., et al. 2021, PSJ, 2, 75

Hubbard, W. B., Hunten, D. M., Dieters, S. W., Hill, K. M., & Watson, R. D. 1988, Natur, 336, 452

Hudson, R. L., Gerakines, P. A., & Moore, M. H. 2014b, Icar, 243, 148

Hudson, R. L., Ferrante, R. F., & Moore, M. H. 2014a, Icar, 228, 276

Hussmann, H., Sohl, F., & Spohn, T. 2006, Icar, 185, 258

Izenberg, et al. 2021, Hopper Missions to Triton and Pluto using a Vehicle with In-Situ Refueling: A White Paper for the Planetary Science and Astrobiology Decadal Survey 2023-2032, NASA https://assets.pubpub.org/wk40adqi/51617915652846.pdf

Kargel, J. S. 1994, EM&P, 67, 101

Kargel, J. S., & Strom, R. G. 1990, Abstract of the Lunar and Planetary Science Conf., 21, 599

King, O., & Fletcher, L. N. 2022, JGRE, 127, e2022JE007323

Krasnopolsky, V. A. 2020, Icar, 335, 113374

Krasnopolsky, V. A., & Cruikshank, D. P. 1999, JGR, 104, 21979

Landis, et al. 2019, International Astronautical Congress (IAC) Conf. (Washington, DC: NASA) https://ntrs.nasa.gov/citations/20190032656 GRC-E-DAA-TN74147

Larkin, J., Barczys, M., Krabbe, A., et al. 2006, NewAR, 50, 362

Laufer, D., Bar-Nun, A., Pat-El, I., & Jacovi, R. 2013, Icar, 222, 73

Lavvas, P., Lellouch, E., Strobel, D. F., et al. 2021, NatAs, 5, 289

Lellouch, E., de Bergh, C., Sicardy, B., Ferron, S., & Kaufl, H.-U. 2010, A&A, 512, L8

Levison, H. F., Morbidelli, A., Van Laerhoven, C., Gomes, R., & Tsiganis, K. 2008, Icar, 196, 258

Malhotra, R. 1995, AJ, 110, 420

Masters, A., Achilleos, N., Agnor, C. B., et al. 2014, P&SS, 104, 108

McCord, T. B. 1966, AJ, 71, 585

McCord, T. B., Hansen, G. B., Buratti, B. J., et al. 2006, P&SS, 54, 1524

McEwen, A. S. 1990, GeoRL, 17, 1765

McKinnon, W. B. 1984, Natur, 311, 355

McKinnon, W. B., Nimmo, F., Wong, T., et al. 2016, Natur, 534, 82

Merlin, F., Lellouch, E., Quirico, E., & Schmitt, B. 2018, Icar, 314, 274

Meza, E., Sicardy, B., Assafin, M., et al. 2019, A&A, 625, A25

Milani, A., Nobili, A. M., & Carpino, M. Icar, 82, 200

Moore, J. M., McKinnon, W. B., Spencer, J. R., et al. 2016, Sci, 351, 1284

Moore, J. M., Haward, A. D., Umurhan, O. M., et al. 2017, Icar, 287, 320

Moore, M. H., Ferrante, R. F., Moore, W. J., & Hudson, R. 2010, APJS, 191, 96

Moore, M. H., & Hudson, R. L. 2003, Icar, 161, 486

Nakamura, R., Sumikawa, S., Ishiguro, M., et al. 2000, PASJ, 52, 551

Neufeld, M. J. 2014, HSNS, 44, 234

Nimmo, F., Orkan, U., Lisse, C. M., et al. 2017, Icar, 287, 12

Nimmo, F., & McKinnon, W. B. 2021, Geodynamics of Pluto ed. S. A. Stern, et al. (Tucson, AZ: Univ. Arizona Press) 89

Oliveira, J. M., Sicardy, B., Gomes-Júnior, A. R., et al. 2022, A&A, 659, A136

Owen, T. C., Roush, T. L., Cruikshank, D. P., et al. 1993, Sci, 261, 745

Prockter, L. M., Nimmo, F., & Pappalardo, R. T. 2005, GRL, 32, L14202

Protopapa, S., Grundy, W. M., Reuter, D. C., et al. 2017, Icar, 287, 218

Protopapa, S., Herbst, T., & Bohnhardt, H. 2007, Msngr, 129, 58

Protopapa, S., Grundy, W. M., Tegler, S. C., & Bergonio, J. M. 2015, Icar, 253, 179

Prusti, T., de Bruijine, J. H. J., Brown, A. G. A., et al. 2016, A&A, 595, A1

Quirico, E., Douté, S., Schmitt, B., et al. 1999, Icar, 139, 159

Rymer, A., Runyon, K., Vertisi, J., et al. 2021, BAAS, 53, 374

Schenk, P., & Jackson, M. P. A. 1993, Geo, 21, 299

Sicardy, B., Widemann, T., Lellouch, E., et al. 2003, Natur, 424, 168

Singer, K. N., White, O. L., Schmitt, B., et al. 2022, NatCo, 13, 1542

Soderblom, L. A., Kieffer, S. W., Becker, T. L., et al. 1990, Sci, 250, 410

Stern, S. A., Bagenal, F., Ennico, K., et al. 2015, Sci, 350, aad1815

Stern, S. A., Kammer, J. A., Gladstone, G. R., et al. 2017, Icar, 287, 47

Stern, S. A., & Grinspoon, D. 2018, Chasing New Horizons: Inside the Epic First Mission to Pluto (New York: Picador) 320

Sulcanese, D., Cioria, C., Kokin, O., et al. 2023, Icar, 392, 115368

Tegler, S. C., Stufflebeam, T. D., Grundy, W. M., et al. 2019, AJ, 158, 8

Telfer, M. W., Parteli, E. J. R., Radebaugh, J., et al. 2018, Sci, 360, 992

Trafton, L. 1984, Icar, 58, 312

Trowbridge, A. J., Melosh, H. J., Steckloff, J. K., & Freed, A. M. 2016, Natur, 534, 79

Tyler, G. L., Sweetnam, D. N., Anderson, J. D., et al. 1989, Sci, 246, 1466

Umurhan, O. M., Howard, A. D., Moore, J. M., et al. 2017, Icar, 287, 301

Weaver, H. A., Buie, M. W., Buratti, B. J., et al. 2016, Sci, 351, aae0030

Young, L. A., Bertrand, T., Trafton, L. M., et al. 2021, The Pluto System after New Horizons, ed. S. A. Stern, et al. (Tucson, AZ: Univ. Arizona Press) 321

Chapter 10

Future Measurement Needs for the Atmospheres of Pluto and Triton

Manuel Scherf, Audrey Vorburger, Peter Wurz and Helmut Lammer

N_2 is the main constituent in Pluto's and Triton's atmospheres and both are believed to have originated in the Kuiper Belt. Significant similarities are hence shared between their atmospheres but some key differences, such as their thermal atmospheric structures, stand out. The reason for this major difference is yet poorly understood as are other parameters related to their atmospheres. Among others, these include their volatile origin and atmospheric evolution, and unknowns related to important model parameters. To resolve these unknowns, various remote and *in situ* measurement needs arise that require future missions. These include ongoing and continuous monitoring of their atmospheres, spectrally mapping their surface volatile ices, and measuring important isotopic and elemental ratios within their atmospheres and surfaces. In this chapter we will introduce some of the key unknowns related to Pluto's and Triton's atmospheres including discussions on observations needed for tackling these unknowns and future pathways for obtaining these measurements. We specifically highlight the need for future *in situ* measurements including a lander on at least one of both bodies, and emphasize the importance of comparative planetology for tackling unknowns related to the atmospheres of Pluto, Triton, and any other Kuiper Belt objects (KBOs).

10.1 Introduction

Early hypotheses suggested that Pluto was an escaped satellite of Neptune but due to Triton's retrograde orbit and its similar composition to Pluto it is now expected that both originated in the Kuiper Belt (Agnor & Hamilton 2006), with Triton being captured by Neptune. The formation of both bodies might therefore have happened within the protosolar nebula (PSN) and not in a circumplanetary disk (such as, e.g., Titan), which may have affected the accretion of their volatiles.

doi:10.1088/2514-3433/ad5278ch10

N_2 is the main gas in Triton's (e.g., Krasnopolsky et al. 1993) and Pluto's (e.g., Gladstone et al. 2016) atmospheres and is considered to be in vapor–pressure equilibrium. CH_4 and CO further comprise the main trace species in these atmospheres. CO was discovered by ground-based observations in 2010 within Triton's atmosphere (Lellouch et al. 2010) and in 2017 in Pluto's together with HCN (Lellouch et al. 2017). These species undergo photochemical reactions that produce C_2H_6, C_2H_4, C_2H_2, HCN, nitriles, other heavy hydrocarbons, as well as complex hazes (e.g., Benne et al. 2022; Gladstone et al. 2016; Krasnopolsky 2019, 2020; Luspay-Kuti et al. 2017; Mandt et al. 2017). In addition, their atmospheres likely include noble gases although no observations have yet confirmed their existence. The non-detection of Ar at Pluto by New Horizons' UV spectrograph Alice, for instance, might indicate an Ar column density as low as 6% compared to CH_4 or even as high as for N_2, simply depending on the implemented eddy diffusion coefficient (see Steffl et al. 2020).

Although various ground-based observations and the flybys of Voyager 2 at Triton and New Horizons at Pluto revealed many similarities in their atmospheres, some key differences were found and important unknowns remain. The atmosphere of Pluto, for example, is significantly cooler than initially expected (e.g., Gladstone et al. 2016; Strobel & Zhu 2017) while Triton's atmosphere temperature can be reproduced relatively well by models (e.g., Krasnopolsky et al. 1993; Strobel & Zhu 2017). Haze formation in Pluto's atmosphere might contribute to cooling in its thermosphere (Zhang et al. 2017) while haze could have differing effects in Triton's atmosphere (e.g., Ohno et al. 2021).

The origin of both bodies' building blocks and the evolution of their atmospheres are relatively unknown as well. Here, future detection of different, currently unknown isotope and elemental ratios (e.g., $^{14}N/^{15}N$, $^{12}C/^{13}C$, and $^{36}Ar/N_2$) in combination with modeling studies will be crucially important (e.g., Glein et al. 2023; Scherf et al. 2020). Space missions to other bodies such as Comet Interceptor (e.g., Snodgrass & Jones 2019), which will measure isotope ratios at a potentially pristine comet, will give further insights into the origin and evolution of Triton, Pluto, and other KBOs. Such missions will be complementary to further ground-, space-, and laboratory-based measurement needs to better understand the atmospheres of Pluto and Triton, their composition, structure, origin, and evolution within the context of comparative planetology.

The unknowns mentioned before are examples hinting toward crucial measurement needs for better understanding their atmospheres. There are, however, further unresolved issues for which future observations will be crucially important. Section 10.2 will briefly overview Pluto's and Triton's atmospheres while Section 10.3 will outline open questions emerging from observations and models, and what kind of measurement needs these entail. Finally, Section 10.4 will discuss potential measurements for resolving them while Section 10.5 will conclude the chapter.

10.2 Brief Overview on Pluto's and Triton's Atmosphere

Information about Pluto's atmosphere mostly stems from Earth-based stellar occultation measurements and the flyby of NASA's New Horizons mission on

2015 July 14. The latter conducted atmospheric experiments with the Radio Experiment (REX; e.g., Hinson et al. 2018, 2017) and Alice instruments (e.g., Young et al. 2018), the Long-Range Reconnaissance Imager (LORRI), and the Multispectral Visible Imaging Camera (MVIC; e.g., Cheng et al. 2017). Together these instruments provided us with high-resolution data on Pluto's atmospheric pressure, temperature profile, its composition, and global haze distribution (Stern et al. 2015; Gladstone et al. 2016). Its atmosphere consists mainly of N_2, with trace amounts of CH_4, CO, HCN, C_2H_6, C_2H_4, C_2H_2, and HNC (e.g., Lellouch et al. 2017, 2022; Stern et al. 2015). The surface pressure is on the order of 1 Pa (Gladstone et al. 2016; Gladstone & Young 2019). Contrary to expectations, Pluto's surface pressure has almost tripled rather than collapsed since then (Sicardy et al. 2016), although it remained constant since ~2016 (Poro et al. 2021; Sicardy et al. 2021) and is expected to reach its peak pressure in the next few years (Meza et al. 2019). Its atmosphere can be divided into a lower, middle, and an upper atmosphere, each of which exhibits a different temperature profile. The lower atmosphere shows a strong positive temperature gradient and, during ingress of New Horizons, ended abruptly at an altitude of ~4 km, which marks the top of a distinctive boundary layer, while no evidence of such a layer was found at its egress (e.g., Gladstone & Young 2019). The middle atmosphere shows a slight negative thermal gradient (Lellouch et al. 2017), while the temperature unexpectedly decreases in the upper atmosphere due to an unknown cooling agent (e.g., Gladstone et al. 2016; Strobel & Zhu 2017).

As for Pluto, the main observations of Triton's atmosphere stem from Earth-based stellar occultations and from one space mission, NASA's Voyager 2 mission, which flew by Triton on 1989 August 25. On Voyager 2, the Radio Science Subsystem (RSS) measured Triton's tenuous atmosphere from which the surface density, pressure, and temperature of Triton's atmosphere were determined (Gurrola 1995). It made four observations with its ultraviolet spectrometer (UVS) through which N_2, CH_4, N, and HCN were identified in its atmosphere (Broadfoot et al. 1989; Herbert & Sandel et al. 1991). The atmospheric pressure at Triton is also on the order of ~1 Pa at the surface (e.g., Gurrola 1995), and it consists mainly of N_2, with trace amounts of CH_4 (Strobel & Summers 1995; Lellouch et al. 2010; Bertrand et al. 2022), and CO (Lellouch et al. 2010). The CH_4 partial pressure was several times higher in 2009 compared to the Voyager 2 flyby, implying that Triton's atmosphere is seasonally variable (Lellouch et al. 2010). Triton's temperature profile mostly resembles Pluto's, with a positive gradient in the lower atmosphere toward ~30 km, a mild negative temperature gradient in the middle atmosphere (Marques Oliveira et al. 2022), and a nearly isothermal temperature gradient at ~90 K in the upper atmosphere (Strobel & Zhu 2017), of which the latter is a clear difference to Pluto. For more details on Triton's and Pluto's atmosphere, we refer to Chapters 5 and 7.

10.3 Open Questions and Related Measurement Needs

10.3.1 Atmospheric Structure and the Potential Role of Haze Production

Prior to the New Horizons flyby, Pluto's upper atmosphere was expected to be significantly warmer and highly extended, with atmospheric loss rates suggested to potentially be hydrodynamic (e.g., Strobel 2008; Zhu & Strobel 2014). However, New Horizons/Alice observed a much more compact atmosphere and colder temperature with 76 ± 16K at ingress and 79 ± 17 K at egress for an altitude of only ~1000 km above the surface (Gladstone et al. 2016). Triton's upper atmosphere, on the other hand, was observed to have a higher temperature of 102 ± 3K (e.g., Strobel & Summers 1995; Strobel & Zhu 2017) during the Voyager 2 flyby, which can be reproduced quite well by atmospheric models if Neptune's magnetic field is accounted for (Strobel & Zhu 2017). Figure 10.1 illustrates the temperature profile of Pluto during the New Horizons flyby (red line; as adopted by Strobel & Zhu 2017) compared with a temperature profile simulated prior to flyby (Zhu & Strobel 2014). The figure further shows a simulated profile for Triton (Strobel & Zhu 2017) based on the Voyager-flyby.

The findings by New Horizons strongly suggest an unknown cooling agent within the atmosphere of Pluto that is likely not present at Triton. Although HCN is supersaturated by seven orders of magnitudes in the upper atmosphere, its cooling is too inefficient to explain the low temperature (Lellouch et al. 2017). Cooling via H_2O was also suggested (Strobel & Zhu 2017), but this would require an even higher degree of supersaturation with mixing ratios of $10^{-5} - 10^{-6}$ above 500 km (Lellouch et al. 2022). Water itself has not been observed in the atmosphere of Pluto although it must be present to some extent due to an influx of interplanetary dust particles (IDPs; e.g., Poppe & Horányi 2018) but its rate of precipitation is unknown as well. For Triton, observations of water influx from Neptune's magnetosphere would be

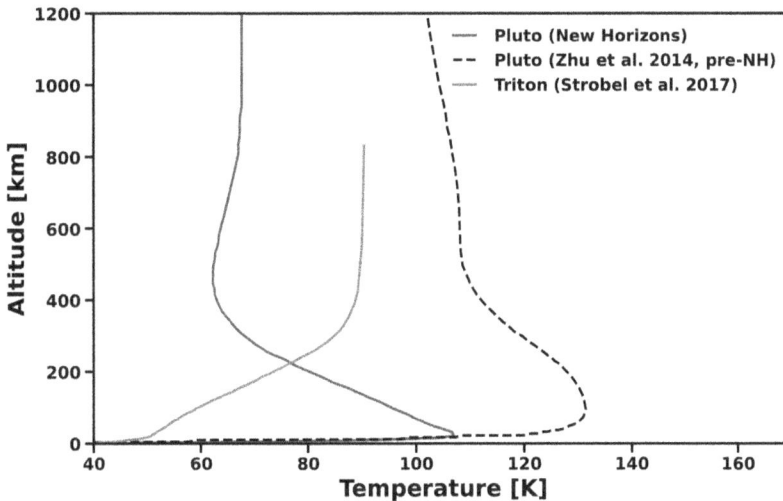

Figure 10.1. Temperature profiles of Pluto (blue and black dashed) and Triton (red). The temperature profile of Pluto (blue) is measured by New Horizons (Gladstone et al. 2016) but slightly updated by Strobel & Zhu (2017) to account for peak and shallow temperatures as measured with ALMA (Lellouch et al. 2017). The black-dashed temperature profile of Pluto is simulated prior to the New Horizons flyby by Zhu & Strobel (2014) but for the assumed solar wind and XUV flux conditions of the New Horizons flyby. The temperature profile of Triton is simulated by Strobel & Zhu (2017), i.e., their "Triton-3 model," which is constrained by the Voyager 2 solar UV occultation data. Data are from Strobel & Zhu (2017) and Zhu & Strobel (2014).

similarly important, as oxygen from H_2O could significantly affect its atmospheric photochemistry (e.g., Benne et al. 2022).

By assuming optical properties of haze similar to Titan, Zhang et al. (2017) proposed cooling and heating in Pluto's upper atmosphere to be controlled by haze instead of gas. However, the similarity between Pluto's and Titan's haze was objected (e.g., Lavvas et al. 2021; Fan et al. 2022). Fan et al. (2022), for example, presented observational evidence for a bimodal distribution of haze particles in Pluto's atmosphere with a small particle population in the size range of ~ 80 nm, which implies radiation timescales to be shorter than collision timescales, and hence heat-transfer between particles and gases to be too inefficient for providing equilibration. Contrasting the observations at Pluto, UVS measurements of Voyager 2 at Triton indicate a rapid redistribution of heat by winds and a much weaker role of haze in Triton's atmosphere (e.g., Lellouch et al. 2022). Although Voyager 2 observations detected haze below an altitude of about 30 km (e.g., Krasnopolsky & Cruikshank 1995), haze formation at Triton is expected to be significantly weaker because of CH_4 being less abundant due to both lower atmospheric temperatures below an altitude of ~ 200 km (Strobel & Zhu 2017) and higher photolysis rates compared to Pluto for most of its orbit (Strobel & Zhu 2017; Zhang et al. 2017).

Several studies (e.g., Fan et al. 2022; Moran et al. 2022) suggest that haze production in Triton's and Pluto's atmosphere may fundamentally change over their orbits due to, e.g., a change in CO/CH_4 mixing ratio (see Chapter 6. for details on haze at Pluto and Triton). If haze particles indeed present the dominant cooling mechanism in Pluto's upper atmosphere, a change in haze chemistry can have significant effects on atmospheric structure. Even if Pluto's atmosphere currently shows small escape rates, such potential season-dependent behavior could lead to a significant variation in loss rates even at present. It is therefore important to monitor haze evolution at Pluto while it recedes from the Sun for understanding the evolution of its atmosphere. Further observations on the evolution of haze at Triton will be similarly important, if one wants to understand atmospheric evolution and the interplay of haze and atmospheric structure at KBOs in general.

10.3.2 The Origin of Their Atmospheres

While there are indications on the origin of Titan's building blocks (e.g., Erkaev et al. 2021; Mandt et al. 2014; Miller et al. 2019), the initial reservoirs from which Pluto and Triton formed are poorly understood (e.g., Glein & Waite 2018; Glein et al. 2023; Scherf et al. 2020). Isotope ratios, their fractionation, and evolution over time are relevant tracers of the building blocks and initial volatile reservoirs out of which icy bodies and their atmospheres formed, with different reservoirs in the solar system being represented by different isotopic values (see, e.g., Scherf et al. 2020). For Pluto, several different reservoirs have been suggested to be potential sources of its N_2 (Glein & Waite 2018; Glein et al. 2023) and it can be expected that similar sources are responsible for the nitrogen inventory of Triton since both bodies originated within the Kuiper Belt (for details on the origin of volatiles, see Chapter 1).

N_2 at Pluto and Triton has likely been accreted from a combination of various volatile reservoirs. Glein (2023) suggests that Pluto's N_2 must have originated to substantial amounts from primordial N_2 and/or organics, while cometary NH_3 as a single source cannot explain the present-day isotope ratio within its vaguely estimated limits. If escape was important, NH_3 contributed $\lesssim 30\%$ and primordial N_2 must have been accreted with a fraction of $\gtrsim 30\%$. Figure 10.2 shows a ternary contour plot by Glein (2023) that illustrates the parameter space for the bulk $^{14}N/^{15}N$ ratio through mixing of three distinct N volatile reservoirs, i.e., primordial N_2, NH_3, and N-bearing organics. Any mixture below the dashed black lines is a potential solution for Pluto's initial N reservoir. While panel (a) does not include isotopic fractionation by atmospheric escape, panel (b) does, illustrating the importance of knowing the history of atmospheric escape. For Triton such studies are currently completely lacking, but one may expect that its initial mixture should be much closer to Pluto than Titan.

To constrain the volatile origin at Pluto and Triton a direct measurement of $^{14}N/^{15}N$ in their atmospheres and surface ices will be needed. By now, however, only theoretical estimates and some lower and upper limits are existing for Pluto while there are none for Triton (see Glein 2023; Krasnopolsky 2020; Lellouch et al. 2022; Mandt et al. 2017). Additionally, reconstructing the evolution of $^{14}N/^{15}N$ with atmospheric and photochemical models will be needed to accurately assess the origin of their volatile reservoirs. For resolving atmospheric escape over time, specifically at Pluto however, it will be crucial to also resolve the conundrum on the unknown cooling agent and how it might have changed in its past. The lack of knowledge about atmospheric evolution and loss rates at Pluto is a key limitation in developing

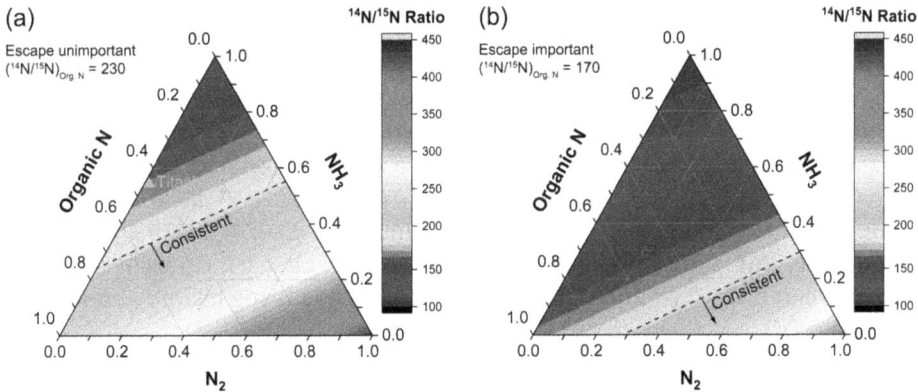

Figure 10.2. A ternary contour plot by Glein (2023) that illustrates the parameter space for the bulk $^{14}N/^{15}N$ ratio through mixing of three distinct nitrogen volatile reservoirs (primordial N_2, NH_3, and N-bearing organics). The three axes show the fractional contribution of each of the three nitrogen reservoirs. Panel (a) illustrates a case for which fractionation through atmospheric escape was not important while panel (b) includes Jeans escape over Pluto's history. The dashed lines show the lower limit for $^{14}N/^{15}N$ as estimated by Glein (2023), i.e., any mixtures below the dashed lines are potential solutions for the origin of N_2 at Pluto. The orange triangle shows a solution for Titan by Miller et al. (2019). Reprinted from Glein (2023). Copyright (2023), with permission from Elsevier.

reliable models on the origin of its volatiles (e.g., Glein & Waite 2018), as this dictates isotopic fractionation over time. This can also be regarded to be true for Triton since it neither has an existing reliable model on its atmospheric evolution, nor is it known whether a cooling agent could have been present in its past.

One also has to account for photochemical fractionation, but the respective fractionation factors are poorly constrained. Measurements of $^{14}N/^{15}N$ in various N-bearing sources, e.g., in HCN, NH_3, and directly in N_2 together with laboratory experiments under Triton/Pluto analog conditions will provide important insights. This, however, also needs a better understanding of photochemistry in the atmospheres of both bodies, specifically by also taking into account the varying solar insulation throughout their orbits since photochemical fractionation is mostly based on observations at Titan for solar minimum conditions (Mandt et al. 2017). Other crucial factors for reconstructing N fractionation is the cycle of condensation and sublimation of N_2 onto and from the surface, and whether N_2 can be lost via aerosol trapping (Mandt et al. 2017).

Other isotope and element ratios also give insights into atmospheric evolution on Triton and Pluto. These include $^{36}Ar/^{38}Ar$, $^{20}Ne/^{22}Ne$, and $^{36}Ar/N_2$ since different non-thermal and thermal escape processes lead to atmospheric fractionation between lighter and heavier isotopes and elements (Erkaev et al. 2021; Mandt et al. 2014). Additionally, $^{36}Ar/N_2$ is important for evaluating the origin of N_2; while a primordial source would have delivered high amounts of Ar, secondary sources would be consistent with a lower value for $^{36}Ar/N_2$ (Glein & Waite 2018). Unfortunately, New Horizons has not been able to measure Ar in Pluto's atmosphere (e.g., Steffl et al. 2020) and neither do any observations or estimates exist for Ar at Triton. Another isotope of Ar, i.e., ^{40}Ar, is an important tracer for tracking a body's interior evolution and outgassing history (Glein & Waite 2018; Glein et al. 2023). If N_2, for example, was produced from N-bearing organic compounds its rocky interior must be hot with temperatures above 350° C (Glein 2023; Miller et al. 2019). In such case, substantial amounts of ^{40}Ar should be in its atmosphere, if degassing occurred later. If it occurred early, we can assume that the N_2 would have been lost from the planet due to early hydrodynamic escape, similar to what has likely happened at Titan (Erkaev et al. 2021). A lack of ^{40}Ar might therefore be a hint toward the source of N_2 being other than N-bearing organics. In such a case, the radiogenic isotope ^{129}Xe should also be deficient in Pluto's atmosphere as both, ^{40}Ar and ^{129}Xe, track a body's outgassing history and efficiency (e.g., Glein 2017).

For Pluto, tholin deposits at Charon's poles (Grundy et al. 2016) can serve as a relevant record for the N_2/CH_4 ratio of escaped gases from Pluto's atmosphere (Glein & Waite 2018). The ratio of N_2/NH_3 would be another important measurement but New Horizons did neither detect NH_3 nor provide an upper limit. Due to the high N_2/NH_3 ratio in the PSN, Pluto and Triton could have accreted N_2 in much larger abundance than NH_3 ice, if they indeed formed in the PSN at temperatures below ~30 K. For comet 67P/Churyumov-Gerasimenko, on the other hand, N_2/NH_3 was found to be 0.13 (Rubin et al. 2015), while a strong enhanced abundance of CO and N_2 was found in the comet R2 PanSTARRS (McKay et al. 2019;

Mousis et al. 2021), indicating a completely different formation history and a potential large spread of N_2/NH_3 ratios within comets, as it was observed also for many other molecules (Biver & Bockelée-Morvan 2019). Further observations of N_2, N_2/CO, N_2/H_2O, and of noble gas isotope ratios such as $^{36}Ar/^{38}Ar$ (Glein et al. 2023) within comets, in particular within pristine comets, will therefore be highly important for understanding the origin and evolution of Pluto, Triton, and KBOs in general.

10.3.3 Atmospheric Pressure Evolution and Volatile Cycling

Pluto and Triton both show significant seasonal variation due to changes in their obliquity and orbits, which also affects the evolution of their atmospheric pressures, mixing ratios, and structures (see Chapter 5 for details). This leads to important changes in the annually averaged insulation as well as in the latitudes that receive the insulation minimum during one year (e.g., Bertrand et al. 2022; Young et al. 2021). For a single orbit of Pluto at the present epoch, models currently suggest a minimum pressure of about $(1.0–5.0) \cdot 10^{-3}$ Pa and maximum pressures only slightly higher than during the New Horizons flyby (e.g., Bertrand et al. 2018).

There is geological evidence of periods of significantly higher (Stern et al. 2017; Telfer et al. 2018) and lower pressures in Pluto's past, however. The latter is in the order of $<10^{-4}$ Pa (Grundy et al. 2018) while its maximum pressure during the last few million years (Myr) could have been as high as several 100 mbar (Stern et al. 2017). Such a thick atmosphere might have also affected atmospheric escape since thermosphere temperature and structure could have been different. A thinner atmosphere, on the other hand, will decrease the amount of N_2 and CH_4 available for haze production while a reduced flux of UV at larger orbital distances will reduce photodissociation of these molecules, which may also affect atmospheric cooling. The upper panel of Figure 10.3 illustrates the maximum and minimum pressure range during Pluto's present orbit while the lower panel shows the same for 0.9 Myr ago, both simulated by Stern et al. (2017). While the maximum pressure during the present epoch within their simulations is about 85 mbar, it even reaches $\sim$760 mbar during the earlier epoch, extreme variations that may affect atmospheric thermal structure and escape over time.

During the Voyager 2 flyby most of Triton's northern hemisphere was hidden from view and only $\sim$60% of its surface was imaged. Even for the observed portion of Triton, the location of surface volatile ices is not known precisely since Voyager 2 did not carry a spectrometer capable of evaluating Triton's surface composition in the near-IR (e.g., Hansen & Castillo-Rogez 2021). Several important issues that affect Triton's volatile cycle and its related atmospheric evolution are therefore still unconstrained or poorly understood, which opens the possibility on Triton's atmosphere collapsing in the future. Simulations by Bertrand et al. (2022) for Triton, however, show an increase in pressure from 1920 until shortly after the summer solstice (around 2005–2010) that is followed by a decrease back to the levels of 1920 in 2080. If a permanent northern ice cap exists, however, Bertrand et al. (2022) suggest surface pressures to be large enough for Triton's atmosphere to remain global.

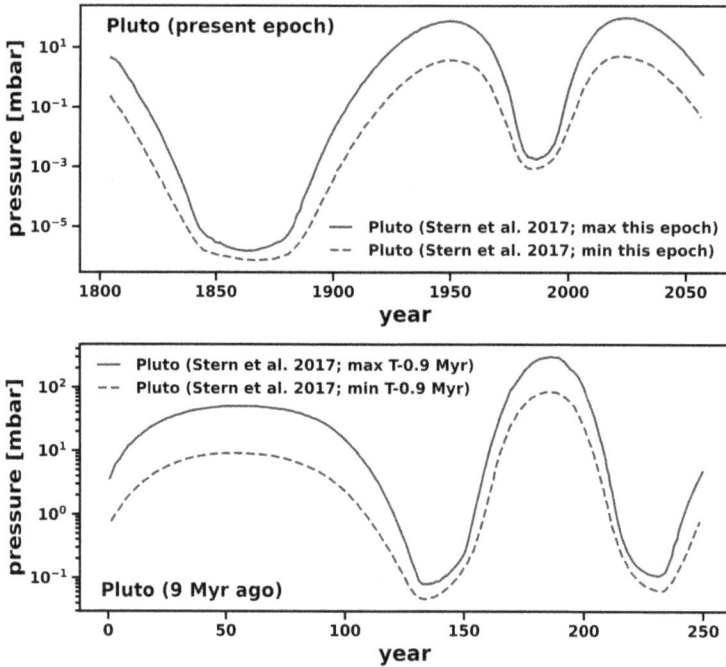

Figure 10.3. The upper panel shows the maximum (solid line) and minimum pressure (dashed line) evolution for Pluto during its entire present orbit while the lower panel shows the maximum and minimum pressure evolution during one orbit of Pluto 0.9 million years ago, both simulated by Stern et al. (2017). Data are from Stern et al. (2017).

For evaluating the maximum and minimum atmospheric pressures of both Pluto and Triton along their orbits, a better spectral mapping of their surfaces will hence be needed. For Pluto, this is specifically important for latitudes south of $\sim 45°$ S to investigate whether an N_2 ice cap exists on its south pole (e.g., Gladstone & Young 2019), and to better understand its topography and albedo, as this may significantly affect its N cycle (Young et al. 2021). For better understanding volatile cycles and atmospheric pressure evolution at Triton spatial and temporal resolution of its surface ices will be needed as well, specifically for the northern hemisphere.

10.3.4 Additional Uncertainties in Model Parameters

Several important parameters have to be considered for modeling the atmospheres of Triton an Pluto that are not or only vaguely known for the conditions present in their atmospheres. These uncertainties can strongly affect simulations, leading either to ambiguous results or even impede understanding specific aspects of their atmospheres. One of these crucial parameters is the eddy diffusion coefficient, K_{zz}. For Pluto, its actual value is not known with K_{zz} ranging between 10^3 (Wong et al. 2017) and 10^4 (Krasnopolsky 2020) to as high as 10^6 (Gladstone et al. 2016; Lellouch et al. 2017; Luspay-Kuti et al. 2017) with the main differences affecting eddy diffusion being the implementation of the CH_4 lower boundary and whether photochemical CH_4 loss is included in the models (Mandt et al. 2021). A value of $K_{zz} \sim 10^6$ cm^{2}s^{-1}, for instance, can reproduce the observed CH_4 profile (Luspay-Kuti et al. 2017; Mandt et al. 2017) but if one includes an upward flux of CH_4 to counteract losses into space, the value must be lower, i.e., between 550 and 4000 cm^{2}s^{-1} (Wong et al. 2017; Young et al. 2018), as pointed out by Gladstone & Young (2019).

For Ar, different eddy diffusion coefficients can lead to highly different results and interpretations of observations. For models with small K_{zz} (e.g., Strobel & Zhu 2017; Young et al. 2018), the non-detection of Ar by New Horizons/Alice is not significant since the density of Ar at Pluto's surface could still be as high as the one of N_2 (Steffl et al. 2020) while this is not the case for high coefficients (e.g., Luspay-Kuti et al. 2017). Also for Triton, a better understanding of the eddy diffusion coefficient, K_{zz}, is relevant. The results by Benne et al. (2022) are strongly impacted by the K_{zz} profile and observations could best be matched with a specific, altitude dependent K_{zz}. Generally, the behavior of the eddy diffusion coefficient and its dependence on density and pressure is poorly known in low-density atmospheres.

A potential cause for an overprediction of CH_3CN at Pluto might be an uncertain sublimation law for the relevant temperatures as these are extrapolated from vapor pressures measured at significantly higher temperatures (see Lellouch et al. 2022). Cross-sections of hydrocarbons (e.g., C_2H_2, C_2H_4, C_2H_6) in various models (e.g., Luspay-Kuti et al. 2017; Mandt et al. 2017; Young et al. 2018) were measured in the laboratory at temperatures around 150 K, which is significantly higher than the temperature range in much of Pluto's and Triton's atmospheres. This might lead to systematic errors in derived abundances of around 10%–20% and could also affect the retrieval of haze abundances around 500–600 km at Pluto (Young et al. 2018).

Photochemical models are strongly dependent on rate coefficients of diverse chemical reactions, which are again not known for such low temperatures. Benne et al. (2022), for instance, found for Triton that these affect key chemical reactions in its atmosphere and are particularly relevant for species such as O(^{3}P), CH_4, and N. It will therefore be a challenge to determine the composition of Triton's atmosphere properly without adequate rate coefficients. We also note that photochemistry may significantly affect fractionation in Pluto's or Triton's atmosphere but only a few measurements are available for isotopolog reaction rates, specifically for such low temperatures.

10.3.5 Some Further Open Questions and Needs

Understanding atmospheric escape is crucial for evaluating atmospheric stability, structure, and evolution. For Pluto and Triton, however, loss rates and their main mechanisms are poorly understood. N_2 and CH_4 Jeans escape rates at Pluto are known within an order of magnitude (e.g., Young et al. 2018) with CH_4 escaping about 500 times stronger than N_2 (Gladstone et al. 2016). On Triton, thermal loss is dominated by H and H_2, and to a lesser extent by N and C (e.g., Krasnopolsky & Cruikshank 1995; Strobel & Zhu 2017). Non-thermal loss rates are even less well established for which additional parameters have to be considered. These include the surrounding magnetic field (i.e., the interplanetary magnetic field in Pluto's and the magnetosphere of Neptune in Triton's case), as well as their particle environments (i.e., the solar wind for Pluto and Neptune's magnetospheric plasma environment for Triton). Although New Horizons/SWAP gave important insights into the solar wind beyond 11 AU (Elliott et al. 2016), it did not carry a magnetometer. Knowledge about Pluto's (potentially induced?) magnetic environment and its interaction with the solar wind is hence scarce.

For Triton orbiting within the magnetosphere of Neptune, the situation is even more complex since it is affected by atmosphere-magnetosphere interactions. In contrast to Pluto, an ionosphere was detected at Triton (Tyler et al. 1989), with its main ions likely being C^+, N^+, H^+, and N_2^+ (Benne et al. 2022). The dominant driver of ionization at Triton was suggested to be particle precipitation from Neptune (e.g., Krasnopolsky & Cruikshank 1995), which might also drive atmospheric chemistry significantly. No *in situ* measurements by Voyager 2, however, were made, which makes the dominant energy input basically unknown. Concerning non-thermal escape, Gu et al. (2021) found that N escapes non-thermally at comparable rates than thermally while O is lost non-thermally 3 times stronger than thermally. For other species, these authors found non-thermal losses to be less significant than thermal ones, but an improved understanding of atmospheric and ionospheric chemistry is needed. Benne et al. (2022), for example, suggest that water influx by IDPs may significantly affect photochemical models as oxygenated species show high rate uncertainties. This may also apply to Pluto and illustrates that laboratory experiments covering Pluto/Triton-like conditions will be another crucial need besides direct observations.

For Pluto, it can also be expected that an ionosphere will actually be present since its peak ionization of N_2 at an altitude of ~ 700 km through XUV is comparable to Triton's. New Horizons/REX was only able to define an upper limit on Pluto's ionospheric electron density of $1000\,e\ cm^{-3}$ and its composition remains largely unknown (e.g., Hinson et al. 2018), although its main ions were suggested to be $C_3H_5^+$ and $C_5H_5^+$ (Krasnopolsky 2020). Pluto's low peak electron density may stem from the high abundance of CH_4 in its atmosphere which might produce an ionosphere dominated by molecular ions (Krasnopolsky 2020). The role of non-thermal escape at Pluto, however, will remain highly uncertain without any knowledge on its ionosphere.

In the lower atmosphere of both bodies, boundary layers were detected that may affect volatile cycling and atmospheric structure. For Pluto, the formation of such layers may be related to sublimation and condensation of N_2 at Sputnik Planitia and its outflowing and inflowing winds (Hinson et al. 2017; Forget et al. 2017). Although the vertical resolution by New Horizons/REX was too low to resolve the vertical temperature gradient in the boundary layer, it would be essential for better understanding energy, momentum, and mass flux over Sputnik Planitia, atmosphere-surface interactions, and volatile cycling (Summers et al. 2021). At Triton, the detection of discrete clouds below 8 km (Yelle et al. 1991) and the appearance of plumes rising toward 8 km (e.g., Lellouch 2018) suggested the presence of a troposphere. However, to confirm its existence, observations of the lower atmosphere will be needed, which cannot be performed through stellar occultations (Marques Oliveira et al. 2022).

10.4 Future Observations

As the last section has illustrated, important complementary measurement needs arise for resolving open questions related to Pluto's and Triton's atmospheres. Some insights can be gained by remote observations, laboratory measurements and theoretical modeling while others depend on *in situ* measurements via a flyby mission, an orbiter or a lander.

10.4.1 Ground-Based Observations

Pluto and Triton are difficult to observe via ground-based facilities, which implies that only sensitive instruments, long-term monitoring, or combined observations can give further insights into their atmospheres. However, some atmospheric species such as CH_4 and CO can already be observed in the near-IR with echelle or cross-dispersed spectrographs and resolutions of $R \sim$ 13,000–70,000 such as the IRTF/CSHELL and VLT/CRIRES+ (see discussion in Buie et al. 2021). For high signal-to-noise (S/N) ratios, as Buie et al. (2021) point out, atmospheric CH_4 could also be detected for lower resolutions in the range of only a few thousand, as likely validated for Triton (Merlin et al. 2018). Monitoring CH_4 and CO with Earth-based facilities such as VLT/CRIRES+ and the upcoming ELT/HIRES instrument can establish the evolution of their mixing ratios, if the total pressure can be derived by other observations such as stellar occultations. Such measurements are key for understanding CH_4-related haze production, volatile cycles, and the related evolution of Pluto's and Triton's atmospheric structure and pressure.

Molecules such as CO, HCN, and HNC can also be observed via its absorption lines (e.g., Lellouch et al. 2017) with large submillimeter interferometers such as ALMA. Depending on the abundance, some isotopic counterparts such as $H^{13}CN$ and $HC^{15}N$ may also be observable, although this has not yet been achieved (see, e.g., Lellouch et al. 2022). A potential detection of $H^{13}CN$ and $HC^{15}N$, however, will be important for retrieving reliable values for $^{12}C/^{13}C$ and $^{14}N/^{15}N$ in HCN. In view of spatial resolution, this can already be provided at present, e.g., for Pluto at an angular resolution of $0.06''$ at a wavelength of 0.87 mm for CO and HCN (Lellouch et al. 2017, 2022) while an additional increase of resolution by about a factor of 3 can in principle still be achieved for substantially longer integration times in the order of $\sim$25–36 hr (Buie et al. 2021; Lellouch et al. 2022). Buie et al. (2021) also note that HCN and HNC are products of the CH_4–N_2 photochemistry, which is coupled to the density, distribution, and microphysics of the haze; this makes monitoring these species an important tool for monitoring and understanding haze chemistry in Pluto's and Triton's atmosphere over time. Moreover, Forget et al. (2021) point out that telescopic observations with ALMA and similar observatories might in principle be able to allow retrieving the magnitude and direction of zonal winds by measuring the Doppler shift of atmospheric CO lines at submillimeter wavelengths.

For spectroscopic observations in the optical and IR, however, ground-based instruments are neither capable of spatially resolving Pluto nor Triton at present. Upcoming large telescopes such as the Thirty Meter Telescope (TMT) and the ELT will collect about 10 times more light than the most powerful optical ground-based telescopes at present (e.g., Lellouch et al. 2022), which will (together with space-based facilities such as *HST* and *JWST*) provide important data sets on the surfaces of Triton and Pluto. These will be able to spectrally resolve some surface features at these bodies; ELT/MICADO, for instance, will be able to provide spatial resolution corresponding to about 250 km at Pluto with a spectral resolution of R=8000 (Buie et al. 2021). Here, we note, however, that Pluto's south polar cap will not be visible from Earth until 2109, i.e., other means of observations will be needed to observe its southern hemisphere.

For Triton, spectroscopic data including surface composition, rotational light curves, and albedo maps will allow constraints on the extension of its northern and southern caps (Bertrand et al. 2022). Mapping the entire surface of Triton would in general be needed for refining volatile transport models and hence simulations on atmospheric evolution. The volatile model by Bertrand et al. (2022), for instance, predicts that N_2 ice sublimation will dominate at the southern cap until about 2025–2030 which makes the southern pole appear darker and enriched in CH_4 ice.

Another important means of ground-based observations are stellar occultation campaigns (e.g., Sicardy 2023). These present an enormous opportunity for studying the atmospheres of Pluto (e.g., Sicardy et al. 2016; Arimatsu et al. 2020; Sicardy et al. 2021; Poro et al. 2021), Triton (e.g., Elliot et al. 2000; Marques Oliveira et al. 2022), and other KBOs (e.g., Arimatsu et al. 2019; Ortiz et al. 2020) with a network of relatively small scale, even mobile telescope facilities. Such campaigns are specifically good for monitoring the evolution of atmospheric pressure and for validating and differentiating between various models of volatile cycles as these make different predictions on their atmospheric evolution (see, e.g., for Pluto, Betrand et al. 2016). For characterizing atmospheric structure via stellar occulta-tions, however, one needs better S/N ratios from larger facilities (with telescopes $\gtrsim 1.0$ m). The probability of bright star occultations for Pluto, however, will decrease since the planet is receding from the galactic plane. Observations of stellar occultation events of fainter stars with high-cadence observations by high speed CMOS cameras on meter-class telescopes will therefore become more important, since these occur more frequently (e.g., Arimatsu et al. 2019, 2020). Mobile telescopes for high-cadence observations will also be essential for future observa-tions (Arimatsu et al. 2019); a strong integration of amateur astronomers and the related organisation of observation campaigns can therefore be a crucial pathway in addition to professional observations. Besides small-class facilities, large ground-based telescopes, specifically TMT and ELT, present opportunities for recording stellar occultations of extremely faint stars, thereby detecting significantly more events. For bright star occultations these observatories would additionally provide an extremely high S/N ratio (Ortiz et al. 2020). Airborne telescopes such as SOFIA can also efficiently be used for observing stellar occultation events (e.g., Person et al. 2021).

Ground-based observations will therefore give important insights into the volatile ice distribution on Pluto and Triton and into their seasonal climate systems, which is crucial for predicting atmospheric evolution. It may also give highly restricted insights into some isotope ratios but information on these remain limited if no other means of observations are applied. In general, remote space missions and in particular *in situ* measurements will be needed for resolving most of the open questions discussed in the last sections. Long-term monitoring, i.e., surface pressure via stellar occultations, surface temperature via thermal radiometry, atmospheric and surface composition via spectroscopy and surface albedo via imaging, through ground- and space-based instrumentation are essential to resolve several of the outstanding questions since various atmospheric characteristics might be interrelated and change while Pluto recedes from its perihelion and Triton moves through its seasons.

10.4.2 *JWST* and Other Remote Space-Based Observations

The *James Webb Space Telescope* (*JWST*) will be able to observe Pluto and Triton in the IR at 0.6–28 μm with high sensitivity through its spectroscopic instruments NIRSpec and MIRI. Crucially, the haze cooling model (Zhang et al. 2017) can be tested via observations with *JWST*/MIRI. This scenario provides large thermal emissions in the mid-IR, which could exceed emission from Pluto's surface for wavelengths below 25 μm by orders of magnitude with its thermal light curve being progressively suppressed between 25.5 and 18.0 μm (Zhang et al. 2017). *JWST*/MIRI has the required sensitivity to detect any thermal emission of haze that was suggested to dominate Pluto's mid-IR spectrum at wavelengths between 5 and 25 μm. Even though Charon's warm emission might contaminate the IR spectrum, Pluto's haze radiation should still exhibit a strong infrared excess below 15 μm (Zhang et al. 2017). Figure 10.4 illustrates the thermal emission of Pluto with and without haze compared with the sensitivity of *JWST*/MIRI (Zhang et al. 2017). Even in the case that the suggestion of a bimodal haze particle size distribution is correct, the efficient radiative cooling of the smaller particles might result in another peak and/or a steeper slope in the mid-IR emission spectrum, which can also be investigated with *JWST*/MIRI (Fan et al. 2022).

$\quad$ *JWST* will likely also be able to observe different CH_3D bands through which D/H ratios at Pluto and Triton could be measured (Grundy et al. 2011) although this will be more likely on Pluto than on Triton due to Triton's 10 times lower CH_4 ice abundance (see Buie et al. 2021). It may also observe ^{13}CO (Buie et al. 2021), which can give key insights into these bodies' $^{13}C/^{12}C$ ratios. *JWST*/NIRSpec will allow observations of methane bands between 1.3 and 2.7 μm, CO at 2.3 and 4.7 μm, and to measure the spatial distribution and state of volatiles such as ice temperatures and N_2–CH_4 ice mixing ratios (Buie et al. 2021). Thermal emission monitoring of the upper atmosphere in the near-IR will further allow to model effects of lower atmosphere changes on the temperature structure of the upper atmosphere (Summers et al. 2021). It can further study surface composition in the 2.5–5.0 μm range with its high sensitivity and good spectral resolution (Buie et al. 2021). Larger, proposed space telescopes in the UV, optical and IR such as the Habitable Worlds Observatory (HWO) and the Large Interferometer For Exoplanets (LIFE) may even

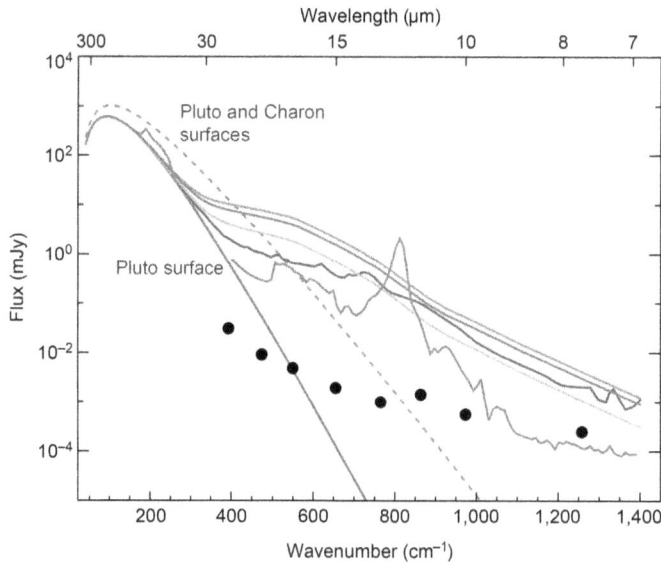

Figure 10.4. IR spectra of Pluto with and without the effect of haze as seen from Earth's distance. The various colored spectra show haze-mediated cooling from different model (for details see Zhang et al. (2017), and references therein). The solid and dashed grey lines illustrate the IR spectrum of Pluto alone and with Charon combined, both without the effect of haze cooling. The black dots show the 10σ faint-source detection-limits in a 10,000s integration for several channels of *JWST*/MIRI. Figure adapted by permission from Springer Nature Customer Service Centre GmbH: Zhang et al. (2017). Copyright 2017.

allow spatial resolutions of Pluto (and Triton) approaching the resolution of New Horizons (Summers et al. 2021).

Besides Pluto, *JWST*/MIRI will also be able to measure Triton's thermal light curve for the first time in the mid-IR. Such observations would strongly constrain the volatile ice distribution at its surface, specifically its temperature, since the thermal light curve and the overall flux level strongly depend on the ice distribution (Bertrand et al. 2022). They would help to discriminate between the different climate scenarios by Bertrand et al. (2022) and to evaluate the related evolution of Triton's surface pressure. Further spectral observations with the *HST* and *JWST* will provide additional complementary data on the volatile distribution of Triton and Pluto.

10.4.3 In Situ Measurements

Some measurement needs for Pluto and Triton can only be achieved through *in situ* measurements, either via a flyby, an orbiter, or even a lander. For Pluto, Howett et al. (2021) propose a mass spectrometer that can identify N_2, CH_4, C_2H_x, and further constituents that were not measured so far, such as H_2. Most of these measurements have to be performed at altitudes above 500 km (Howett et al. 2021) since saturation may affect the results of such measurements, an altitude that is consistent with the predicted density peaks of photochemical products and ions as produced through absorption of solar XUV (Krasnopolsky 2020). However, for gathering reliable data on isotopes and isotopologs, a lander will be more convenient or at least an accurate determination of eddy diffusion as a function of altitude will be needed.

The viable instruments suggested by Howett et al. (2021) include a UV spectrometer, a near-IR spectrometer, and a radio science instrument for sounding Pluto's atmosphere to derive its structure and composition. It also covers a mass spectrometer that measures neutrals at densities >1.0 cm^{-3}, a fluxgate magnetometer, and a plasma spectrometer for measuring composition, energies, angular and spatial distributions, as well as densities of pickup, solar wind, and ionosphere ions. This would allow us to measure the interaction between the solar wind and Pluto's atmosphere, any existing bow shock and magnetotail, as well as Pluto's ionosphere and potential magnetic field (Howett et al. 2021). An orbiter with such an instrument suite would hence be able to probe the dynamics of Pluto's neutral and charged particle environment, essential for better understanding the thermal balance within its atmosphere and the evolution of atmospheric loss, specifically, the role of non-thermal escape. If such an orbiter is deployed in the next decades, it would give—in addition to observations by New Horizons—insights into the long-term atmospheric evolution along Pluto's orbit. A similar suite of instruments could perform similar tasks at Triton. The proposed TRIDENT mission would have similarly been equipped with plasma and high-resolution IR spectrometers with spectral ranges up to $5\,\mu$m (e.g., Prockter et al. 2019). Together with a magnetometer, such instrumentation would be essential to study the complex interaction of Neptune's magnetosphere and plasma environment with Triton's atmosphere. In addition, observations in the UV would help to resolve the different ice condensation models suggested by Ohno et al. (2021) since the optical depth of ice aggregates increases with decreasing wavelengths for $\lambda < 0.15\,\mu$m while it is invariant for their ice ball scenario.

An orbiter or a flyby mission at Pluto and Triton could further measure the impact of IDPs with a dust analyzer, specifically for studying the related influx of H_2O but also of minor species such as CO_2, CO, and H_2O_2. This would be important for further investigating the hypothesis of water or other minor species contributing to cooling within Pluto's upper atmosphere (see Strobel & Zhu 2017), but also for generally deriving the cosmic IDP influx. For Triton, observations of dust and water influx from Neptune's magnetosphere would be similarly important, as oxygen from H_2O could significantly affect photochemistry in its atmosphere (Benne et al. 2022).

Flyby or orbiter missions could also perform high-resolution measurements of the near-surface regions that are important for volatile transport and for energy and momentum exchange with the surface. This would provide the boundary-layer constraints needed for developing high-resolution climate models and for understanding volatile transport (e.g., Summers et al. 2021). For Pluto, mapping the entire surface would be key for investigating N_2 and CH_4 ice deposits specifically in the southern hemisphere. For investigating the evolving atmospheric circulation and thermal structure of Pluto's (or Triton's) atmosphere, Forget et al. (2021) suggest a submillimeter heterodyne instrument on an orbiter that resolves CO line emissions, could make Doppler wind measurements at the limb, and provides temporal and spatial monitoring of atmospheric circulations. For Triton, a flyby mission or orbiter could image parts in the northern hemisphere that were not imaged by Voyager 2 and could study its atmospheric structure and the interaction of its ionosphere with Neptune's magnetosphere. A simple imaging IR spectrometer that maps Triton's surface (e.g., Hansen & Castillo-Rogez 2021) could inform on the distribution of ices, cryovolcanic processes and the existence of a permanent northern ice cap.

Our knowledge about Triton would also benefit significantly by a Neptune flagship mission that would likely yield a scientific output comparable to Cassini/ Huygens and that would be a milestone for solar system exploration (e.g., Fletcher et al. 2020). Multiple flybys at Triton could perform spectral mapping of its surface and the related volatile distribution, and study variations in Triton's atmosphere and ionosphere. A combination of an orbiter with subsequent missions would be complementary since they enable studying atmospheric long-term changes under different environmental conditions shaped by the changing system architecture (Hansen & Castillo-Rogez 2021).

In the case of Pluto, we emphasize that Charon can provide a rich source of information on Pluto's atmospheric evolution. The tholin regions at Charon, for example, should be spectroscopically investigated for spectral line signatures of nitriles. If these are present, they may have been formed by photolysis of CH_4–N_2 mixtures (Wong et al. 2015), indicating strong escape of both species from Pluto in the very recent past (Tucker et al. 2015). Observing the composition of Charon's poles within 2.5–5.0 μm (Howett et al. 2021) would therefore help to gain key insights into the evolution of Pluto's atmosphere. Observing CH_3D or other D and H containing molecules not only at Pluto (e.g., with *JWST*) but also on Charon where CH_4 could be cold trapped at the poles during winter season (but evaporates during other seasons; Grundy et al. 2016) would give unique insights into atmospheric escape at Pluto since potential differences in D/H on both bodies may be a result of the more easily escaping H being preferentially deposited onto Charon.

Finally, a lander will likely be fundamental for measuring various isotopes, isotopologs, and their respective ratios in atmosphere and surface reservoirs. Due to strong diffusive fractionation in their atmospheres, because of their low homopause levels below ~12 km at Pluto (Young et al. 2018) and around ~30 km at Triton (Krasnopolsky 2019), it will be difficult to extrapolate measurements from their upper to their lower atmospheres or bulk compositions. Isotopic observations of N_2 and other N-bearing compounds, as well as gaining information about $^{36}Ar/N_2$ (see also Glein & Waite 2018; Glein 2023), with a mass spectrometer at the surface would therefore be highly favorable. These measurements are key for understanding the origin and evolution of Pluto's and Triton's volatile reservoirs and for reconstructing isotopic fractionation through atmospheric escape and photochemistry. Observations of $^{36}Ar/^{38}Ar$ and $^{20}Ne/^{22}Ne$ would give additional insights into the evolution of atmospheric escape while complementary data such as D/H and $^{12}C/^{13}C$ will help to understand how N_2 relates to other volatiles such as CH_4 and CO (e.g., Glein 2023). Atmospheric observations of the radiogenic isotopes ^{40}Ar and ^{129}Xe might similarly be feasible at the surface only.

For measuring haze particles at Triton with a mass spectrometer, a lander will be needed as well since it is only present within the first ~30 km of its atmosphere. For this, Moran et al. (2022) emphasize the need of mass spectrometers reaching out to as high as $\geqslant$450 amu, and of near-IR spectrometers covering wavelengths up to at least 6.5 μm; these characteristics are needed for accurately probing the importance of large and complex molecules with significant carbon–oxygen compounds for haze formation at Triton. A lander mission to at least Pluto (or Triton) will hence provide a swath of rich data that cannot only be applied to Triton (or Pluto) but to a certain extent to KBOs in general, as they will likely share similar origins and evolutions. Even a singular mission to either Pluto or Triton will therefore illuminate the origin and history of a multitude of KBOs and the entire outer solar system making a strong point for investigating future possibilities on such missions.

10.4.4 Laboratory Measurements and Experiments

As discussed in Section 10.3.4, atmosphere models of Pluto and Triton suffer from the fact that several important model parameters are poorly known. Specific laboratory measurements can therefore significantly advance the field. Condensation and sublimation under icy bodies' physical conditions are needed, specifically for better understanding isotope fractionation on bodies whose atmosphere is shaped by equilibrium vapor pressure. For species in Pluto's and Triton's atmospheres, however, the saturation vapor pressures have to be extrapolated from measurements at temperatures that are higher than their respective temperature range, thereby introducing an additional uncertainty into the results. Young et al. (2021) further point out that experiments on ice mixtures will be needed to study the behavior of N_2, CO, and CH_4 under disequilibrium at low-temperature conditions to test whether CH_4-rich ices can be formed for potentially explaining the observed mixing ratio in their atmospheres.

Cross-sections of hydrocarbons, which are crucial to model their profiles within Pluto's, and to a lesser extent within Triton's, atmosphere were measured in laboratory experiments (see Young et al. 2018) at 150 K, 140 K, and 150 K for C_2H_2, C_2H_4, and C_2H_6, respectively, which is significantly warmer than the temperatures in the atmospheres of Pluto and Triton. Although uncertainties of these high-temperature cross-sections might only be 10%–20% (Summers et al. 2021), the hydrocarbon profiles are essential for understanding atmospheric chemistry. Reaction rates at temperatures around 100 K and lower also have to be extrapolated from laboratory measurements at higher temperatures. Performing such experiments at the atmospheric temperatures prevalent at Pluto and Triton will be of crucial importance for improving photochemical models since high uncertainties in reaction rates of important species such as atomic nitrogen or CH_4 (see, e.g., Benne et al. 2022) prevent an accurate determination of atmospheric composition and profiles. In addition, future laboratory studies at low temperatures of the UV and optical properties of hydrocarbon ices are needed to accurately distinguish between different haze formation models (Ohno et al. 2021). For better understanding aerosol formation, desorption energies and contact angles of hydrocarbon ices on tholin are further properties that have to be studied through laboratory measurement, as currently the only available data stem from CH_4 and C_2H_6 (see Ohno et al. 2021).

For Triton, Moran et al. (2022) point out that laboratory work on haze formation for conditions prevalent in its upper atmosphere will be important to disentangle the roles of CH_4 and CO in haze formation, as its upper atmosphere is basically CH_4 free. This could also help resolving the question on whether photochemical haze or ice condensates are more relevant for the production of haze in Triton's atmosphere. There might also be differences in haze production in the ionosphere and lower atmosphere, which can also be studied by such experiments (Moran et al. 2022).

To accurately model isotopic fractionation, laboratory measurements of ^{14}N–^{15}N fractionation associated with HCN condensation and aerosol trapping under varying XUV irradiation for Pluto and Triton atmosphere conditions are needed. Finally, it has to again be emphasized that eddy diffusion, as well as molecular diffusion, for the conditions present in Pluto's and Triton's atmosphere are not known precisely. Molecular diffusion is only measured in the laboratory for temperatures much higher than the ones at Pluto and Triton although its inherited uncertainty was estimated to be less than 7% (Plessis et al. 2015). For the eddy diffusion coefficient, K_{zz}, Mandt et al. (2021) point out that collaborative efforts between the various modeling groups will be needed for assessing the differences in K_{zz}.

10.4.5 Comparative Planetology

For better understanding the atmosphere of Pluto, one can also probe Triton and vice versa. Isotope and element ratios at these bodies, e.g., $^{14}N/^{15}N$, D/H, $^{12}C/^{13}C$, and $^{36}Ar/^{14}N$, may have been similar initially. However, present-day variations within these ratios will, along with atmospheric and photochemical models, identify the potential differences in their evolutionary paths over the last few billion years. Comparing results from both, but also from Pluto and Triton with Titan, will be key for understanding the evolution of KBOs in general.

Investigating the differences in their atmospheric thermal structure will also give important insights into the drivers of this difference. Triton, for instance, is likely dominated by a $\sim$8 km deep cold troposphere that could be related to a lower mixing ratio of CH_4 at Triton and, hence, to less haze in its atmosphere (see, e.g., Forget et al. 2021). Such a difference might fundamentally alter the atmospheric structure, temperature, and dynamics in both satellites' atmospheres. Comparative studies on haze formation, condensation, and sticking of aerosols at Pluto and Triton are, hence, important for generally understanding haze in planetary atmospheres (see, e.g., Mandt et al. 2021).

Besides investigating similarities and differences between Pluto and Triton, it will also be fundamental to gather data from other solar system bodies. Future atmospheric observations of Titan, remote and *in situ*, will be of high value for understanding similarities and differences between Saturn's major satellite and KBOs. While remote or even *in situ* observations of other KBOs would be of high relevance but difficult to achieve, investigating comets will help in deriving the origin and evolution of Pluto's and Triton's atmospheres. It will hence be crucial to proceed measuring isotope ratios and element abundances within various comets. ESA's upcoming mission Comet Interceptor (e.g., Snodgrass & Jones 2019) will therefore be a milestone mission also for understanding the origin of Pluto's and Triton's atmospheres. Any *in situ* space mission or remote telescope that will probe isotope ratios within the atmospheres of the ice giants can further the understanding of the origin of nitrogen and other volatiles at the icy bodies.

In addition, one should also emphasize the importance of space missions for observing the characteristics of the solar wind at larger heliospheric distances and the flux and composition of IDPs for better assessing the interaction of Pluto and Triton with the interplanetary medium. A reconstruction of the Sun's history will finally be crucial for modeling atmospheric evolution at Pluto, Triton, and other KBOs, which can best be achieved by modeling the atmospheres of several bodies within a single framework under the same assumed solar evolution. Isotope measurements of any solar system planet will therefore help increasing our understanding of Pluto's and Triton's atmospheric evolution.

10.5 Conclusion

Pluto and Triton are both likely KBOs with their atmospheres sharing significant similarities and important differences. One of the major differences lies in the thermal structure of their atmospheres with Pluto's being significantly colder than Triton's. The physical reason for this difference is presently poorly understood as are other important atmospheric parameters. The dominant cooling process within Pluto's atmosphere was suggested to either be cooling via haze or the influx of H_2O. These differing scenarios can be distinguished by future *JWST*/MIRI measurements since Pluto's thermal emission in the mid-IR could exceed its surface emission by orders of magnitude, if the haze cooling scenario is correct. It will therefore also be key to establish comparative models of haze formation in both bodies' atmospheres for better understanding their thermal differences and for reconstructing their atmospheric evolution. Observing the volatile distribution at the surface of Pluto and Triton will be another key measurement for reconstructing not only their seasonal volatile cycles but also the evolution of their atmospheric structure and pressure. A continuous monitoring of their atmospheres via stellar occultation and ground-based UV and near-IR observations will be additionally needed for evaluating molecular mixing ratios and atmospheric pressure over time. In the future, however, *in situ* measurements of isotope ratios in N_2- and N-bearing ices, in noble gases, and of radiogenic isotopes with mass spectrometers in the atmospheres of Pluto and Triton and, specifically, at their surfaces with future landers will be essential for evaluating the origin of both bodies' volatiles. In addition, laboratory experiments to probe conditions in their cold atmospheres will help to refine atmospheric and photochemical models for validating past and upcoming atmospheric observations. Finally, we also emphasize the important role of comparative planetology. A future mission to one of these bodies will significantly enhance our knowledge on other KBOs and on the evolution of the outer solar system in general. Observations of isotope ratios within pristine comets will be further milestone measurements for understanding the atmospheres of Pluto and Triton.

References

Agnor, C. B., & Hamilton, D. P. 2006, Natur, 441, 192

Arimatsu, K., Hashimoto, G. L., Kagitani, M., et al. 2020, A&A, 638, L5

Arimatsu, K., Ohsawa, R., Hashimoto, G. L., et al. 2019, AJ, 158, 236

Arimatsu, K., Tsumura, K., Usui, F., et al. 2019, NatAs, 3, 301

Benne, B., Dobrijevic, M., Cavalié, T., Loison, J. C., & Hickson, K. M. 2022, A&A, 667, A169

Bertrand, T., & Forget, F. 2016, Natur, 540, 86

Bertrand, T., Forget, F., Umurhan, O. M., et al. 2018, Icar, 309, 277

Bertrand, T., Lellouch, E., Holler, B. J., et al. 2022, Icar, 373, 114764

Biver, N., & Bockelée-Morvan, D. 2019, ESC, 3, 1550

Broadfoot, A. L., Atreya, S. K., Bertaux, J. L., et al. 1989, Sci, 246, 1459

Buie, M. W., Hofgartner, J. D., Bray, V. J., & Lellouch, E. 2021, The Pluto System After New Horizons, ed. S. A. Stern, et al. (Tuscon, AZ: Univ. Arizona Press) 569

Cheng, A. F., Summers, M. E., Gladstone, G. R., et al. 2017, Icar, 290, 112

Elliot, J. L., Person, M. J., McDonald, S. W., et al. 2000, Icar, 148, 347

Elliott, H. A., McComas, D. J., Valek, P., et al. 2016, ApJS, 223, 19

Erkaev, N. V., Scherf, M., Thaller, S. E., et al. 2021, MNRAS, 500, 2020

Fan, S., Gao, P., Zhang, X., et al. 2022, NatCo, 13, 240

Fletcher, L. N., Helled, R., Roussos, E., et al. 2020, P&SS, 191, 105030

Forget, F., Bertrand, T., Hinson, D., & Toigo, A. 2021, The Pluto System After New Horizons, ed. S. A. Stern, J. M. Moore, W. M. Grundy, L. A. Young, & R. P. Binzel (Tuscon, AZ: Univ. Arizona Press)

Forget, F., Bertrand, T., Vangvichith, M., et al. 2017, Icar, 287, 54

Gladstone, G. R., Stern, S. A., Ennico, K., et al. 2016, Sci, 351, aad8866

Gladstone, G. R., & Young, L. A. 2019, AREPS, 47, 119

Glein, C. R. 2017, Icar, 293, 231

Glein, C. R. 2023, Icar, 404, 115651

Glein, C. R., Waite, J. H., et al. 2018, Icar, 313, 79

Grundy, W. M., Bertrand, T., Binzel, R. P., et al. 2018, Icar, 314, 232

Grundy, W. M., Cruikshank, D. P., Gladstone, G. R., et al. 2016, Natur, 539, 65

Grundy, W. M., Morrison, S. J., Bovyn, M. J., Tegler, S. C., & Cornelison, D. M. 2011, Icar, 212, 941

Gu, H., Cui, J., Niu, D. D., et al. 2021, A&A, 650, A130

Gurrola, E. M. 1995, PhD thesis, Stanford University

Hansen, C. J., Castillo-Rogez, J., et al. 2021, PSJ, 2, 137

Herbert, F., Sandel, B. R., et al. 1991, JGR, 96, 19241

Hinson, D. P., Linscott, I. R., Strobel, D. F., et al. 2018, Icar, 307, 17

Hinson, D. P., Linscott, I. R., Young, L. A., et al. 2017, Icar, 290, 96

Howett, C. J. A., Robbins, S. J., Holler, B. J., et al. 2021, PSJ, 2, 75

Krasnopolsky, V. A. 2019, Spectroscopy and Photochemistry of Planetary Atmospheres and Ionospheres. Mars, Venus, Titan, Triton and Pluto (Cambridge: Cambridge Univ. Press)

Krasnopolsky, V. A. 2020, Icar, 335, 113374

Krasnopolsky, V. A. 2020, P&SS, 192, 105044

Krasnopolsky, V. A., & Cruikshank, D. P. 1995, JGR, 100, 21271

Krasnopolsky, V. A., Sandel, B. R., Herbert, F., & Vervack, R. J. 1993, JGR, 98, 3065

Lavvas, P., Lellouch, E., Strobel, D. F., et al. 2021, NatAs, 5, 289

Lellouch, E. 2018, Handbook of Exoplanets, Vol. 47, ed. H. J. Deeg, & J. A. Belmonte (Berlin: Springer)

Lellouch, E., Butler, B., Moreno, R., et al. 2022, Icar, 372, 114722

Lellouch, E., de Bergh, C., Sicardy, B., Ferron, S., & Käufl, H. U. 2010, A&A, 512, L8

Lellouch, E., Gurwell, M., Butler, B., et al. 2017, Icar, 286, 289

Luspay-Kuti, A., Mandt, K., Jessup, K.-L., et al. 2017, MNRAS, 472, 104

Mandt, K. E., Luspay-Kuti, A., Cheng, A., Jessup, K. L., & Gao, P. 2021, The Pluto System After New Horizons, ed. S. A. Stern, J. M. Moore, W. M. Grundy, L. A. Young, & R. P. Binzel (Tuscon, AZ: Univ. Arizona Press) 279

Mandt, K. E., Mousis, O., Lunine, J., & Gautier, D. 2014, ApL, 788, L24

Mandt, K., Luspay-Kuti, A., Hamel, M., et al. 2017, MNRAS, 472, 118

Marques Oliveira, J., Sicardy, B., Gomes-Júnior, A. R., et al. 2022, A&A, 659, A136

McKay, A., DiSanti, M. A., Kelley, M. S. P., et al. 2019, AJ, 158, 128

Merlin, F., Lellouch, E., Quirico, E., & Schmitt, B. 2018, Icar, 314, 274

Meza, E., Sicardy, B., Assafin, M., et al. 2019, A&A, 625, A42

Miller, K. E., Glein, C. R. Jr, & Waite, J. H. 2019, ApJ, 871, 59

Moran, S. E., Hörst, S. M., He, C., et al. 2022, JGRE, 127, e06984

Mousis, O., Aguichine, A., Bouquet, A., et al. 2021, PSJ, 2, 72

Ohno, K., Zhang, X., Tazaki, R., & Okuzumi, S. 2021, ApJ, 912, 37

Ortiz, J.L., B. Sicardy, J.I.B. Camargo, P. Santos-Sanz, and F. Braga-Ribas. 2020. The Trans-Neptunian Solar System, eds. Prialnik, M.A. Barucci, and L. Young, (Amsterdam: Elsevier) 413

Person, M. J., Bosh, A. S., Zuluaga, C. A., et al. 2021, Icar, 356, 113572

Plessis, S., McDougall, D., Mandt, K., Greathouse, T., & Luspay-Kuti, A. 2015, P&SS, 117, 377

Poppe, A. R., & Horányi, M. 2018, A&A, 617, L5

Poro, A., Ahangarani Farahani, F., Bahraminasr, M., et al. 2021, A&A, 653, L7

Prockter, L. M., Mitchell, K. L., Howett, C. J. A., et al. 2019, 50th Annual Lunar and Planetary Science Conf., (Houston, TX: Lunar and Planetary Science Institute) 3188

Rubin, M., Altwegg, K., Balsiger, H., et al. 2015, Sci, 348, 232

Scherf, M., Lammer, H., Erkaev, N.V., et al. 2020, SSRv, 216, 123

Sicardy, B. 2023, CRPhy, 23, 213

Sicardy, B., Ashok, N. M., Tej, A., et al. 2021, ApL, 923, L31

Sicardy, B., Talbot, J., Meza, E., et al. 2016, ApL, 819, L38

Snodgrass, C., & Jones, G. H. 2019, NatCo, 10, 5418

Steffl, A. J., Young, L. A., Strobel, D. F., et al. 2020, AJ, 159, 274

Stern, S. A., Bagenal, F., Ennico, K., et al. 2015, Sci, 350, aad1815

Stern, S. A., Binzel, R. P., Earle, A. M., et al. 2017, Icar, 287, 47

Strobel, D. F. 2008, Icar, 193, 612

Strobel, D. F., & Summers, M. E. 1995, Neptune and Triton (Tuscon, AZ: Univ. Arizona Press) 1107

Strobel, D. F., & Zhu, X. 2017, Icar, 291, 55

Summers, M. E., Young, L. A., Gladstone, G. R., & Person, M. J. 2021, The Pluto System After New Horizons ed. S. A. Stern, et al. (Tuscon, AZ: Univ. Arizona Press) 257

Telfer, M. W., Parteli, E. J. R., Radebaugh, J., et al. 2018, Sci, 360, 992

Tucker, O. J., Johnson, R. E., & Young, L. A. 2015, Icar, 246, 291

Tyler, G. L., Sweetnam, D. N., Anderson, J. D., et al. 1989, Sci, 246, 1466

Wong, M. L., Fan, S., Gao, P., et al. 2017, Icar, 287, 110

Wong, M. L., Yung, Y. L., Gladstone, G. R., et al. 2015, Icar, 246, 192

Yelle, R. V., Lunine, J. I., & Hunten, D. M. 1991, Icar, 89, 347

Young, L. A., et al. 2021, The Pluto System After New Horizons, ed. S. A. Stern, J. M. Moore, W. M. Grundy, L. A. Young, & R. P. Binzel (Tuscon, AZ: Univ. Arizona Press) 321

Young, L. A., Kammer, J. A., Steffl, A. J., et al. 2018, Icar, 300, 174

Zhang, X., Strobel, D. F., & Imanaka, H. 2017, Natur, 551, 352

Zhu, X., Strobel, F., & Erwin, J. T. 2014, Icar, 228, 301

Chapter 11

Planning for Long-Lived Missions

Janet Vertesi, Matthew J Bietz, Marisa Cohn, Stephanie Jordan and David Reinecke

11.1 Introduction

Forty years of Voyager, 20 years of Cassini, 10 years of Curiosity, and a decade-long cruise for New Horizons to reach its primary target. With the exception of the latter, most space science missions were not designed with excessive longevity in mind, despite their impressive age (Fox et al. 2013). Spacecraft teams instead tend to improvise extensively in the moment, rely on charismatic leaders, hope for serendipitous circumstances, and scramble for survival amid successive financial crises. As a result, little systematic attention has been given to the supporting infrastructures and human organizations necessary to sustain space science over the long term, even as next-generation missions contemplate very long lifetimes.

While no mission planner has a crystal ball, there are ways to plan for long-term success that avoid reinventing the wheel. A mission to the Ice Giants, their moons, and beyond will do well to heed the lessons learned from studies of successful long-term scientific investigations in the planetary sciences and elsewhere to plan for a long-lived future. This contribution reviews longevity-oriented considerations for large sociotechnical science teams that prospective missions can build into their plans from the ground up.

This paper draws on the authors' two decades of qualitative and historical studies among the planetary science community, including ethnographic immersion among several mission teams and interviews with dozens of participating scientists and engineers at multiple levels of mission hierarchy. We bring into the conversation relevant literature from sociology of science, history of technology, and computer-supported cooperative work. We draw on our advisory capacity roles alongside scientists, engineers, and project managers developing long-lived teams, including an Interstellar Probe and prospective missions to Europa, Io, and Neptune. Among our co-authors are ethnographers who have been embedded in scientific fields of collaboration at scale, including neuroscience, genetics, and oceanography. Across these domains we have conducted semi-structured interviews with leading scientists, administrators, project managers, and historians of science on the challenges of very long duration science in order to generate insights for long-term teams.

We present findings incorporating lessons from (a) past studies of spaceflight (Conway 2015; McCurdy 2003; Mirmalek 2020; Vaughan 1996; Vertesi 2020); (b) big science, including long-term experiments like particle accelerators or observatories (Knorr Cetina 1999; Patrick McCray 2000; Galison & Hevly 1992); and (c) the sociological literature on team science (Cooke et al. 2015) and science and technology studies (Latour 1992; Leigh Star 1999). Building upon this evolving body of practice, we introduce a sociotechnical approach to thinking about spacecraft teams, their technical tools, and their participants. We propose that long-duration science excels when mission planners consider experimental longevity from the outset rather than after the fact. We therefore document challenges and strategies for success concerning: (a) multi-generational team structure; (b) cultures of knowledge management; (c) aging hardware, software, and data sets; (d) sustainable funding models; and (e) workforce diversity and inclusion. Our goal is to be comprehensive but not exhaustive, offering a summary of and an introduction to a rich body of work in each domain.

Our findings and recommendations build upon prior co-authored white papers, technical reports, and proceedings in which we have advanced previous articulations of these ideas (Fountain et al. 2020; Reinecke & Brandt 2021; Reinecke & Vertesi 2021; Rymer et al. 2021; Vertesi 2019). We are grateful to our many coauthors and interlocutors in the space sciences—especially Glen Fountain, Stamatios Krimigis, Abi Rymer, Pontus Brandt, and Kathy Mandt—for their sustained engagement with our investigations and our findings.

11.1.1 Take a Sociotechnical Approach

Prior studies of long-lived craft examine primarily technical recommendations for longevity, redundancy, and system resilience. In contrast, our contribution centers *social* and *sociotechnical* considerations. By *social*, we refer to the organizational, interactional, and human elements of spacecraft and scientific teams. Social considerations include questions like: Who is in charge? How will our team make decisions? How do we teach newcomers? Many social questions involve institutions and politics, such as: How can we ensure we stay funded at the necessary level throughout development and exploration?

Sociotechnical considerations add an additional layer of complexity to social or technical thinking alone. This requires imbricating both the human and the mechanical sides of any technological enterprise (Latour 1992), demonstrating how they interact and influence each other and cannot be considered without one another (Leonardi 2012; Leigh Star 1999). On the one hand, the human organizations and institutions that sustain and animate our technologies play a tremendous role in how we imagine, build, and use the tools at our disposal. On the other hand, behind every technical decision about operational modes or implementation strategies are human organizations, relationships, and necessities we must contend with as well.

Thinking sociotechnically brings a shift in perspective from one that places technology in the drivers' seat, to one in which societal and technical arguments go hand in hand. An example of a sociotechnical question is: How do we ensure that an evolving list of team members—including some who have not yet been born—can keep working effectively with aging hardware and software over time? This requires thinking of the technical capacities of an aging apparatus alongside questions like: Who will take care of these systems? How will they be trained? What status will their skills have? Who will pay for them? And how do we attract newcomers to take inspiration in our tools, care for them, or prioritize them in our national best interest?

Spacecraft are sociotechnical systems par excellence. Unlike tools in the private sector like social media systems or manufacturing equipment, planetary scientists and space engineers work together to build the bespoke machinery that they will ultimately use to explore scientific phenomena. Scientific, technical, and political considerations (such as which launch vehicle to purchase or how little money to spend) weave together in the phased development of a spacecraft to produce the final product (Latour 1992). Spacecraft teams therefore develop complex, emergent, even intimate relationships with their technologies (Vertesi 2020). This means that the effects of a team upon their spacecraft (and the effects of the spacecraft upon their team) cannot be so easily disentangled. Additionally, the sociotechnical view points out where and how technological and human considerations may conflict: for instance, in the temporalities of spacecraft versus people. A single spacecraft lifetime may span across multiple full or partial careers for the people involved in the mission. Managing this tension is of utmost concern for long-lived mission teams. In the coming sections, we detail how to lay these social, organizational, and technical considerations out in advance, in order to proactively design a mission, a team, a ground system, and a craft to achieve goals over decadal scales.

11.2 Build a Multi-Generational Team

11.2.1 The Current Model Is Limited

Most NASA missions begin by selecting one or more PI's with an associated science team. The Principal Investigator (PI) position is considered a lifetime appointment, at least until the end of the lifetime of the spacecraft, and sometimes beyond to cover data processing. This convention in the planetary sciences has several downstream effects. One is a persistent limitation against newcomers' promotions into positions of authority, another is the seeming impossibility of considering mission feasibility without its associated leader.

This has promoted an emphasis on charisma and on certain personal character-istics associated with the position of PI. Such leaders must endure throughout many years of the slings and arrows of fortune and must move mountains across institutional boundaries, often with limited funds. They are promised a series of rewards if they can do so successfully. However, the ability to act in a brash or strident manner contrasts with stereotypical behavior for women and people of color, producing backlash (Rudman & Fairchild 2004). An additional cultural element of this view of a PI-ship is that such difficulties earn PIs the right to produce papers based on the instruments that they craft and see through to completion. This model hampers open access to ongoing research or to a diverse group of scientists. When leadership qualities are narrowly scoped or defined based on a response to institutional limitations, mission teams may suffer from gendered, racialized, and ageist bias accordingly.

This model causes serious problems for long-term missions. First, the charismatic form of leadership does not tend to endure far beyond the leader's death: A robust organization must be put in place that includes succession planning. Succession is a complex organizational task, so it is unsurprising that planetary science has shied away from this problem. The Voyager team was even chastised by its review panel for not planning for the deaths of its PI's and Project Scientist. Merely transferring the baton after death implies a patriarchal, oligarchic, organization, which can be subject to political chaos once transfer ultimately takes place (Weber 1968; Grusky 1960). For the sake of the longevity of the craft, we cannot rely merely on the continued longevity of the humans who command it.

Second, this format does not allow young people to move up other than to acquire and achieve their own PI-ships of other missions, where their attention is divided. Third, this form of leadership does not encourage other personal qualities necessary for leading different mission phases or to encourage newcomers (Traweek 1988). It shies away from resolving some of the institutional issues associated with longevity, such as funding, by placing all of the burden on a single PI's shoulders. Fourth, as will be discussed below, this format of leadership and organizational development discourages diverse voices entering and rising through the ranks. This leaves mission personnel increasingly at odds with the demographic reality and prioritization of its stakeholders and funding agency. Finally, the idea that the reward is in the data at the end is moot for a mission that takes over 30 years to reach its target. A PI who is suitably senior in their career at time of inception may or may not be alive to see the mission's conclusion.

Fortunately, this is not the only way to be a PI. NSF grants, NIH research centers, observatories, and even hospitals feature other ways of doing scientific investigations alongside complex machinery (Shrum et al. 2007; Patrick McCray 2000; Borgman et al. 2016; Galison & Hevly 1992). In many cases, tools are shared and opportunities for observations or data collection are administered by a granting committee. In cases where an instrument is too complex to share beyond a bounded community (for instance with particle physics accelerators or detectors), there are opportunities for contributing scientists to lead sub-units and inquiries of their own. NIH-sponsored research centers feature entire PI-ships associated with data management and synthesis, apart from the management of instrumentation. There are many models to choose from that generate success along several different axes (Balakrishnan et al. 2011). Optimizing for long-term missions requires selecting one that enables career paths through the ranks.

11.2.2 Adopt a Long-Term Structure

Long-lived organizations must choose a form optimized for longevity. A streamlined bureaucratic–hierarchical structure is best suited for this task, despite its shortcomings

(Weber 1968). This is an organization in which jobs and offices (bureaus) are well-defined and scoped, there are opportunities for advancement given jobs well done, people in positions of authority are not coterminous with their position, and communication flows upward while decision-making authority generally flows downward. Bureaucracy in the late twentieth century has become associated with inefficiency due to middle-managerial bloat, and it can encourage stovepiping between units. However, this form is not only efficient, it enables long-term turnover alongside institutional continuity.

Notably, these kinds of structures may be especially effective for remote teaming considerations due to their reliance upon documentation and clarity of hand-offs instead of participation and co-presence (Yang et al. 2022). They also provide opportunities for advancement. A well-trained military, for instance, presents opportunities for newcomers to move up from one level to the next. Someone who starts off as a Lieutenant can eventually become a four-star General, not just because of whom they know or how long they have been around, but through precisely specified achievements. More recently, engineering organizations have embraced a matrix form (Davis & Lawrence 1977), a dual hierarchy which enables teams to coordinate among units and to make decisions close to the problem at hand through cross-functional teaming. The Cassini science team adopted an effective matrix form (Vertesi 2020), organizing participating scientists into both instrument teams and thematic working groups responsible for allocating observations. While matrices can introduce managerial confusion, they enable the strengths of bureaucratic hierarchy alongside a certain degree of integrative cross-coordination.

Whether a hierarchy or a matrix, long-lived mission teams must conceptualize the PI-ship in successive waves, with clear hand-offs between these eras. It is not necessary for a single person to have leadership of the organization or decision-making authority vested in them for the entire duration of all phases. Instead, a mission oriented toward longevity may consider breaking up the PI position into successive waves or phases. There are certain phases that are best associated with conceptualization, construction, maintenance, operations, or data analysis: Such phases can readily be distinguished throughout the lifetime of a mission and may be better suited to one or another individual skillset. Alongside succession, such missions must have clear deputy positions that are shared by several individuals, to train the next PI waiting in the wings for the next phase, and to encourage young people to try out leadership positions and to move up the chain. Instead of appointing a beloved acolyte to be the next PI at random, clear characteristics associated with the job and an outline of responsibilities must be made clear from the outset in order to encourage a diversity of applications and fair consideration of response. Without such clearly defined roles and characteristics, PIs as appraisers will readily fall into the pattern of selecting based on "fit" and existing privilege instead of based on merit and diversity of contribution (Rivera 2015).

11.2.3 Invest in Successful Succession

Teams must practice for successful succession planning. It is inadvisable to have one PI for 25 years covering the entire duration of development, cruise and the primary mission phase, only then to hand over to a newcomer upon retirement. At that phase, such a position may be seen as lesser or unable to give newcomers the skills they need to move up in their careers as well. It also does not give an opportunity for the team to practice passing the torch. With clear structure and clear goals, such a transition can take place effectively and efficiently, with minimal disruption. The prior PI may even join an Advisory Council of sorts for the mission, to ensure that their respected wisdom is still available and accessible for the next generation working on the mission. Such attention to succession and focus away from the singular, heroic PI can make organizations even more resilient in the face of exogenous shocks, including changes in NASA administration, funding paradigms, political shifts, or pandemics (Hannan & Freeman 1984; Haveman et al. 2001; Reinecke 2021). Teams must practice for successful succession planning. It is inadvisable to have one PI for 25 years covering the entire duration of development, cruise and the primary mission phase, only then to hand over to a newcomer upon retirement. At that phase, such a position may be seen as lesser or unable to give newcomers the skills they need to move up in their careers as well. It also does not give an opportunity for the team to practice passing the torch. With clear structure and clear goals, such a transition can take place effectively and efficiently, with minimal disruption. The prior PI may even join an Advisory Council of sorts for the mission, to ensure that their respected wisdom is still available and accessible for the next generation working on the mission. Such attention to succession and focus away from the singular, heroic PI can make organizations even more resilient in the face of exogenous shocks, including changes in NASA administration, funding paradigms, political shifts, or pandemics (Hannan & Freeman 1984; Haveman et al. 2001; Reinecke 2021).

To operate over a long period of time with a solid succession plan in place requires bringing multiple generations of scientists and engineers together in a workplace. This is of increasing concern around the world as the Baby Boom generation works longer, the Silent Generation remains in the workplace, while Generation X, Millennials, and Generation Z together occupy a variety of positions in the workplace (Cekada 2012; Flinchbaugh et al. 2018). These generations grew up in different formative periods and bring diverse approaches and considerations to the workplace for how work ought to be done and how their person must be treated with respect (Twenge 2023). It is advisable that long-term missions invest in some of the training and knowledge acquisition necessary to work constructively, instead of obstructively, across these generational divides. There are many ways to harness the intelligence, ingenuity, experience, and spirit of each generation and what they bring to the table.

Beyond the PI-level, teams can use this approach to mentorship, deputy positions, and intergenerational mobility to structure their science team across its components. This can include those with responsibilities for data management, cleaning, and security. It can include cross-functional teams such as interdisciplinary working groups that bring people together from multiple mission stakeholder groups. For instance, the Europa Clipper mission features thematic working groups with co-chairs that rotate every two to three years. These co-chairships pair a senior scientist with a mid-career or early-career scientist. Placing these scientists into collaborative leadership roles enables them to transmit information and knowledge and gives young scientists the opportunity to try out leadership positions and to show their worth. It also provides externally recognized criteria of leadership to enable their advancement in their careers beyond the mission, and it exposes more senior scientists to changing scientific and social priorities that can better inform their practice. As teams become more diverse at the junior level, this practice enables the broader science team to benefit from asking a more diverse range of questions accordingly (Wullum Nielsen et al. 2018). In sum, thinking multi-generationally in advance can set a mission up for long-term success by structuring appropriately and investing in newcomers. The next step is to consider how to facilitate knowledge transfer through team cultural practices.

11.3 Transmit Knowledge Effectively to Newcomers

There is a lot to learn and a lot to know on spacecraft mission teams. Each instrument has its quirks. Calibration is never straightforward. Decisions taken early on have ripple effects. Knowledge transfer and knowledge management are therefore essential characteristics of longevity-oriented organizations. An entire field of knowledge management has sprung up in order to encourage the correct documentation and findability of the materials necessary to work alongside complex socio-technical systems (Linde 2009).

Knowledge management impacts questions of repair and maintenance, ongoing decision-making about scientific targeting, and problem-solving as the mission progresses. It can also arise in accident investigation board reports as critical omissions, silences, or gaps in institutional knowledge that imperil spacecraft success, especially when there is a lot of team turnover. It is one thing for a group working with a short-lived craft to maintain knowledge of precisely which operational modes matter, how a piece of hardware got made, which software patch workarounds are already installed, and the history of safe modes and their exit. But when the technical errors can occur years, decades, or even generations apart, such knowledge can easily be lost.

In addition, lifetime issues are precisely those that surface in engineering projects that are around long enough to produce novelty through aging. Anomalies of this kind, while rare, can surface throughout a mission and are harder to diagnose without the maintenance of oral history in addition to good "documentation." Engineers who inherit a system without deep knowledge of its past may misdiagnose a problem or upload commands that contradict prior upgrades or lessons learned. Fortunately, there are ways to resolve these problems by applying sociological insights.

11.3.1 Embrace Tacit Knowledge

In bureaucratic organizations, as in large software projects, documentation is essential (Weber 1968; Cohn 2019). Documentation enables newcomers to move into a position and acquire (to a large extent) the necessary context for work. Still, evidence shows that documentation is not enough. Unlike Motor Vehicle Commission offices (a classic bureaucratic environment in which documentation serves a recording purpose and in which new employees can get quickly up to speed), spacecraft teams are "high-context" organizations (Hall 1990) wherein hands-on experience, mentorship, storytelling, and culture are essential for information transmission over the long term.

Tacit knowledge is the sociological term for the kind of knowledge that cannot be put into writing but which requires hands-on experience, physical demonstration, and extended mentorship to acquire. Classic examples are teeth-brushing or riding a bicycle (Polanyi, 1996; Fountain 2021). High-context and embodied tasks involve a considerable degree of tacit knowledge. Sociologist of science Harry Collins noted the importance of tacit knowledge in the sciences when he observed lab technicians struggling to replicate another lab's experiment (Collins 1985). An experimentalist from the initial lab had to visit and demonstrate how they had tweaked the instrument just so, providing the high-context knowledge necessary to get the instrument working. These details could not have been written up in any documentation, no matter how thorough. Tacit knowledge has since been shown to be an essential part of scientific and engineering training. It is transmitted through degree programs, internships, peer-to-peer communication, and mentorship experiences. Particle accelerator experiments especially emphasize this crucial aspect of knowledge transfer: like spacecraft, scientists, and technicians work together over a long period of time with the same instrument and facility (Doing 2004; Knorr Cetina 1999; Traweek 1988). The prevalence of high-context, tacit instrumental knowledge that can only be transmitted through interactive learning highlights the priority role of mentorship and story-telling for any long-lived scientific team (Linde 2009).

11.3.2 Structure Mentorship

Mentorship is already an important element of planetary science, but often occurs in an ad hoc, unstructured way. Every spacecraft team learns the quirks of their tools, from communicating with a Voyager probe that is described as "a little bit deaf," to working creatively with Galileo's downlink limitations, to driving Spirit backwards to accommodate its broken wheel. With several months of exposure and work alongside more seasoned members, even newcomers can develop approximately the same feeling for the spacecraft as those who have worked alongside it for a long time (Vertesi 2015). Formalizing this otherwise informal knowledge transfer is an effective way to maintain a functioning instrument across generations.

Mentorship can also promote diversity within teams. A strong mentor from a majority group can support a young minority in a subordinate network to make their way up in an organization (Burt 1998; McDonald 2011). Additionally, a strong mentor from a minority group can aid in navigating and defining productive strategies for culturally specific issues that arise during their term on a project. While informal interactions are important for the success of any organization (Meyer & Rowan 1977), when left entirely to informal mechanisms classic status hierarchies will emerge (Freeman 1992; McDonald 2011; Rivera 2015). Mentorship programs must therefore be formally structured, instructed, and guided to guard against reproducing inequities.

Mentorship ties typically relay both information and opportunities for advancement (McDonald 2011). This exchange goes both ways—"reverse mentorship" from junior to senior can be just as important for bringing fresh ideas into the field (Jordan & Sorell 2019). If planetary science missions can move toward a culture in which teams expect the keys to an instrument to be handed over to the next generation, projects can then use mentorship constructively. This includes allowing newcomers to play important roles, and keeping knowledge of instrumentation alive beyond the lifetimes of singular individuals on the team.

11.3.3 Facilitate Culture

Mentorship and tacit knowledge transmission form part of a team's culture. Culture is not something that is good or bad, or a hegemonic national identity: It is the communication substrate for teams. In flat hierarchies, culture is essential for the communication of information and generating a sense of membership (Kunda 2006). Small groups like science teams enable their collaborations through shared cultural ways of talking, telling stories, making decisions, and solving problems (Vertesi 2020). In the sciences, culture even enables discovery (Knorr Cetina 1999; Vertesi 2020).

Culture is often emergent and can therefore seem slippery and difficult to control. However, there are certain features that enable long-term efforts instead of hampering them. For example, rituals and shared objects of attention are powerful ways to bring teams together across time and distance (Collins 2004; Vertesi 2015). Daily or weekly planning meetings or Project Science team meetings are frequently ritualized, especially if they have the same cadence or shared focus. The Mars Exploration Rover team, for example, nurtured a powerful relationship with their robots through their shared meetings and ritualized interactions. Scientists and engineers alike even impersonated the rovers with specialized movements that brought the robots' experiences into their own bodies on Earth: a form of reverse anthropomorphism called technomorphism (Vertesi 2015). This technomorphism relayed deep, tacit, embodied knowledge about the robots and their capabilities to all members of the team, whether newcomers or old hands. In another ritual that benefitted a science team's collaboration, the Europa Clipper team deployed purpose-built monoliths and prizes that referenced the Moon's role in *2001: A Space Odyssey*. As shared cultural symbols, they provided humor as well as a mark of membership with the team shared by multiple generations of scientists.

Story-telling and intergenerational care are essential to the Interstellar Probe team even in its development phase (McNutt et al. 2019). With a planned 50-year development and cruise, those making decisions during planning will never witness the craft reach its destination. The team invested in videos and oral histories even before the mission received approval, to capture the voices and the ideas in the room that will ultimately shape the mission. They also developed shared visualizations to demonstrate the generational handoff alongside the spacecraft's trajectory, or pictures that envision the probe as a prized heirloom to hand down to the next generation of scientists. Collaborations in social science also produce insights and documentation that can further support future transformation and support. An intergenerational culture focused on knowledge transmission, mentorship, and a shared sense of belonging, can contribute to a mission's long-term strategic imperatives even with a changing roster of personnel over time.

11.4 Anticipate Aging Systems

Spacecraft teams are well accustomed to working with hardware that is significantly out of date. For hardware engineers or operators, it can even be a point of pride (although, notably, this is not the case for software engineers). Even as long-lived spacecraft are designed to last beyond their primary missions, hardware in space decays, software on earth upgrades, and new operating systems are brought online. While system resilience is a core concern in spacecraft technology development, the sociotechnical nature of questions of maintenance and repair demands special attention in long-lived projects (Ribes & Jackson 2013). Creative solutions to problems of breakdown over a distance are part of the thrill of operating spacecraft in the outer solar system; it also connects the maintenance interests of engineers across deep space and deep ocean contexts (Steinhardt 2016; Cohn 2017). In these cases, maintenance and repair at-distance must be thoughtful, adaptable, and sustainable.

Insights from sociotechnical studies of infrastructure, especially the recent turn toward repair, breakdown, and maintenance, are therefore of particular interest (Rosner & Ames 2014; Henke & Sims 2020; Steinhardt 2016). Such insights reveal special challenges for long-lived missions. They must contend with the fact that, while the science continues to be cutting edge long into the life of the mission, the engineering systems that support them may no longer inspire the same level of enthusiasm among young engineering talent. Certain aged components, especially in the software realm, will not support career mobility among a next generation. But with the shifting climate favoring sustainability, there is an opportunity for long-lived missions to recapture imagination around long-term futures of computation in times where an appreciation for design within limits and finite resources is increasing.

11.4.1 Lifetime Issues in Hardware and Software

Technological systems degrade over time, and missions must be prepared to deal with hardware and software failures. Long-lived spacecraft face many "lifetime issues" where hardware wears out over time (Cohn 2017). Engineers who build spacecraft imbue them with system redundancy and backup systems to accommodate failure, like low gain antennas and backup memory systems, such that a single problem will not entirely cripple the spacecraft. The New Horizons mission team reportedly bought up as many computers from the auction site eBay as possible in order to provide redundancy for their aging ground system. If one computer went down, an identical aged machine could be swapped in to take its place.

Yet this is only way that technology ages. Early literature on large scale technical systems identified an aspect of big systems called a "reverse salient" (Hughes 1999). Named after a formation observed in a military advance, when a piece of the army falls behind and the rest surges ahead, lagging system elements must be kept running while surrounding tools continually march forward to new file formats, programming languages, and advancements in artificial intelligence (Ribes & Finholt 2007; Cohn 2016). Long-lived missions will inevitably be faced with reverse salients in their hardware and software systems. Due to the nature of working with remote irretrievable hardware, resolving failures and reverse salients through replacement or upgrades is often impossible. Instead, missions must find ways to manage aging systems and keep them operational *within an evolving environment*. This requires continual attention to interfaces and specific techniques for working alongside aging systems.

Managing the reverse salient at tremendous distance suggests that multi-generational spacecraft teams would do well to invest in technologies of *emulation*. Instead of relying on online marketplaces to find 20-year old computers, emulators recreate older operational and run-time environments so that legacy software can be run on contemporary equipment as virtual machines. Emulation as a Service (EaaS) systems can assist in maintaining access to aging systems.[1] File formats that were once popular and are no longer usable in current software will require patches, emulators, API integrators, or other forms of upgrade to continually enable their access. Note that paradigmatic shifts in how computational systems are built, such as from procedural to object-oriented paradigms, may render emulation untenable. So, this must be one among many strategies.

Operations teams in long-lived missions frequently demonstrate significant innovation and creativity in figuring out how to keep aging systems running (Cohn 2017). Engineers implement various workarounds and kluges to ensure spacecraft operations can continue. These improvisatory fixes carry risk in the long term through the accrual of "technical debt." Technical debt is a powerful metaphor to understand how software built towards deadlines or other constraints compromises on decisions that will support longevity, both in terms of evolvability and maintainability (Yang et al. 2023). When an engineer is forced to use a sub-optimal solution (like a workaround), that decision accrues a kind of debt, which has both principal and interest. The principal could be repaid by "refactoring the quick and dirty design to a better solution" (Kruchten et al. 2013). The "interest" on the debt is the extra work that will result in the future to manage this sub-optimal solution. While these compromises in commercial software are often due to market deadlines, they can also arise due to the urgent software or fiscal needs leading up to launch, deferring the design and development of software needed for later phases to be developed "along the way." Technical debt in complex systems is considered inevitable in many ways, especially over the long term, since changes in requirements can arise from new functional needs as well as policy or political changes (see below).

---

[1] For instance, https://www.softwarepreservationnetwork.org/ (Accessed June 15, 2023).

Combined with the concept of the reverse salient, technical debt can powerfully inform our understanding of long-lived systems and their recurring issues. On the one hand, system components that are dependent on obsolete software languages or formats will require keeping older software interoperable with newer systems. On the other hand, managing software change over the long durée will encounter increasing forms of such technical debt. The longer a mission goes on, the more software it is likely to accrue in the form of workarounds to continue pushing maintainability forward. Such software accrual is often not possible to manage entirely within controlled review processes, as some of it adheres more to quality and efficiency of on-Earth work practices. While the costs of new system development are obviously lower during operations, it is still necessary to pay off the accrued technical debt and the compounding interest that builds up in the form of systems that are increasingly fragile and difficult to maintain.

11.4.2 Sustain Human Infrastructure

Long-lived missions must attend to longevity of the "human infrastructure" that supports their hardware and software (Lee et al. 2006). As Cohn observes, "lifetime issues surface … the ways that both development and operations are bound up in the co-histories of people and objects over time" (Cohn 2017). The same succession planning necessary for science teams is also important for operations and systems engineering teams. If new generations of engineers lack the necessary skills and local knowledge to maintain older languages and increasingly complex and aging systems, missions will find that their systems fail without appropriate training, mentorship, and structures to support intergenerational mobility. For example, the Voyager team still transmits code to the spacecraft in FORTRAN'77, but young engineers and computer scientists are no longer trained in this language so options for new recruits are thin. Engineers also need to understand the mission and system history, as system anomalies may arise not only from new problems but also when something already known about spacecraft or its operational environment is forgotten (Cohn 2017). Missions must consider the career trajectories of its engineers, as the career mobility of those working to maintain increasingly obsolescent components of the technical infrastructure will also become increasingly limited over time.

Later phases of an infrastructure's lifetime are never subject to the same levels of funding and support as the initial building phase–yet ongoing support and maintenance are no less essential to the continued successful operation of the spacecraft and conduct of science. Teams should engage emulation experts and library and information scientists early on in system development. Such groups can bring "long now" thinking to the table, considering proposed materials to collect prior to acquisition or ways to create operational bubbles around reverse salients. The earlier such groups are involved, the more strategic decisions can be made about which kinds of tools and techniques are necessary for continued mission success. Planning for longevity should also include considerations of the costs of migrating to new software systems when emulation to extend life of software is no longer possible. Migrating a system that has been ongoing for a decade can come with unforeseen costs as sedimented infrastructures become increasingly invisible (Leigh Star 1999). Changes in software infrastructures in the later phases of missions must be allocated substantial resources rather than being viewed as sunk costs.

11.4.3 Accommodate Data (R)evolutions

Data maintenance and access is well studied among large scale scientific collaborations in neuroscience, geology, astronomy, and planetary science, demonstrating the challenges of long-term storage, processing, and access. Social scientists have identified several issues regarding the inevitable changes in the technical environments in space and on earth in which data is stored, transferred, and accessed. One problem concerns bringing multiple data sets together across different communities, or changes to data sharing cultures over time (Birnholtz & Bietz 2003; Bietz & Lee 2009; Vertesi 2020; Borgman et al. 2016). Another concerns agreement upon the structures and systems that are necessary for storage and access that can appease all involved groups (Bietz & Lee 2009; Ribes & Finholt 2007). As the data repository grows, continual challenges arise with respect to funding, open access, and trust.

Collecting usable data over the long term requires acknowledging that change is inevitable: changes to the data itself, to questions asked, repositories, file formats, metadata, research communities, access assumptions, and more. Studies of environmental science, oceanography, and AIDS research indicate how data collected early on is often out-of-phase with data collected later, and how instrumentation is frequently repurposed to address new questions (Ribes & Jackson 2013; Ribes & Polk 2015; Steinhardt 2016). Initial categorizations—used for metadata categories, to establish domains for inquiry, and render data searchable or visualizable—are subject to change over the long term (Ribes & Polk 2015). This is true in space as well. Cassini's discovery of the plumes at Enceladus and its investigation of Saturn's aurora necessitated changes in the use of its instrument suite accordingly (Vertesi 2020). End-of-mission calibrations of Cassini's Imaging Science Subsystem not only account for the instrument's decay over time but also provide new calibration values that "differ considerable from those previously reported" (Knowles et al. 2020). MER's Mini-TES suffered dust contamination and degrading throughout its life, such that the same calibration tools and mechanisms could not be used across the lifetime of the data set. The team even continued to collect Mini-TES data when the instrument was damaged in expectation that a later calibration fix might render the data usable. Evolving calibration mechanisms, including those developed in the wake of instrument fault, must also be archived.

Planetary scientists are used to thinking about data questions in the context of requirements levied by the NASA Planetary Data System: What must we deliver such that the community can use our data? Which resources will become available through a Data Analysis Program to support investigation? Best practices indicate, however, that these questions are best considered during the instrument development phase. Among PI-led missions, New Horizons had synergistic instrumentation that developed joint data products, while MAVEN used overlap among spectral bands or regional coverage among its instruments for cross-calibration purposes. On strategic missions, the Europa Clipper team established working relations among instrument PI's in Phase A to ensure that each instrument's data set would eventually synergize with the rest of the payload suite. Because Level One science required multi-instrumental investigations of Europa, instruments developed synergistic data collection and visualization support from the outset of development.

Data are themselves sociotechnical artifacts and can play a role in the cultural issues outlined above. Successfully interpreting data often requires high context, tacit knowledge about the design of the instrument, how the instrument has aged, and how it is operated. Giving a young scientist responsibility for a data set (e.g., validating and calibrating data before release) can help them to develop skills and create opportunities for training and mentorship (Birnholtz & Bietz 2003). Data circulation is also important to mission structure and culture. Both broader scientific disciplines and local mission teams develop "data cultures"—sets of norms around appropriate uses of data, authorship, acknowledgement, and ownership (Poirier & Costelloe-Kuehn 2019; Vertesi 2020).

Data calibration, metadata tags, and quality control are necessary to ensure that data streams remain consistently usable despite aging systems (Höltgen & Williamson 2023). These processes are also sociotechnical. Alongside knowledge of technical concerns, the ability to identify subtle deviations, to ascertain categories, or distinguish signal from noise is a social practice and a learned technique. For instance, calibration processes frequently differ from instrument to instrument, with standards rooted in different disciplinary or community commitments. Since data management frequently involves collaboration across multiple subdivisions or subprojects, communication practices must be well defined across instrumentation to avoid missing, mischaracterizing, or misinterpreting signals.

In sum, developing a usable long-term research infrastructure means accommodating the fact that the systems involved and the data collected will be subject to change. Ensuring appropriate funding for system maintenance, repair, upgrades, and flexibility for continual data analysis programs that re-investigates data categorization is essential for long-lived craft. Teams should propose techniques to produce long-term plans for data stewardship that include clear expectations shared across the instrument suite (Bietz & Lee 2009); that share data in ways consistent with the evolving operational and scientific considerations of the mission (Vertesi & Dourish 2011); that enable flexible categories within and across data sets Ribes & Polk 2015; and that make data work and maintenance work a valued part of the funding for collaboration from the outset (Lee et al. 2006). Confronting these questions early in mission development will aid tremendously in the successful resolution of these well-known maintenance challenges over the long term.

11.5 Sustain Support

11.5.1 The Time Inconsistency Problem in Planetary Science

Assuming that a spacecraft operates as intended, the biggest risk to long-term space science missions is rarely technical: its political. Support for space science is historically an uncertain endeavor given downward budget pressure and precarious electorate support that challenges long-lived commitments. Changes in administrations, agencies, or Congress can also radically reorient science policy and funding priorities. Congressional cycles typically require all missions to continually renew or extend their funding on a short-term basis irrespective of their technical or scientific time horizons (Broniatowski & Weigel 2008). Analogous to the internal problem of succession, long-duration missions must also sustain the support of successive generations of administrators, politicians, and, ultimately, the public at large to survive.

Spacecraft teams observe a special version of the "time inconsistency problem," noted in public policy and in macroeconomics, wherein promises made under one set of conditions may later change in response to new parameters and present downstream effects on interdependent projects and economies (Lynch & Zauberman 2006). Certainly, to operate over the very long term requires matching fiscal commitments from funding agencies or other resource-granting organizations. Yet mission leaders know that a commitment from external funders today is no guarantee of a commitment tomorrow. Additionally, innovative spacecraft systems rarely follow a linear function for funding requests throughout their development cycle. The need to return to Congress yearly for funds and re-encounter this time inconsistency problem when the life-cycle of development requires more investment produces a situation that we call "funding asynchrony" (Reinecke 2021). In this scenario, the money necessary for key stages of spacecraft development can readily be withheld or cut back in response to the political climate of the moment. This results in drastic impacts upon teams that have long-term effects.

Importantly, the political realities of the time inconsistency problem and funding asynchrony combine to disincentivize true cost savings for mission teams, undermining a team's ability to continually request necessary resources. For example, time-inconsistency incentivizes decision-making that produces short-term gains with long-term pains for mission teams. Cassini's loss of their scan platform during a recession in the 1990s kept the mission alive but produced an operationally complex, cumbersome, and ultimately very expensive system. As another example, to pass through the short-term gates of congressional fiscal oversight, NASA has repeatedly developed systems aimed at encouraging more accurate funding requests, while raising the bar for political acceptability of large-scale projects. These measures incentivize inaccuracies in planning or reporting, over-charging on subcontracts, even cutting corners on technical safety—to name only a few observed downstream effects.

Importantly, then, missions face political requirements and fiscal requirements for support; in the current environment, these requirements are tightly coupled, yet often produce counter-incentives and operate at odds to each other. Our studies have revealed local strategies used by missions to enable political stability for their projects. They have also revealed the missing pieces in budgeting that lead to frequent overruns due to these counter-incentives, despite best intentions oriented toward cost savings.

11.5.2 Sustainable Support

To counteract time inconsistency and funding asynchrony and sustain long-term sciences, we have observed six common practices among teams that enable continual support. This support generates legitimacy—a generalized perception that an entity's actions are desirable, proper, or appropriate—which is essential to survival during times of organizational failure, economic uncertainty, or political reversal. Note that these strategies enable sustainability but are not necessarily cost effective, for reasons we describe below.

First, missions must engage with other external audiences like the wider space science community or the public, whose support bestows legitimacy and funding upon the project. A strong endorsement from the Space Science Board, letters from the Planetary Society, or priority in a Decadal Survey can help a mission avoid cancellation. Similarly, arresting images from a physics mission's add-on camera (Hansen et al. 2017) or from a costly space telescope can smooth many tensions. Stakeholder engagement, however, requires clear metrics that are easy to reach in both the short and long term. The Cassini extended-extended-mission, for instance, faced its first evaluation immediately after the 2009 market crash, when it was only 1 year into its proposed plan to detect seasonal variation over 7 years. Despite a healthy spacecraft and a robust plan for observations over the coming years, this initial check-up lacked easily visible metrics for stakeholders to gage productivity.

Second, funding agencies can extend time horizons and review cycles to match the tempo of scientific activity. The NSF's Long-Term Ecological Research Network funds investigators on renewable 6-year grants (the NSF's longest such funding duration) and reviews the entire research network only once a decade (Willig & Walker 2016). A slower funding pace avoids disrupting long-term scientific operations needed to collect data over extended temporal and spatial scales (Vanderbilt & Gaiser 2017). Third, projects can seek diversified funding sources to prevent an over-reliance on a sole funding source. For example, in the world of high-energy particle accelerators, international participation and public–private partnerships are now the norm to get new projects off the ground (Westfall 2010). NASA often engages with international partners such as ESA, wherein costs for instrumentation or hardware are provided by another space agency. A Memorandum of Understanding typically guides such partnerships and can provide some shelter from political shifts, even if it may also increase the collaborative costs of working effectively across borders and cultures.

Fourth, projects can also seek to secure a legal mandate that insulates their support from future reprioritization. The US's desire to maintain scientific diplomatic ties and honor international agreements with its European partners helped to save Cassini after the CRAF cancellation in the early 1990s. Relatedly, the Europa Clipper mission was upheld in multiple Decadal surveys and written into NASA appropriation bills, providing a measure of political stability and funding to the project over the nascent and vulnerable period of its development. Fifth, a champion in Congress like Senator Barbara Mikulski or Congressman John Culbertson can ensure that missions such as Hubble, *JWST*, or Europa Clipper make it to the launch pad. Note that these bedfellows may come with additional constraints. For instance, the Congressional legislation that enabled Clipper's continuation under austerity required holding open three different launch farings in the design process, effectively increasing development cost.

11.5.3 The People Equation

Effective ways to counter the time inconsistency problem are not always cost-effective, despite apparent gains. Small wonder, then, that large-scale missions repeatedly eclipse their budgets as they seek support over the long term (Large Mission Study Report 2020). Sociotechnical thinking helps to redress this problem. After all, historical budgetary data indicates that the vast majority—between two-thirds and four-fifths—of money spent on spacecraft mission teams goes to *people*, not hardware. Akin to the "Rocket Equation" which determines the relationship between spacecraft mass and propulsion necessary to escape orbit, we suggest there is also a "People Equation" which indicates a similarly exponential relationship between spacecraft personnel and cost (Reinecke & Vertesi 2022).

The People Equation demonstrates an exponentially increasing labor cost relative to the sociotechnical complexity of a mission. But this is not how missions are budgeted. Budgets are instead premised upon component parts, each stove-piped for traceability and institutional accountability. Each unit appears to have its own cost and a project office is responsible for reserves in case of coordination costs. Yet the more institutionally complex or tightly coupled a mission becomes, the more additional personnel are required to smooth interfaces, resolve emergent problems, track requirements, test hardware, and conduct reviews. These personnel are not factored into modeling and predictive tools, although they are visible in forensic investigation.

A focus on the connective glue of people instead of upon proposed technological parts offers a different approach to costing missions. It also demonstrates how measures that appear to improve a mission's support over time can increase costs in the long term. Consider how science teams search for partnerships in which instrumentation or payload elements can be provided at a discount to manage costs and keep proposals low. This includes working with prior flight spares, contracting a private sector partner, collaborating with a foreign agency, or adhering to fixed price contracts. From the perspective of the People Equation, each of these options comes with additional interface personnel, risk, and cost risk. Contractors pass costs along to their shareholders or other purchasers, but their firm fixed price contracts do not allow for flexibility to respond to ripple effects of hardware changes to the evolving craft. Adding too many multi-lateral and multi-party relationships can severely overconstrain the tradespace as a mission takes shape. Flight spare payload elements provided at discount are inflexible and do not assist cost estimators in developing accurate models for future flight. Each of these cases requires human investment in communication, oversight, and management.

While the People Equation shows an increase in labor cost associated with complexity, investment in people and human relationships can make a significant difference for resolving these issues and de-escalating cost. For instance, international partnerships that invest in a strong foundation of friendship and multi-lateralism, as in the Cassini-Huygens mission, build on relationships for success; fractures in communication and coordination can jeopardize timely launch, as in Mars InSight. Intercultural communication can smooth interaction and risk penetration in multi-institutional teams too.

In adopting one or more of the above-mentioned strategies oriented at long-term support, long-lived missions must pay attention to the People Equation up front. Teams should look out for the following pitfalls: If the projected cost of FTE's in a budget is not equivalent to *at least* two-thirds of a mission's proposed cost, if the budget is focused on hardware components and not the interfaces between them, or if the organization has become unwieldy and complex due to too many partners. In such cases, the mission will undoubtedly run over projected cost, jeopardizing continued support in an asynchronous funding environment and exacerbating the same political instability that they had hoped to resolve.

11.6 Invest in Meaningful Partnerships

11.6.1 Diversity and Inclusion

The planetary sciences have invested in the diversification of their ranks to a limited degree. Studies show that current levels of minority success in the field are largely limited to white women, who comprise under 30% of active science teams (Rathbun et al. 2018). Other minorities—including racial and sexual minorities, neurodiverse or differently abled individuals—are at extremely low levels of representation. Current studies indicate that the climate in planetary science does not support these minorities' advancement in the field (Rathbun & Rivera-Valentín 2021).

Planetary science missions must improve these demographics. A more diverse team is better able to spot and solve problems as well as to bring creativity and innovation to the table (Nielsen et al. 2018). Diversity is also necessary to ensure continued support from an increasingly diverse demographic of stakeholders. The American taxpayer, by 2050, will not be predominantly white, male, or "neuro-typical," and will expect its publicly funded scientists to follow their lead (Rivera-Valentin et al. 2020; Rathbun 2017).

Issues of diversity and retention are often discussed as a problem of a "leaky pipeline," where even when people from diverse backgrounds are better represented in a scientific discipline, their ability to complete degrees and be promoted to high levels remains low. The institutions that train planetary scientists and space engineers have a long way to go in terms of creating equitable environments for training (Prescod-Weinstein 2021; Riley et al. 2014). One way that mission teams can support this work is by funding fellowships and internships that expose young scientists and engineers to diversified teams. Another way to craft intentional partnerships is "thick bridges" with minority-serving institutions: Teams must welcome participants from minority institutions as full participants, set real and complex problems for solving while providing active mentorship, pay for internships (with an appreciation that brilliant scholars subject to socioeconomic disadvantage cannot work or travel for free), and enable reverse mentorship opportunities such that team members can learn from newcomers. Reverse mentorship is important, but teams cannot expect newcomers from diverse backgrounds to take up the work of diversification, inclusion, and training (Shim 2021). Funded professional training can better assist in this regard. Last, activism is increasingly named as a part and parcel of scientific experience: as both a coping mechanism and a strategy for better stewarding diverse scientists across roles, particularly early career scientists. Often this activism is in pursuit of safety for minoritized participants. This trend tells us that new infrastructures must include more assurances of a sustainable life in science including grievance policies that take seriously harm (which falls disproportionately on minoritized persons in science) and community-building opportunities to allow for culturally sensitive support to promulgate.

Thinking sociotechnically, long-lived missions can address the problem of diversity and support marginalized pathways through the pipeline by contending with how technical systems evolve steadily but unevenly over the time. Recalling reverse salience and technical debt, we already know that certain technical components will obsolesce more readily than others; still others will be less amenable to upgrade, migration, or emulation. The individuals working to maintain these systems will also be hampered in their career progress if they are not also given opportunities to upskill to current systems (Mason et al. 2023). Technical obsolescence and specialization can also produce differential impacts on marginalized groups. As the scientific recognition and compensation for critical engineering work diminishes due to obsolescence, this work may become less attractive, leading to male attrition and the feminization or racialization of certain forms of labor (Vertesi 2020; Ensmenger 2010; Wingfield & Alston 2014). The result is a pattern where minoritized or marginalized members of the space science community disproportionately fill longer term maintenance roles, which are considered less important or skilled than other roles at the forefront of discovery.

At the same time, long-term mission opportunities can also be very attractive to typically disadvantaged groups, particularly those with familial obligations, or for whom maintaining advancement within the profession entails higher levels of social and cognitive load. For these individuals, a long-term mission produces the sustainable life in science that runs in contrast to the short-term precarity and churn. As such, providing options for longer-term, more stable employment can be a boon to increasing diversity in space science missions. This is only the case, however, if knowledge transmission, succession, and mentoring are closely attuned to the differential impacts of obsolescence on opportunities for career advancement and security among marginalized groups.

11.6.2 Encountering Differences

Diversity is not just a question of adding a few minorities to a team: Inclusion is necessary for retention (Ahmed 2012). Team members must be fully integrated as participating, active scientists, lest they be subject to the considerable problems of tokenism (Shim 2021; Zimmer 1988). Among science and engineering teams, the benefits of diversity are only apparent when minority members are fully integrated into the organization (Smith-Doerr et al. 2017). Training and support is necessary to develop an environment that is inclusive and equitable, including management of unconscious bias, bystander intervention, and intercultural communication in diverse workplaces. Such training is not simply a mechanism for overcoming visible differences. Spacecraft teams must form working partnerships across invisible divides too, including institutional ones. Investment in these forms of training can help avoid the problems associated with miscommunication over remote tools and distances. Without such training, team members are likely to make attribution errors, in which a minority or distant participant is unjustly considered to be less competent than local, trusted team members, with negative effects on their scientific or technical reputations (Yang et al. 2022).

These forms of social training are complemented by proactively diversifying the intellectual training of the field to include voices and discoveries from under-represented groups. Countless studies in the past decade across education, gender, and science and technology studies have reaffirmed that inclusion and retention are only bolstered by representation in the academic canon. Promoting reading groups that emphasize cultural and historical underpinnings of scientific goals have been shown to elevate group cohesion and shared understanding. These jointly cultural and intellectual practices provide opportunities for finding common ground as well as identifying cross-cultural challenges that can then be discussed and problematized outside of scenarios in which science work is in action.

Prior studies of spacecraft collaborations indicate that even at the institutional level, cultural differences exist that challenge effective communication. These cultures, especially in engineering organizations, concern the ways in which risk is identified, managed, tracked, constrained, and resolved (Vaughan 1996). When two organizations with divergent risk cultures attempt to collaborate, each group's attempt to assuage the other can instead inflame concern; or, at worst, miscommunications across divides can have disastrous consequences. Effective communication across divides, with recognition of the differences between the groups' articulation of problems, can enable better understanding, more trust, productive conflict, and deeper penetration into those persistent scientific and technical issues that require resolution.

11.6.3 An Aging Workforce

No amount of long-term thinking or organizational overhead can ensure the success of a long-term mission unless the project can also provide adequate career opportunities for its team members over the life course. These include not only opportunities for advancement, training, or doing exciting science, but also recognition of the need for career flexibility and sustainability as team members transition between different critical life points (Tomlinson et al. 2018). This is especially essential as participation varies from full-time personnel, to a regular 10%–20% of a person's time, to a more formal institute akin to the Hubble's Space Telescope Science Institute. Teams can consider how best to match evolving participation levels with changes in personnel over time, for instance as individuals transition from newcomer, to participant, to leader, and eventually, to mentor. Minus these opportunities for advancement, or if these opportunities are disproportionately hoarded, potential team members in the pipeline may be unwilling to join the project. This can impact the diversity of the team over the long term, as described above.

Compounding these issues is the nature of scientific work on long-duration space science missions. From a staffing perspective, there are real challenges in attracting people to a mission in which expected payoffs are uncertain, emerge slowly, or require enormous personal sacrifices. Funding realities can preclude substantial numbers of full-time employees dedicated to a single mission over long periods. In all likelihood, long-lived spaceflight missions will not be the primary focus for most investigators or engineers for much of their careers. As such, assembling a viable career will depend on the wider set of opportunities in space science that may be adjacent or complementary to long-duration missions.

Rather than force human careers to adapt to the cadence of robotic space exploration, project leaders might rethink careers through the lens of the life course. STEM (science, technology, engineering and medicine) jobs are notoriously difficult to combine with family life and these difficulties are exacerbated as one moves up the hierarchy of skill and authority in science teams (Tomlinson et al. 2018). Retaining a diverse and competent scientific workforce over the long term likely requires corresponding flexibility to the life transitions all team members experience as they age.

11.7 Conclusion

Any planned mission to the outer solar system should benefit from the experience of prior missions and sister sciences to plan for its future and to integrate sociotechnical thinking into its mission organization, structure, culture, and strategic planning. Whether considering an optimal structure for succession, the integration of minority team members, keeping up hardware and software, or maintaining stakeholder support, a sociotechnical approach offers new insights and opportunities for success. Even in space, our robots do not operate in a vaccuum: Technologies do not work without the people who bring them to life and interpret their discoveries. Considering the people alongside the machineries of exploration and the scientific questions involved can enable long-term mission success. We offer these insights for implementation from the earliest stages of mission development through its successful completion.

References

National Aeronautics and Space Administration, 2020, (Washington, DC: NASA) Large Mission Study Report

Ahmed, S. 2012, On Being Included: Racism and Diversity in Institutional Life (Durham: Duke Univ. Press)

Balakrishnan, A. D., Kiesler, S., Cummings, J. N., & Zadeh, R. 2011, Proc. CSCW 2011, (New York: ACM) 523

Bietz, M. J., & Lee, C. P. 2009, Proc. ECSCW, (London: Springer) 243

Birnholtz, J. P., & Bietz, M. J. 2003, Proc. GROUP'03, (New York: ACM) 339

Borgman, C. L., Darch, P. T., Sands, A. S., & Golshan, M. S. 2016, Proc. of the 79th ASIS&T Annual Meeting Vol. 53, (New York: ACM) 00055

Broniatowski, D. A., & Weigel, A. L. 2008, SpPol, 24, 148

Burt, R. S. 1998, RS, 10, 5

Cekada, T. L. 2012, PS, 57, 40

Cohn, M. 2016, Proc. CSCW 2016, (New York: ACM) 1509

Cohn, M. 2017, Continent, 6, 4

Cohn, M. L. 2019, DigitalSTS: A Field Guide for Science & Technology Studies, ed. J. Vertesi, & D. Ribes (Princeton, NJ: Princeton Univ. Press) 423

Collins, H. M. 1985, Changing Order: Replication and Induction in Scientific Practice (London: Sage)

Collins, R. 2004, Interaction Ritual Chains (Princeton, NJ: Princeton Univ. Press)

Conway, E. M. 2015, Exploration and Engineering: The Jet Propulsion Laboratory and the Quest for Mars (Baltimore, MD: Johns Hopkins Univ. Press)

Cooke, N. J., & Hilton, M. L.US National Research Council 2015, Enhancing the Effectiveness of Team Science, (Washington, DC: The National Academies Press)

Davis, S. M., & Lawrence, P. R. 1977, Matrix (Reading, MA: Addison-Wesley)

Doing, P. 2004, SSS, 34, 2004

Ensmenger, N. 2010, The Computer Boys Take Over: Computers, Programmers, and the Politics of Technical Expertise (Cambridge, MA: MIT Press)

Flinchbaugh, C., Valenzuela, M. A., & Li, P. 2018, JMO, 24, 517

Fountain, G. H., Clay, S., Whitley, S., et al. 2020, AGUFM, 2020, SH017

Fountain, G. H. 2021, Challenges & Strategies for Supporting a Very Long Duration Mission. *AIAA Scitech 2021 Forum.* (Reston, VA: AIAA)

Fox, G., Salazar, R., Habib-Agahi, H., & Dubos, G. F. 2013, IEEE Aerospace Conference, (Piscataway, NJ: IEEE) 1

Freeman, J. 1992, BJS, 17, 151

Galison, P., & Hevly, B. 1992, Big Science: The Growth of Large-Scale Research (Stanford, CA: Stanford Univ, Press)

Grusky, O. 1960, SF, 39, 105

Hall, E. T. 1990, The Silent Language (New York: Anchor Books)

Hannan, M. T., & Freeman, J. 1984, ASR, 49, 149

Hansen, C. J., Caplinger, M. A., Ingersoll, A., et al. 2017, SSRv, 213, 475

Haveman, H. A., Russo, M. V., & Meyer, A. D. 2001, OS, 12, 253

Henke, C., & Sims, B. 2020, Repairing Infrastructures: The Maintenance of Materiality and Power (Cambridge, MA: MIT Press)

Hughes, T. P. 1999, The Evolution of Large Technological Systems (202,; New York: Routeledge)

Höltgen, B., & Williamson, R. C. 2023, 2023 ACM Conf. Fairness, Accountability, and Transparency, (New York: ACM) 1124

Mason, J. P., Begbie, R.G., Bowen, M., Caspi, A., Chamberlin, P.C., et al. 2023, BAAS, 55, 268

Jordan, J., & Sorell, M. 2019, Why Reverse Mentoring Works and How to Do It Right, Harvard Business Review, https://hbr.org/2019/10/why-reverse-mentoring-works-and-how-to-do-it-right

Knorr Cetina, K. 1999, Epistemic Cultures: How the Sciences Make Knowledge (Cambridge, MA: Harvard Univ. Press)

Knowles, B., West, R., Helfenstein, P., et al. 2020, P&SS, 185, 104898

Kruchten, P., Nord, R. L., Ozkaya, I., & Falessi, D. 2013, SIGSOFT Softw. Eng. Notes, 38, 51

Kunda, G. 2006, Engineering Culture: Control and Commitment in a High-Tech Corporation (Philadelphia, PA: Temple Univ. Press)

Latour, B. 1992, Where Are the Missing Masses? The Sociology of a Few Mundane Artifacts (Cambridge, MA: MIT Press) 225

Lee, C. P., Dourish, P., & Mark, G. 2006, Proc. CSCW'07, (New York: ACM) 483

Leonardi, P. M. 2012, Materiality, Sociomateriality, and Socio-Technical Systems: What Do These Terms Mean? How Are They Different? Do We Need Them? (1st edn; Oxford: Oxford Univ. Press)

Linde, C. 2009, Working the Past: Narrative and Institutional Memory (Oxford: Oxford Univ. Press)

Lynch, J. G., & Zauberman, G. 2006, J. Public Policy Mark, 25, 67

McCray, W. P. 2000 *SSS 30 685*

McCurdy, H. E. 2003, Faster, Better, Cheaper: Low-Cost Innovation in the U.S. Space Program (Baltimore, MD: Johns Hopkins Univ. Press)

McDonald, S. 2011, SN, 33, 317

McNutt, R. L., et al. 2019, AcAau, 162, 284

Meyer, J. W., & Rowan, B. 1977, AmJS, 83, 340

Mirmalek, Z. 2020, Making Time on Mars (Cambridge, MA: MIT Press)

Nielsen, M.W., Bloch, C. W., & Schiebinger, L. 2018, NatHB, 2, 726

Poirier, L., & Costelloe-Kuehn, B. 2019, DatSJ, 18, 1

Polanyi, M. 1966, The Tacit Dimension (Chicago, IL: Univ. Chicago Press)

Prescod-Weinstein, C. 2021, The Disordered Cosmos: A Journey into Dark Matter, Spacetime, and Dreams Deferred (New York: Bold Type Books)

Rathbun, J. A., et al. 2017, Planetary Science Vision 2050 Workshop (Houston, TX: LPI) 8079

Rathbun, J. A., et al. 2018, 49th Lunar and Planetary Sci. Conf. Abstract #2668 http://www.lpi.usra.edu/meetings/lpsc2018/pdf/2668.pdf

Rathbun, J., Rivera-Valentín, E. G., Keane, J. T., Lynch, K., Diniega, S., Quick, L. C., Richey, C., Vertesi, J., Tucker, O. J., & Brooks, S. M. 2021, BAAS, 53, 435

Reinecke, D. 2021, SSS, 51, 750

Reinecke, D., Brandt, P. C., Fountain, G. H., Rymer, A. M., & Vertesi, J. A. 2021, Planetary Science and Astrobiology Decadal Survey 2023–2032, Vol 53, 428

Reinecke, D., & Vertesi, J. 2021, BAAS, 53, 06

Reinecke, D., & Vertesi, J. 2022, 53rd Lunar and Planetary Science Conf. (Houston, TX: LPI) 2474

Ribes, D., & Finholt, T. A. 2007, Proc. GROUP 2007, (New York: ACM) 229

Ribes, D., & Jackson, S. J. 2013, Data Bite Man: The Work of Sustaining a Long-Term Study (Cambridge, MA: MIT Press) 147

Ribes, D., & Polk, J. B. 2015, SSS, 45, 214

Riley, D., Slaton, A. E., & Pawley, A. L. 2014, Social Justice and Inclusion: Women and Minorities in Engineering (Cambridge: Cambridge Univ. Press) 335

Rivera, L. A. 2015, Pedigree: How Elite Students Get Elite Jobs (Princeton, NJ: Princeton Univ. Press)

Rivera-Valentin, E., et al. 2020, BAAS, 52, 502.07

Rosner, D. K., & Ames, M. 2014, Proc. 17th ACM CSCW Conf., (New York: ACM) 319

Rudman, L. A., & Fairchild, K. 2004, JPSP, 87, 157

Rymer, A. M., et al. 2021, PSJ, 2, 184

Shim, J. 2021, ERS, 44, 1115

Shrum, W., Genuth, J., & Chompalov, I. 2007, Structures of Scientific Collaboration (Cambridge, MA: MIT Press)

Smith-Doerr, L., Alegria, S. N., & Sacco, T. 2017, ESTS, 3, 139

Leigh Star, S. 1999, ABS, 43, 377

Steinhardt, S. B. 2016, Breaking Down While Building Up: Design and Decline in Emerging Infrastructures (New York: ACM) 2198

Tomlinson, J., Baird, M., Berg, P., & Cooper, R. 2018, HR, 71, 4

Traweek, S. 1988, Beamtimes and Lifetimes: The World of High Energy Physicists (Cambridge, MA: Harvard Univ. Press)

Twenge, J. M. 2023, Generations: The Real Differences Between Gen Z, Millennials, Gen X, Boomers, and Silents and What They Mean for America's Future (New York: Atria Books)

Vanderbilt, K., & Gaiser, E. 2017, Ecosphere, 8, e01697

Vaughan, D. 1996, The Challenger Launch Decision: Risky Technology, Culture, and Deviance at NASA (Chicago, IL: Univ. Chicago Press)

Vertesi, J. 2015, Seeing Like a Rover: How Robots, Teams, and Images Craft Knowledge of Mars (Chicago, IL: Univ. Chicago Press)

Vertesi, J. 2019, AGUFM Vol. 2019, SH54A-03

Vertesi, J. 2020, Shaping Science: Organizations, Decisions, and Culture on NASA's Teams (Chicago, IL: Univ. Chicago Press)

Vertesi, J., & Dourish, P. 2011, CSCW '11: Proc. of the ACM 2011 Conf. on Computer Supported Cooperative Work 533 (New York: ACM)

Weber, M. 1968, Economy and Society: An Outline of Interpretive Sociology (New York: Bedminster Press)

Westfall, C. 2010, HSNS, 40, 350

Willig, M., & Walker, L. 2016, Ecological Research: Changing the Nature of Scientists (New York: Oxford Univ. Press)

Wingfield, A. H., & Alston, R. S. 2014, ABS, 58, 274

Yang, C.-L., Yamashita, N., Kuzuoka, H., Wang, H.-C., & Foong, E. 2022, Proc. of the ACM on Human-Computer Interaction, Vol. 6 (New York: ACM) 44

Yang, Y., Verma, D., & Anton, P. S. 2023, Syst. Eng., 26, 590

Zimmer, L. 1988, SP, 35, 64

Chapter 12

Pluto and Triton: Open Questions and Decadal Linkages

S Alan Stern, Candace J Hansen, William B McKinnon, Carol Paty, Louise Prockter
and Leslie A Young

12.1 Introduction

The purpose of this chapter is to explore the key open questions regarding Pluto and Triton, and how answering these open questions responds to the 2023 US National Academies' Decadal Strategy for Planetary Science and Astrobiology (henceforth, "the Decadal" or OWL (National Academies 2023). We proceed as follows.

Section 12.2 focuses on open questions regarding a series of key areas for the future study of Pluto and Triton. These include their interiors, their surfaces/geologies, their atmospheres, their ionospheric, magnetospheric and radiation environments, their origins, their ocean worlds connections, and their comparative planetology. Section 12.3 then discusses the relevance of these questions to OWL, and the needed missions to these worlds that are required to address the open questions described in Section 12.1. Section 12.4 provides a brief summary.

12.2 Open Questions

12.2.1 Origin

Pluto and Triton are both considered to be representatives of large Kuiper belt worlds, albeit ones with radically different histories. The Kuiper belt, or more broadly speaking, the trans-Neptunian population of Trans-Neptunian Objects (TNOs), is thought to have been, for the most part, emplaced as part of a dynamical instability involving the giant planets early in solar system history (Nesvorný 2018; Morbidelli & Nesvorný 2020). Recent work has postulated that this instability was quite early, in which a closely-spaced resonant chain amongst Jupiter, Saturn, Uranus, and Neptune—and one or more additional ice giants—emerged from the dissipating protosolar nebula $\sim$5 Myr after the formation of the first solar system solids (the calcium-aluminum inclusions) at 4.567 Ga (Amelin et al. 2010).

Deprived of the stabilizing influence of nebular gas, the giant planets both underwent an orbital reconfiguration, ejecting one or more ice giants from the solar system, but also destabilizing a remnant disk of planetesimals orbiting beyond Neptune's original orbital position (Nesvorný 2018; Liu et al. 2022). Further outward migration of Neptune was driven by dynamical scattering of the TNOs. Numerous such bodies moved in and out of mean-motion resonances with Neptune, while others were scattered into the Oort cloud, ejected from the solar system, or collided with the giant planets or their satellites. Broadly speaking, Pluto is one of these surviving resonant bodies, now in the 3:2 Neptune mean-motion resonance, while Triton (the larger of the pair) is viewed as a Kuiper Belt (KB) planet captured by Neptune. As described in Chapter 1, Triton and Pluto share many similarities in terms of bulk density and surface composition; a major line of evidence linking the two, but there remain many major open questions regarding their origin, the sequence of events, and implications. We summarize these next.

- *When and where did Pluto form? When was it emplaced into the 3:2 resonance? What was its dynamical history?*

Figure 12.1 illustrates possible stages in Pluto's formation. In the beginning in the outer protosolar nebula, small particles undergoing collective aerodynamic aggregation (the most well-studied being the streaming instability) formed initial planetesimals of order 100 km across (Morbidelli & Nesvorný 2020). A combination of conventional hierarchical planetesimal accretion and (while nebular gas persisted) gas-drag assisted pebble accretion then built proto-Pluto (and indeed, thousands of proto-Plutos; Nesvorný 2018; Nesvorný & Vokrouhlický 2016; but see Kaib et al. 2024). Insofar as we can tell, proto-Pluto could have formed anywhere in the ancient or ancestral trans-Neptunian belt, somewhere between $\sim$20 and 30 AU (the outer boundary a requirement to halt Neptune's migration, e.g., Nesvorný 2018). Its subsequent orbital evolution during the instability noted above could have been quite chaotic (in terms of orbital elements), before ending up in the 3:2 mean-motion resonance with Neptune and migrating further outward in lockstep with Neptune. This early epoch is of some interest because of the implied time Pluto would have spent much closer to the Sun (for perhaps 10s of Myr), which has implications for the sources and evolution of its volatiles and atmosphere.

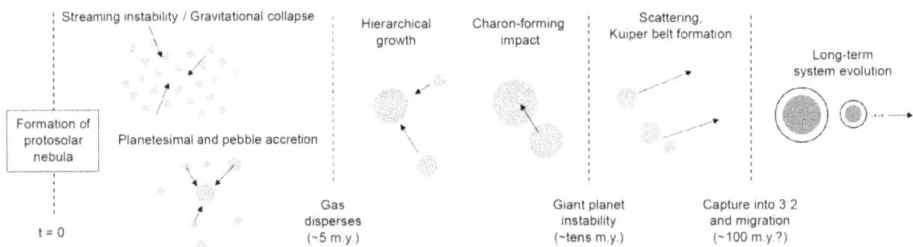

Figure 12.1. Possible stages in the origin and evolution of the Pluto system. Figure from McKinnon et al. (2021) © 2021 The Arizona Board of Regents. Reprinted by permission of the University of Arizona Press.

- *When and how was Charon formed? In the context of the giant impact scenario, what are the constraints on the impact parameters and internal state of the precursor bodies?* Pluto has a relatively massive satellite, and its formation is a key clue to the state of the trans-Neptunian disk when Pluto formed. Many KBOs are binaries (>10%; Noll et al. 2020), and the leading theory for the formation of large KBO binaries (or large KBOs with satellites) is giant impact (see the review by Canup et al. 2021, and references therein). Thus, when we discuss the accretion of proto-Pluto we really mean the accretion of two protoplanets of large and comparable mass. Numerical models of Charon-forming giant impacts implicate both relatively gentle collisions (low approach speeds at "infinity") and partially differentiated conditions (separation of ice from rock, the latter forming a core) for both protoplanets. The low speed is necessary to get a big mass in permanent orbit (the Charon/Pluto mass ratio is 1/8, an order of magnitude larger than the Moon/Earth mass ratio), whereas partial differentiation means that Charon can be as similarly rock-rich as Pluto. Low impact speeds point toward an earlier, less dynamically-excited era in the ancestral Kuiper belt, prior to the early giant planet dynamical instability. Nevertheless, it is important to explore giant impacts with greater numerical fidelity, and over a broader range of impact speeds, masses, internal states, and equations-of-state to really be sure (Arakawa et al. 2019).

- *What are the effects of the giant impact on the initial states and early evolutions of Pluto and Charon? To what extent does the small satellite system constrain the impact conditions and early evolution of Pluto-Charon?* It is of great geophysical interest as to the effects of giant impacts on Pluto and Charon, and KBOs generally. Even for partially differentiated precursors, giant impact may push the bodies toward full differentiation and even internal ocean formation. The internal state then further influences the tidal evolution of Pluto and Charon to the dual synchronous state it occupies today, with different possible thermal implications. The lack of compressional tectonic features on Pluto's surface and the spectacular extensional faulting (rifting) of Charon's surface (as well as more modest, late rifting on Pluto) are all consistent with a "hot" initial thermal state for these bodies (e.g., Bierson et al. 2020), as well as early freezing and global expansion on Charon (Moore et al. 2016; Moore & McKinnon 2021). No geological evidence (fossil bulge, tectonic pattern) has been identified to date for the tidal evolution of the Pluto-Charon binary, so the relevance of this early epoch to the evolution of these bodies remains an open question. The origin of Pluto's small outer satellites is another unsolved puzzle. Spectrally they are made of water ice, but their bulk densities are as yet poorly constrained. They are close to but not in mean-motion resonances with Charon, and no satisfactory model of their orbital evolution has been produced to date (Canup et al. 2021, and references therein). The characteristics of the satellites seem to point toward a giant impact origin between bodies that are at least partially differentiated (icy surfaces) followed by considerable outward tidal evolution, but alternative scenarios that invoke neither also exist (e.g., Bromley & Kenyon 2020).

- *When and how was Triton captured? Is exchange capture from a massive binary plausible, given our present understanding of the Kuiper belt?* Early proposals had Triton captured from heliocentric orbit by energy loss through collision with a regular satellite of Neptune (now destroyed) (Goldreich et al. 1989) or via gas drag in a precursor protosatellite disk around Neptune (McKinnon et al. 1995). An arguably more likely scenario is, however, "exchange capture," in which a binary KBO encounters Neptune deep in the planet's Hill sphere (gravitational sphere of influence), which disrupts the binary, with one half of the pair ending up in an eccentric orbit around Neptune and the other half escaping back to heliocentric orbit (Agnor & Hamilton 2006). Given that binaries are common in the Kuiper belt, such encounters (as opposed to collisions, or encounters at a special time) may have been likely. Exchange reaction capture of Triton is, however, not favored during the dynamic instability described above, nor later, as encounter velocities would have been too high (Vokrouhlický et al. 2008), so an earlier capture (or a later, low probability event; Nogueira et al. 2011) is implied. Early capture also mitigates against scattering and loss of Neptune's distant, irregular satellites during Triton's subsequent tidal evolution toward its present circular orbit (Ćuk & Gladman 2005). Neptune likely acquired these satellites during the dynamic instability (Nesvorný 2018). Nevertheless, it is not at all obvious that binaries with components the scale of Triton or greater were common, or are an expected, direct outcome of planetesimal formation by the streaming instability or similar processes (Johansen et al. 2015; Morbidelli & Nesvorný 2020).

- *What are the implications of Triton's capture for its early evolution and ongoing activity?* Triton's geological vigor is likely tied to tidal heating, as with other active icy satellites (e.g., Europa and Enceladus), but Triton is not being tidally heated today by eccentricity tides. Substantial heating would have occurred as its post-capture orbit circularized, as noted above. The degree to which tides circularized Triton's post-capture orbit, as opposed to collisional interactions with Neptune's original regular satellites (Ćuk & Gladman 2005), and when this occurred are major open questions. The possibility of profound internal heating (global melting) and thermochemical processing (Shock & McKinnon 1993; McKinnon et al. 1995) is undeniable, but relating any of this directly to Triton's very young, lightly cratered surface (Schenk & Zahnle 2007) has proved elusive. While more data from Triton would no doubt be helpful, the riddle of Triton's ongoing geological activity has likely been solved (at least in broad outline) by the recognition (actually, re-recognition) of the existence and potential importance of obliquity tides (Jankowski et al. 1989; Nimmo & Spencer 2015). But as always, the devil is in the details, and estimates of the true strength of this obliquity tidal heating are uncertain and require further work.

12.3 Interior/Ocean

Three of the most pressing, unanswered questions about Pluto and Triton concern their interiors: *What are their bulk compositions and radial structures? Are or were oceans ever present in their interiors?* The current state of knowledge for the formation and bulk composition of both Pluto and Triton was reviewed in Chapter 1 and the interior was reviewed in Chapter 2. As noted in these chapters the current evidence for oceans on both bodies is circumstantial, inferred from limited flyby observations and theoretical arguments. Few constraints are available for the initial thermal states, radioactive isotope abundances, ice viscosities, silicate and ice-shell porosity, organic mass fractions, interior ammonia and other volatile ice abundances, and the extent of any hydrothermal alteration of either body.

The lack of knowledge of Pluto's interior structure and initial conditions make any inferences about its thermal evolution speculative. Pluto is not currently experiencing significant tidal heating. However, early in its history it likely contained sufficient radioactive elements to result in partial melting of the interior (McKinnon et al. 1997). We caution though, that large uncertainties remain about the temperature at which melting would have been initiated, the effects of ice grain sizes, and the influence of any fine silicate or other grains present in the ice, and therefore the predicted nature of any ocean that formed this way.

Pluto's bulk density is similar to that of many other icy dwarf planets and Triton, consistent with an interior mostly composed of about one-third water ice and two-thirds silicates (McKinnon et al. 2017). However, large uncertainties in the density of its core yield a wide range of potential layer thicknesses. If a substantial, non-silicate component such as carbon is present, the planet's rock fraction could be substantially lower (McKinnon et al. 2021), and any interior ocean consequently thinner. Most models predict a present-day ocean up to 130 km thick lying ~100–200 km beneath the ice-rich mantle (see Nimmo & McKinnon 2021 and references therein). The stability of such an ocean depends on largely unknown parameters such as the viscosity of the overlying ice shell, which determines whether heat is lost by convection or conduction, and the influence of salts and volatiles such as ammonia within the ocean and ice shell. Hydrothermal reactions are predicted between the rocky core, water, and any organic material present, and are expected to lead to a salty ocean (Zolotov 2007). However, core permeability is another key unknown in these scenarios.

A differentiated interior has been inferred for Pluto from other observations, including the lack of contractional geological features in Pluto's water-ice bedrock (McKinnon et al. 1997), a lack of a fossil tidal bulge (Robuchon & Nimmo 2011), and the similarity between Pluto's and Charon's rock fractions (Canup 2010). The proposed reorientation of Sputnik Planitia basin to Pluto's equator by true polar wander (e.g., Keane et al. 2016) may well require a persistent thinned and rigid subsurface ice shell overlying an ocean. Such conditions could be met if a layer of insulating clathrates lie atop the ocean, preventing both viscous relaxation of the basin and freezing of the ocean (Kamata et al. 2019). Candidate cryovolcanic features on Pluto's surface include reddish materials and coatings that exhibit the spectral signatures of water ice with an ammoniated compound (Dalle Ore et al. 2019; White et al. 2021 and references therein). These are associated with some of Pluto's youngest extensional features, indicating that they may have tapped into a subsurface aqueous reservoir, possibly an ocean. Such eruptions could result from over-pressurization in a slowly freezing ammonia-water ocean. However, if clathrates are present at the top of the ocean, the ocean could be effectively thermally isolated from the surface, so how and whether such cryolavas originate in a subsurface liquid reservoir remains to be investigated. More information is needed about Pluto's global geology, the communication between subsurface liquid reservoirs and the surface, and the role clathrates play in its interior (figure 12.2).

Following its capture about Neptune, Triton would have undergone significant tidal heating resulting in differentiation into a rocky core and water-ice mantle (e.g., McKinnon 1984; Gaeman et al. 2012). However, it is not known whether present-day Triton's shell is in hydrostatic equilibrium and decoupled from the interior. Confirmation of an ocean could have major implications for understanding alternative pathways toward the conditions necessary for Triton as a habitable world.

Triton's high inclination with respect to Neptune results in obliquity tides (Jankowski et al. 1989), which potentially set up resonant flows in liquid layers (Tyler 2009). Heat fluxes in excess of 10 mW m^{-2} are predicted due to tidal dissipation in any subsurface liquid due to turbulence, and an ocean ~150 km thick is expected to persist beneath an ice shell of a similar thickness (Nimmo & Spencer 2015). The effects of obliquity tidal heating on Triton's interior evolution, heat flow and surface geology, and comparisons of the effects of obliquity vs. eccentricity tidal heating all remain to be thoroughly explored.

Little is known about the physical state and composition of Triton's hypothesized ocean, and few constraints exist on its salinity. The volatile composition of that ocean may be cometary in nature (Shock & McKinnon 1993), and salts are expected to comprise a few weight percent if the ocean is thick (Neveu et al. 2017). A near-surface, liquid-water ocean could provide the thermal energy needed to drive the geological processes that refresh Triton's extremely young surface, such as cryovolcanic eruptions, cryomagmatic intrusions, and tectonic disruption; however, an understanding of such processes on Triton, and how they may be facilitated by

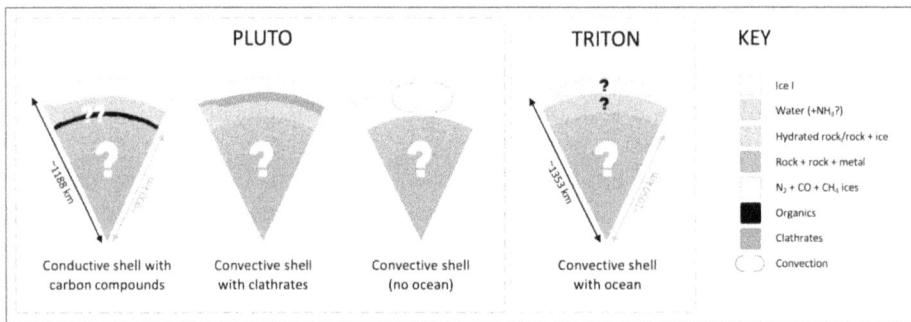

Figure 12.2. Schematic view of three hypothesized interiors for Pluto (Nimmo & McKinnon 2021) and a predicted interior of Triton (Nimmo & Spencer 2015). Interfaces are approximately to scale; yellow basin on Pluto represents Sputnik Planitia and is exaggerated for clarity. Question marks denote interfaces which are inferred, and for which more information is needed to determine the correct interior structure (modified from Nimmo & McKinnon 2021).

antifreeze agents such as ammonia in the ice, remain to be determined. A detailed discussion of methods for future ocean detection is provided in Chapter 8.

12.4 Ocean World/Astrobiology

For many years after Voyager 2 flew by Neptune, Triton languished as an enigmatic body; mysterious, yet strangely, without urgency for a return mission. The Galileo mission revealed Jupiter's moon Europa to be an "ocean world" with a subsurface layer of salty liquid water, but this too seemed to be something of a curiosity. Then Saturn's moon Enceladus was discovered by the Cassini spacecraft to be spewing water vapor from its interior; a second water ocean world in the outer solar system. Cassini executed multiple flybys and established that Enceladus' liquid water interior is likely to be habitable.

As a result of these discoveries, the planetary community quickly realized that our concept of a Goldilocks habitable zone around stars where liquid water on the surface could host life was a very narrow, terrestrio-centric concept. The water interiors of icy moons and dwarf planets are also able to provide the conditions for life to flourish regardless of their heliocentric distance. New analyses showed that many moons and dwarf planets could be ocean worlds, including both Pluto and Triton, with habitable internal liquid water layers. We review the potential for detecting oceans in Chapter 8. Beyond detection, our focus is also turning to access these oceans. With thick, icy crusts, Ganymede's and Callisto's oceans will remain out of reach for a long time. But does Triton's youthful surface imply that its interior ocean is in communication with its surface, as Pluto's appears to be (e.g., Cruikshank et al. 2021)?

The questions we ask today are whether or not there is life anywhere else in our solar system besides Earth. What conditions are required for life to appear and evolve? The easiest place to search for life right now is Enceladus, helpfully spewing the contents of its subsurface ocean into space. If we do not find life in Enceladus's ocean, we will ask *why not*? If we do find life, we will ask *where else*? And are all oceans hospitable to life? Once the Europa Clipper mission's reconnaissance is underway, we will have a much better idea as to where to send a future mission to Europa (potentially involving a lander), asking the same questions.

Bodies formed in the outer solar system are of particular interest for ocean world science because of their rich chemistry, including nitrogen compounds, which may be much less abundant in, e.g., the Galilean satellites. To address these questions, the groundwork must be laid now:

- *Are both Pluto and Triton in fact ocean worlds?*
- *How thick is the crust of ice separating the surface from the liquid on each?*
- *Is there communication between their surface and the subsurface oceans?*
  - Do Triton's cryovolcanic terrains have cracks/faults/conduits allowing material from the ocean to reach the surface?
  - Do the hypothesized thermal diapirs that may transport fluids between Triton's ocean and its surface introduce chemical gradients in the ice shell and create habitable niches (Ruiz et al. 2007)?

      ○ What volatile reservoir is accessed by Triton's plumes, and how far beneath the surface does it lie?

      ○ Did the linear fissure adjacent to Pluto's Hardie crater fill the crater with a watery-cryolava?

- *Do the oceans provide habitable environments and host life?* Liquid water, organic compounds, and chemical energy are generally accepted to be necessary ingredients for life (e.g., Cockell et al. 2016; Hendrix et al. 2019). Both Triton and Pluto are known to have C, H, O and N, but ascertaining the presence of P and S awaits us.

- *Are the hazes produced by photochemistry of CH_4 generated in the tenuous atmospheres of Pluto and Triton, precipitating out of the atmosphere, possibly contributing organics to their potential subsurface oceans?* Is there ionospheric production of organic compounds, which could include materials of high chemical potential?

Confirmation of a rich, organic chemistry environment and ice-shell transport processes connecting the proposed ocean to its surface would place Triton and Pluto among the highest value targets in the search for life, adding to the diversity of potentially habitable worlds, all the way out at 49 AU (Pluto at aphelion).

12.5 Surface/Geology

As pointed out above, both Triton and Pluto are suspected ocean worlds, and the surface geology of both shows signs of geologically recent active surface processes. For a detailed review of surface geology and cryovolcanism see Chapters 3 and 4. Questions that remain include: What is the energy source required to erase craters? Are the volatile inventories of both worlds similar? Why does Pluto have mountains and Triton not? For both bodies, these and related questions are summarized by this one: How does the interplay of tidal dissipation, heat transfer, tectonics, cryovolcanism, diapirism, and surface-atmosphere interactions drive resurfacing on Triton and Pluto? Other important questions also remain open, including:

What is the global signature of the stress history of Triton and Pluto? The evolutionary histories of Triton and Pluto diverged dramatically after the capture of Triton into orbit around Neptune. Both Triton and Pluto have experienced tidal stresses, e.g., Pluto as it synchronized with Charon, and Triton in its high obliquity orbit. But Voyager and New Horizons gave us high-resolution images of only one hemisphere of each body, so we do not have the global tectonic signatures to map these stresses and the differing outcomes.

How has cryovolcanism sculpted the surface feature morphologies? The high-resolution imaging available for one hemisphere of each body shows exotic terrain and hints at a variety of cryovolcanic processes responsible for geologic features. The crustal materials of Pluto and Triton are similar: water ice and (for Triton) CO_2 ice, form the bedrock. Inventories of more volatile ices (CH_4, N_2, CO) are likely similar to first order. The main difference in our state of knowledge is that New

Horizons was able to map the composition of Pluto's surface units, but Voyager did not have this capability so the same kind of data is not available for Triton. Hence, the composition of Triton's surface units and potential linkage to ocean chemistry represents a critical gap in our understanding of Triton's cryolava.

Cryovolcanism is implicated on both worlds as a key process for forming surface features. Many of the questions we have regarding surface processes have to do with how their various ices deform and flow at the temperatures of the outer solar system. Assuming that an adequate internal energy source for endogenic cryovolcanism is available, or that sufficient compounds such as NH_3, CH_3OH, and/or salts can lower the melting temperature, the questions we can pose are many, for example:

At Pluto:

- Is Wright Mons cryovolcanic?
- Are Pluto's tall mountains cryovolcanic?
- Are plumes the source of recent mantling deposits of ammonia compounds mixed with water ice near Virgil Fossae?

At Triton:

- Do guttae result from sublimation erosion or surface collapse over hot spots?
- Is cantaloupe terrain formed by diapirism?
- Are walled plains formed by the eruption of cryomagma?
- Are Triton's plumes solar-driven or endogenic?
- What do the sites and timings of plume occurrence tell us about the energetics and relevant processes?

In their distant realm of the solar system, materials respond to vastly different environmental conditions such as temperature, pressure and gravity, compared to terrestrial experience. For example, New Horizons showed evidence of solid-state convection in the thick N_2 ice filling Sputnik Planitia, and valleys cut by nitrogen glaciers (figure 12.3).

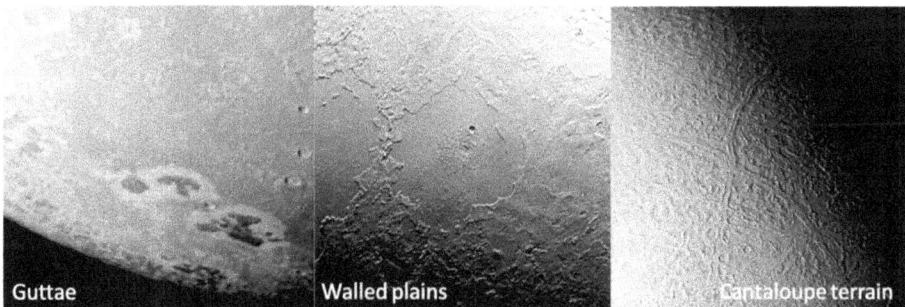

Figure 12.3. Three examples of Triton's possible cryovolcanic terrains are shown here. The guttae may have been formed by viscous lava flow. The walled plains may be collapsed calderas. The cantaloupe terrain may be the expression of subsurface, compositionally-driven convection. Figure from McKinnon et al. (2021) © 2021 The Arizona Board of Regents. Reprinted by permission of the University of Arizona Press.

We also point to the following important questions:

- *What role do ices play in the development of the surfaces we see today?* Both bodies show the interplay of climatic and geologic processes. On Pluto, where compositional mapping of surface units is available, it is possible, for example, to describe how CH_4-rich ice forms bladed terrain. The expected seasonal migration of N_2 ice shows up in northern mid-latitude spectra, along with less volatile CH_4 ice. Pluto's rugged terrain modulates the cold traps for its volatiles, and Sputnik Planitia maintains its nitrogen atmosphere well beyond perihelion. Mysteries remain however: Sputnik Planitia has only 0.3%–0.5% CH_4 in the N_2 ice, while thermal equilibrium models predict values an order of magnitude more. Why hasn't this evolved toward equilibrium? Is this a question of timescales?
- Triton, which similarly to Pluto has an N_2 atmosphere in volatile equilibrium with surface frosts, has very subdued topography. Theoretical models should be adequate to explain its polar caps, and yet simple models fail dramatically to correlate surface ice composition with albedo boundaries. Basic questions include:
  - Where are Triton's volatiles located on its surface?
  - What drives their distribution? Are the approximately latitudinal color bands imaged by Voyager consistent with frost on the move seasonally?
  - Are there permanent N_2 polar caps/deposits at both poles? Or conversely, just one in the southern hemisphere (e.g., Bertrand et al. 2022)?
  - Since the time of the Voyager flyby, the subsolar point has passed through the southern summer solstice and is approaching the equator. How much mass has been transferred into the northern polar region?
- *How has the climate history affected surface evolution?* On both bodies we believe long-term climate changes have influenced surface processes. On Pluto, for example, the questions include:
  - How did its dunes form?
  - Was there a time when environmental conditions could have allowed liquid N_2 to flow across the surface and carve dendritic channels and valleys?
  - And on Triton, what is the nature of the bright southern hemisphere polar terrain?

12.6 Atmosphere

The current state of knowledge for the atmospheres is covered in Chapters 5, 6, and 7. Both Triton and Pluto have N_2-dominated atmospheres; each also has CO and CH_4 as minor species, and all of these species also exist as surface ices on the two bodies. Triton has much less CH_4 than Pluto, both in its atmosphere and on its surface. This fundamental difference affects atmospheric temperatures, stability, vertical mixing, photochemistry, and haze production, and leads to specific open questions. These include:

- *What is the CH_4 mixing ratio at the surfaces of Pluto and Triton, and how does it vary with latitude, altitude, ice composition and temperature, time of day,* *surface pressure, and horizontal mixing?* The Voyager solar EUV occultation measured Triton's CH_4 to within a scale height of the surface (Herbert & Sandel 1991) at summer and winter locations. The derived number density was consistent with a mixing ratio of $\sim$0.01% (Herbert & Sandel 1991), which is similar to the vapor pressure expected for a system in vapor-solid–solid equilibrium at 38 K (Tan & Kargel 2018). Pluto's CH_4 was too opaque near the surface in the EUV to measure directly from the New Horizons solar occultation; this was described in Young et al. (2018), who extrapolated downward from 80 km to derive mixing ratios at the surface of 0.28%–0.35%, much higher than Triton's, and much higher than expected from vapor-pressure over the $\sim$37 K N_2-rich ices. Pluto, unlike Triton, has copious areas of CH_4-rich ices that must be warmer than the N_2-rich ices (or else they would be N_2 condensation sites), and which must display elevated CH_4 vapor pressures. The expectation is that Pluto's CH_4 mixing ratio at the surface will depend on many factors, both global and local. If Triton's CH_4 mixing ratio is only controlled by the vapor pressure over N_2-rich ices, it may be much more uniform.

- *What is the vertical CH_4 profile, and how does it relate to vertical mixing, upward flux of CH_4, and photochemical destruction of CH_4?* On Triton, CH_4 appears well-mixed at the lowest levels (Herbert & Sandel 1991), perhaps due to terrain-induced turbulence. The scale height of CH_4 decreases with altitude on Triton, more quickly at the summer site than at the winter, indicative of photochemical destruction (Strobel et al. 1990). On Pluto, the ingress and egress CH_4 profiles are very similar over the altitudes of reliable measurements, 80–1200 km (Young et al. 2018). Photochemistry does not affect Pluto's CH_4 profile much, which is instead subject to diffusive separation, so that CH_4 becomes the dominant species above $\sim$1400 km (Young et al. 2018). On Pluto, the vertical profile of CH_4 transitions from diffusive separation to well-mixed somewhere below 80 km, where the CH_4 mixing ratio was not well measured by New Horizons. Therefore, this transition, and the inferred CH_4 mixing ratio at the surface, is model dependent. The vertical dynamics is not yet fully understood, and depends on the eddy diffusion, and also on the vertical flux of CH_4 at the surface (Luspay-Kuti et al. 2017; Young et al. 2018; Krasnopolsky 2020).

- *What is the nature and effect of hazes?* On both Triton and Pluto, hazes were seen in EUV occultations (Herbert & Sandel 1991; Young et al. 2018) and imaging (Hillier et al. 1991; Cheng et al. 2017). Triton, but not Pluto, also had distinct plumes emanating from its surface (Hansen et al. 1990). Pluto, but not Triton, had haze extending to high altitudes. The source of Triton's plumes is still an area of active investigation (Hofgartner et al. 2022). How will different seasons affect the locations of Triton's plumes? Triton's hazes were formed by condensation, but Pluto was observed to have hazes to higher altitudes where condensation was ineffective, and its hazes appear to be due to both Titan-like molecular growth and Triton-like condensation (Lavvas et al. 2021). Since the formation path depends on the high-altitude CH_4

abundance, how much do the haze properties change with season and latitude? On Pluto, calculations indicate that haze is more important than gases for the radiative heating and cooling of the atmosphere (Zhang et al. 2017), given the limited measurements of haze extinction coefficients at sub-mm wavelengths. How effectively do Pluto and Triton's hazes cool? Is haze important for the energetics of Triton's atmosphere? On both Triton and Pluto, how do the haze particles affect the surface after they settle? And how does Pluto's surface remain heterogeneous after billions of years of haze particles settling onto the surface (Cheng et al. 2017, Grundy et al. 2018)? In a related topic, Triton, but not Pluto, also displayed distinct plumes emanating from its surface (Hansen et al. 1990); the source of these plumes is still an area of active investigation (Hofgartner et al. 2022). How will different seasons affect the locations of Triton's plumes, and do the plume particles affect either the atmospheric energetics or surface properties?

- *What controls global circulation, climate, and wave activity?* On both Triton and Pluto, gaseous N_2 is in vapor-pressure equilibrium with the surface N_2 ice. Is the subsurface thermal inertia high enough to keep some areas free of N_2 ice (Young et al. 2021; Bertrand et al. 2022)? How does thermal inertia affect the global pressure over an entire orbit? How does the annual climate change with changing obliquity? As N_2 is transported from areas of high to low insolation, how do the zonal winds react? What atmospheric waves are produced by the daily cycle of sublimation and condensation, and what is the role of waves in transporting momentum and energy within the atmosphere?

- *How important are exogenic sources of material?* Interplanetary dust delivers H_2O to both Pluto and Triton (Poppe & Horányi 2018). How does this affect radiative cooling (Strobel & Zhu 2017)? How does water affect atmospheric and ionospheric chemistry? Can dust serve as condensation nuclei for the formation of hazes?

12.7 Magnetospheric and Radiation Environments

While Triton and Pluto may share a similar origin, their divergent evolutionary paths have taken them to exist in completely different radiation environments. Their interactions with their environments are reviewed in Chapter 7. Pluto resides in the solar wind, albeit far from the Sun at an average of 39 AU where the plasma density is low ($\sim$0.002–0.025 cm^{-3}) and the Interplanetary Magnetic Field (IMF) is weak (0.08–0.3 nT), but the flow speed remains similar to that near the Earth ($\sim$400 Km s^{-1}; Table 1 in Bagenal et al. 2016). This places the interaction of the solar wind with Pluto in a significantly supersonic and super-Alfvénic regime. Conversely, Triton's orbit about Neptune (itself on average 30 AU from the Sun) places it well within the ice giant's magnetosphere, where it experiences periodic magnetic waves of 5–8 nT in amplitude, plasma densities of 0.03–0.11 cm^{-3} and a relative flow speed of 43 km s^{-1}, placing the interaction thoroughly in a subsonic and sub-Alfvénic regime (Table 1 in Liuzzo et al. 2021 and Table A2 in Cochrane et al. 2022).

The New Horizons Alice ultraviolet spectrograph detected the production of ions in Pluto's atmosphere during the New Horizons encounter (Steffl et al. 2020). However, the ionosphere was below the detection threshold for direct observation via radio occultation, which was consistent with modeling estimates that placed the peak at the terminator at $\sim 1 \times 10^3$ cm^{-3} (Hinson et al. 2018). Conversely, Triton's ionosphere was observed to be incredibly well defined with a peak electron density of $2-5 \times 10^4$ cm^{-3} (Tyler et al. 1989); much higher than anticipated from the weak solar UV flux. This indicates a non-solar source mechanism, perhaps from Triton's atmosphere interacting with energetic particles in Neptune's magnetosphere. However, without direct particle observations close to Triton, the available energy flux remains unknown. Another unknown factor is how the presence of an induced magnetic field from a potential subsurface ocean (described below) at Triton might structure and/or accelerate the flux of local magnetospheric plasma, or potentially shield its ionosphere from loss due to ion pick-up.

Some of the key questions that arise from these distinctive environments relates to the formation of the ionospheres of each body.

- *What causes this intense ionosphere at Triton? How might Triton's interaction with Neptune's magnetosphere be controlled by an induced magnetic field? Why are the ionospheres of Triton and Pluto so different?*

It is interesting to note the role that Triton might play in Neptune's magnetospheric system. N$^+$ and H$^+$ ions were detected at Triton's orbital distance by Voyager 2 (Sittler & Hartle 1996), indicating Triton may provide a source of plasma to the magnetosphere. However, the amount of material sourced this way was difficult to ascertain given that Triton was not close to the location that Voyager 2 crossed its orbital location. Triton's orbit also appears to sculpt the radiation belts of Neptune (Mauk et al. 1991), potentially due to a putative neutral cloud, but direct observations of the process do not exist.

- *Does Triton play an important role in driving magnetospheric dynamics and establishing aurora at Neptune like the important plasma sources provided by Enceladus at Saturn and Io at Jupiter? Or does Triton's elliptical and highly inclined orbit combined with Neptune's highly asymmetric magnetosphere reduce the longevity and impact of plasma sourced from Triton?*
- *Does the orbit of this large moon (and associated neutrals) also structure Neptune's radiation belts, and if so, then via what processes?*

While Pluto and Charon do not reside in a large planetary magnetosphere, they are both interacting with the solar wind and with one another, as evidenced by the deposition of material on Charon sourced from Pluto (e.g., Stern et al. 2015; Grundy et al. 2016; Gladstone et al. 2016). However, while hybrid-kinetic simulations have modeled the solar wind interaction with Pluto and captured some of its highly asymmetric properties observed by SWAP (Feyerabend et al. 2017; Barnes et al. 2019), they are still limited by the lack of direct observations of the IMF during the New Horizons encounter. Comprehensive modeling of the solar wind interaction across both Pluto and Charon has not yet occurred post-New Horizons, leaving open a range of questions relating to how this binary system interacts with the solar wind and with one another. Similarly, there are few observations of the IMF and solar wind conditions at such large heliocentric distances to indicate the range of upstream conditions experienced by these bodies. For example, during the New Horizons encounter, the solar wind density was significantly enhanced from the anticipated mean and range derived from Voyager 2 observations, likely due to a well-timed interplanetary shock (Bagenal et al. 2016).

- *What are the mean properties and variability of the IMF and solar wind at Pluto and how do these structure local interactions between Pluto and Charon? How has this interaction affected the exchange of material over time between these bodies?*

12.8 Comparative Planetology of Dwarf Planets

Pluto-Charon and Triton illustrate just two examples of the possible planetology of dwarf planets (DPs). Nominally, a DP is a body sufficiently large that it has achieved hydrostatic equilibrium. The International Astronomical Union (IAU) recognizes five such worlds at present: Pluto, Eris, Haumea, and Makemake in the Kuiper belt, and Ceres, the largest asteroid. However, it is clear from a geophysical perspective that there are dozens of likely DPs (bodies the scale of Charon or larger) in the Kuiper belt alone, as well as Ceres in the asteroid belt, and that Triton should be considered a DP as well.

A given DP will represent a combination of systematic and contingent properties and evolution (see Moore & McKinnon 2021 for greater details). Systematic properties include those related to size and rock content (which governs, among other things, the amount of radiogenic heating available and the likelihood of differentiation). The rock content of DPs for which densities are known is relatively important, on the order of 2/3 of the total mass (and even greater for Eris). Volatile ice compositions also appear consistent for most DPs, presumably reflecting the low condensation temperatures across the primordial planetesimal disk that sourced the Kuiper belt. Ammoniated compounds have been detected on Pluto, Charon, and Orcus (Cruikshank et al. 2021; Barucci et al. 2021). CH_4 ice is widespread among DPs, N_2 ice is important on Pluto, Triton, and Eris, and inferred to be present on other bodies. CO ice exists on Pluto and Triton, and CO_2 is an important surface ice on Triton (so far undetected on Pluto, but James Webb Space Telescope (JWST) spectra, sensitive to the CO_2 fundamental, may be telling). The relative proportions of the various volatile ices are strongly biased by stability against sublimation loss (Schaller & Brown 2007), but for the largest bodies (e.g., Pluto, Eris, and Triton) endogenic, geologic processes likely exert a strong control as well and result in important regional differences across a given DP.

Contingent effects include impacts of course, and the largest impacts have pronounced, outsized effects. A key example would be the formation of Sputnik basin on Pluto, but none would be greater than the Charon-forming impact. Most DPs have detected satellites, likely all the result of giant impacts. Haumea is another outlier, in that a giant impact has apparently exposed its deeper, water ice mantle, and created both water-ice-rich satellites and a collisional family. Haumea even possesses a ring (Ortiz et al. 2017). *How many other DPs have the exposed ice mantles, collisional families, or rings?* Subsequent tidal evolution may also be important for global heat budgets. Triton is a special case in terms of tidal evolution, but represents an important end member: *what is the outcome of massive tidal heating of a DP KBO?* Tidal heating is an important determinant of the sustainability of internal oceans, but the threshold DP size for which this is important is poorly defined. *Can we expect surface effects from the long-term thermochemical evolution of the largest rocky cores?*

Ultimately, understanding the comparative planetology of DPs will depend on future spacecraft observations. The most relevant, from a US Decadal Survey point of view, would be a mission to Triton or the Neptune system. Spacecraft study of the regular moons of Uranus, though distinct from heliocentric DPs such as Pluto and Triton in terms of origin, may also be relevant in terms of composition, structure and history; *n.b.* the similarities between Uranus' Ariel and Charon. A return to the Pluto system, or a first reconnaissance of one or more other major known DPs, would be of particularly great scientific value. And many more Pluto-scale DPs likely still exist. It has long been recognized that many (on the order of a thousand) Pluto-scale bodies must have originally formed beyond Neptune, in order to make the Charon-forming impact and Triton's capture likely (e.g., Stern 1991). In support of this, Nesvorný & Vokrouhlický (2016) argued on dynamical grounds that the original trans-Neptunian planetesimal disk must have contained of order 1,000–4,000 Pluto-scale bodies, so that Neptune's outward orbital migration would have been sufficiently grainy (or jittery), so that the various mean-motion resonances with Neptune would not be over-populated. From numerical models, the efficiency of emplacement into the Kuiper belt during the instability is on the order of 10^{-3} (and greater into the scattered disk; Morbidelli & Nesvorný 2020). If so, anywhere from a handful to dozens of Pluto-scale worlds await discovery as telescopic surveys reach farther and farther into the deep outer solar system.

12.9 Decadal Relevance and Response

12.9.1 Decadal Relevance

As this chapter demonstrated in the section just above, the important open questions relating to the study of Pluto and Triton are both extensive and diverse, spanning virtually every aspect of planetology, from interiors to ocean words, to surfaces, atmospheres, magnetospheres, and origins. These open questions also bear on many key aspects of comparative planetology today.

OWL recognized the significance and import of this in a wide range of ways, including through the following priority questions and sub-questions:
- Q2.3b What were the roles of giant impact and capture in the outer solar system for the origin of primordial satellites and planetary rings?

- Q2.4b Did the giant planets of the early solar system migrate, and if so, how far, and what was the effect of this migration on other outer solar system bodies?
- Q2.4c Was there a global instability among the giant planets, and if so, when did it occur? Were any major planets ejected (lost) during this instability?
- Q2.5 How did processes in the early outer solar system produce the structure and composition (surface and interior) of Pluto and Trans-Neptunian Objects?
- Q2.5a When and how did trans-Neptunian objects and cometary bodies form?
- Q2.5b How many trans-Neptunian objects formed, and what were their initial size distribution(s)?
- Q2.5c How prevalent were giant impacts in the early trans-Neptunian belt?
- Q2.5d What were the relative proportions of ices, rock, and organic materials accreted by small objects (comets, TNOs, moons) in the outer solar system?
- Q2.5e During accretion, (how) did the interiors of outer solar system moons and dwarf planets transition from homogeneous to layered?
- Q2.6a How did the dynamical structure of the trans-Neptunian belt originate?
- Q3.1b What were the mechanisms of accretion from planetesimals to larger bodies?
- Q4.1b How has collisional and dynamical evolution affected small body populations now found in stable reservoirs within the inner and outer solar systems?
- Q4.2a What small body populations dominated the early bombardment of worlds in the inner and outer solar systems?
- Q4.2f What is the current impact flux on planetary worlds, and has the flux changed substantially over the last several billions of years?
- Q4.3a How did the earliest and/or largest impact events influence the physical evolution of solar system worlds?
- Q4.3b How do impacts affect surface and near-surface properties of solar system worlds?
- Q6.2d How does orbital forcing, including obliquity and eccentricity changes, govern climate change and surface volatile redistribution on extraterrestrial bodies with volatile cycles like modern Mars, Triton, and Pluto?
- Q6.6c What photochemical pathways take methane to more complex hydrocarbons, including hazes, in the atmospheres of Titan, Triton, Pluto, and possibly other Kuiper Belt Objects?
- Q6.6f What processes control the formation and composition of clouds in the atmospheres of Venus, Mars, Titan, Triton, and Pluto?

The OWL also set objectives relevant to Pluto and Triton, including these examples:
- Investigate the properties of subsurface water or magma oceans and melt reservoirs within Europa, Io, Titan, Enceladus, Triton and the Uranian Moons via electromagnetic sounding (active/passive) or induction, or geodetic measurements from orbiting or landed spacecraft.

- Investigate and classify tectonic and cryovolcanic activity and landforms on icy bodies (Europa, Enceladus, Triton and Titan) to determine properties such as location of deforming regions and the thickness of deforming layers via global high-resolution imaging and topography, and via constraining the ages of cryovolcanic units/structures.
- Study the properties of Triton's vapor-pressure atmosphere, including why/how it is different from Pluto, by measuring the distribution of surface ices, as well as atmospheric pressure and temperature as a function of altitude.
- Investigate the source of Triton's plumes and their contribution to its atmosphere by measuring the composition of plume material.
- Assess the past and present geologic activity of the Uranian satellites, Triton, and large Centaurs and trans-Neptunian objects with moon or ring systems—including their cratering record, tectonic, and cryovolcanic activity—and understand how and why they differ from satellites in other systems by imaging their surfaces with resolution, coverage, and spectral range that is at least comparable to those of the Galilean satellites and mid-sized moons of Saturn.

12.10 Mapping of Open Questions to Decadal Science

The OWL priority questions and sub-questions above include those that refer to either Triton or Pluto, or both specifically, and those that are more generally related (e.g., concerning trans-Neptunian objects). However, exploring Pluto and Triton is valuable for addressing aspects of most of the twelve fundamental questions outlined in OWL as we show in Table 12.1, where we link the topical aspects of future Pluto/Triton studies covered above to their decadal questions.

Table 12.1. Mapping Open Questions to Decadal Science

Topical Area	Decadal Science Areas
Interior	1.1c; 2.5c,d,e; 4.3
Atmospheres	1.1b; 6.1d; 6.2d; 6.3a,c,d; 6.4b,c,d,e; 6.5a; 6.6c,d,f ; 12.6c,d
Surface geology	2.3; 2.4; 2.5; 2.6; 4.1; 4.2; 4.3; 4.4; 5.3; 5.4; 5.5; 5.6; 6.2; 6.4; 8.2; 8.3; 12.5b
Magnetosphere/radiation environment	6.2; 6.3; 6.4; 6.5; 7.4; 8.3; 8.4
Origin	1.3c; 2.3b; 2.4b,c; 2.5; 2.6a; 3.1b; 12.1a; 12.2b; 12.4a; 12.8
Dwarf planet comparative planetology	2.5d,e; 4.1b; 4.2a,f; 4.3a,b; 6.2d; 6.6c,f
Ocean worlds/astrobiology	2.5, 5.1, 5.2, 5.3, 5.4, 6.2, 6.6, 10.1, 10.2, 10.3, 10.4, 10.5, 10.6, 10.7, 11.3

12.11 Mission Needs for Pluto and Triton to Address Science and Open Questions

The Pluto and Triton missions needed to advance critical open questions in Pluto and Triton science, and also to advance the goals of the 2020s Planetary OWL, fall into three common categories: flybys, orbiters, and landers/probes.

Although flyby missions have value, orbiters are the logical and more comprehensive next steps following the initial exploration of these bodies by the New Horizons and Voyager 2 flybys, respectively. A Pluto Orbiter has most recently been studied in detail by an OWL precursor study funded by NASA (Howett et al. 2020). Similarly recent Triton work has been done in the course of Neptune orbiters making numerous Triton flybys (Rymer et al. 2021; see also Hansen et al. 2021). Both of these missions have extensive OWL-relevant scientific rationale, science traceability to payload, technical feasibility, mission and spacecraft/payload design, and costing work already done. Each would require flagship class (multi-$B) expenditures.

As a part of OWL, the Triton Ocean World Surveyor mission concept was developed to address important Triton science objectives within a lower, New Frontiers cost cap. The mission featured a Neptune orbiter with multiple close and distant Triton flybys. Significant new science results were found to be achievable, addressing 9 of the 12 top level questions identified in the OWL (Hansen-Koharcheck and Fielhauer et al. 2021) in this lower cost implementation than a Triton orbiter.

An even less expensive next step for Triton would be a hyperbolic, single flyby, as developed extensively for the 2019 Trident Discovery proposal (Frazier et al. 2020), though we point out that similar missions could be flown with different payload priorities. One variant of a future single flyby Triton mission would be a flyby of Triton as a part of a gravity assist by Neptune targeting another dwarf planet and multiple small KBOs (similar to the Calypso study, Martin et al. 2021). We note that, importantly, any such mission to explore other KB dwarf planets, would also indirectly advance Pluto and Triton science through important comparative planetology studies of the surfaces, atmosphere, and interiors of such worlds, an obviously heterogeneous population with much to teach us about the origin and evolution of these bodies as an ensemble.

Landers on Pluto and Triton are premature at this time, particularly from an engineering design standpoint, and should be deprioritized in our view until after further orbiter or even flyby reconnaissance better refines both the science to be performed and the engineering requirements of such landers.

However, we do point out that simple entry probe/penetrators/crash landers (akin to missions like Huygens, LCROSS, DART, and Deep Impact) could make important atmospheric/ionospheric and high-resolution surface measurements as relatively inexpensive adjuncts to flyby and orbiter reconnaissance missions, and should be closely examined when future flybys or orbiters are further studied/proposed.

Table 12.2 summarizes these mission types.

Table 12.2. Future Pluto and Triton Missions Comparisons

	Pluto	Triton
Discovery/New frontiers class flybys	Not scientifically competitive except with high-altitude atmospheric entry sampling	Scientifically valuable except with high-altitude atmospheric entry sampling
New frontiers or flagship orbiters (Including Neptune orbiters)	Best value	Best value
Landers	Currently premature	Currently premature
Entry probes	Study as adjunct to flybys and orbiters	Study as adjunct to flybys and orbiters

References

Agnor, C. B., & Hamilton, D. P. 2006, Natur, 441, 192

Amelin, Y., Kaltenbach, A., Iizuka, T., et al. 2010, E&PSL, 300, 343

Arakawa, S., Hyodo, R., & Genda, H. 2019, NatAs, 3, 802

Bagenal, F., Horányi, M., McComas, D. J., et al. 2016, Sci, 351, aad9045

Barnes, N. P., Delamere, P. A., Strobel, D. F., et al. 2019, JGRA, 124, 1568

Barucci, M. A., Dalle Ore, C., & Fornasier, S. 2021, The Pluto System After New Horizons, ed. S. A. Stern, J. M. Moore, W. M. Grundy, L. A. Young, & R. P. Binzel (Tucson, AZ: Univ. of Arizona Press) 21

Bertrand, T., Lellouch, E., Holler, B. J., et al. 2022, Icar, 373, 114764

Bierson, C. J., Nimmo, F., & Stern, S. A. 2020, NatGe, 13, 468

Bromley, B. C., & Kenyon, S. J. 2020, AJ, 160, 85

Canup, R. M. 2010, AJ, 141, 35

Canup, R. M., Kratter, K. M., & Neveu, M. 2021, The Pluto System After New Horizons, ed. S. A. Stern, J. M. Moore, W. M. Grundy, L. A. Young, & R. P. Binzel (Tucson, AZ: Univ. of Arizona Press) 475

Cheng, A. F., Summers, M. E., Gladstone, G. R., et al. 2017, Icar, 290, 112

Cochrane, C. J., Persinger, R. R., Vance, S. D., et al. 2022, E&SS, 9, e2021EA002034

Cockell, C. S., Bush, T., Bryce, C., et al. 2016, AsBio, 16, 89

Cruikshank, D. P., Grundy, W. M., Protopapa, S., Schmitt, B., & Linscott, I. R. 2021, in The Pluto System After New Horizons, ed. S. A. Stern, J. M. Moore, W. M. Grundy, L. A. Young, & R. P. Binzel 165 (Tucson, AZ: Univ. of Arizona)

Ćuk, M., & Gladman., B. R. 2005, ApJL., 626, L113

Dalle Ore, C. M., Cruikshank, D. P., Protopapa, S., et al. 2019, SciA, 5, eaav5731

Feyerabend, M., Liuzzo, L., Simon, S., Motschmann, U., et al. 2017, JGRA, 122, 10

Frazier, W., Bearden, D., Mitchell, K. L., et al. 2020, 2020 IEEE Aerospace Conf. (Piscataway, NJ: IEEE) 1

Gaeman, J., Hier-Majumder, S., & Roberts, J. H. 2012, Icar, 220, 339

Gladstone, G. R., Stern, S. A., Ennico, K., et al. 2016, Sci, 351, aad8866

Goldreich, P., Murray, N., Longaretti, P. Y., & Banfield, D. 1989, Sci, 245, 500

Grundy, W. M., Binzel, R. P., Buratti, B. J., et al. 2016, Sci, 351, aad9189

Grundy, W. M., Bertrand, T., Binzel, R. P., et al. 2018, Icar, 314, 232

Hansen, C. J., McEwen, A. S., Ingersoll, A. P., & Terrile, R. J. 1990, Sci, 250, 421

Hansen, C. J., Castillo-Rogez, J., Grundy, W., et al. 2021, PSJ, 2, 137

Hansen-Koharcheck, C., Fielhauer, K. B., Martin, E., et al. 2021, (Columbia, MD: Applied Physics Laboratory) Study Report https://tinyurl.com/2p88fx4f

Herbert, F., & Sandel, B. R. 1991, JGRS, 96, 19241

Howett, C. J. A., Robbins, S., Fielhauer, K., Apland, C., et al. 2020, Outer Planets Assessment Group (Fall 2020) (Houston, TX: LPI) 6032

Hendrix, A. R., Hurford, T. A., Barge, L. M., et al. 2019, AsBio, 19, 1

Hillier, J., Helfenstein, P., Verbiscer, A., & Veverka, J. 1991, JGRS, 96, 19203

Hinson, D. P., Linscott, I. R., Strobel, D. F., et al. 2018, Icar, 307, 17

Hofgartner, J. D., Birch, S. P. D., Castillo, J., et al. 2022, Icar, 375, 114835

Jankowski, D. G., Chyba, C.F., & Nicholson, P. D. 1989, Icar, 80, 211

Johansen, A., Mac Low, M.-M., Lacerda, P., & Bizzaro, M. 2015, SciA, 1, e1500109

Kamata, S., Nimmo, F., Sekine, Y., et al. 2019, NatGe, 12, 407

Keane, J. T., Matsuyama, I., Kamata, S., & Steckloff, J. K. 2016, Natur, 540, 90

Kaib, N. A., Parsells, A., Grimm, S., Quarles, B., & Clement, M. S. 2024, Icar, 415, 116057

Krasnopolsky, V. A. 2020, Icar, 335, 113374

Lavvas, P., Lellouch, E., Strobel, D. F., et al. 2021, NatAs, 5, 289

Liu, B., Raymond, S. N., & Jacobson, S. A. 2022, Natur, 604, 643

Liuzzo, L., Paty, C., Cochrane, C., et al. 2021, JGRA, 126, e2021JA029740

Luspay-Kuti, A., Mandt, K., Jessup, K.-L., et al. 2017, MNRAS, 472, 104

Martin, E., Bottke, W., Hersman, C. B., et al. 2021, (Columbia, MD: Applied Physics Laboratory) Study report available at https://tinyurl.com/2p88fx4f

Mauk, B. H., Keath, E. P., Kane, M., et al. 1991, JGRA, 96, 19061

McKinnon, W. B. 1984, Natur, 311, 355

McKinnon, W. B., Lunine, J. I., & Banfield, D. 1995, Neptune and Triton, ed. D. P. Cruikshank (Tucson, AZ: Univ. Arizona Press) 807

McKinnon, W. B., Simonelli, D. P., & Schubert, G. 1997, Pluto and Charon ed. S. A. Stern, & D. J. Tholen (Tucson, AZ: Univ. Arizona Press) 295

McKinnon, W. B., Stern, S. A., Weaver, H. A., et al. 2017, Icar, 287, 2

McKinnon, W. B., Glein, C. R., Bertrand, T., & Rhoden, A. R. 2021, In The Pluto System After New Horizons, ed. S. A. Stern, J. M. Moore, W. M. Grundy, L. A. Young, & R. P. Binzel 507 (Tucson, AZ: Univ. of Arizona)

Moore, J. M., & McKinnon, W. B. 2021, AREPS, 49, 173

Moore, J. M., McKinnon, W. B., Spencer, J. R., et al. 2016, Sci, 351, 1284

Morbidelli, A., & Nesvorný, D. 2020, in The Trans-Neptunian Solar System, ed. D. Prialnik, et al. 25 (Amsterdam: Elsevier)

National Academies of Sciences, Engineering, and Medicine 2023, Origins,Worlds, and Life: A Decadal Strategy for Planetary Science and Astrobiology 2023–2032 (Washington, DC: The National Academies Press)

Nesvorný, D. 2018, ARA&A, 56, 137

Nesvorný, D., & Vokrouhlický, D. 2016, AJ, 825, 94

Neveu, M., Desch, S. J., & Castillo-Rogez, J. C. 2017, GeCoA, 212, 324

Nimmo, F., & McKinnon, W. B. 2021, The Pluto System After New Horizons (Tuscon, AZ: Univ. Arizona Press) 89

Nimmo, F., & Spencer, J. R. 2015, Icar, 246, 2

Nogueira, E., Brasser, R., & Gomes, R. 2011, Icar, 214, 113

Noll, K. S., Grundy, W. M., Nesvorný, D., & Thirouin, A. 2020, The Trans-Neptunian Solar System ed. D. Prialnik, A. Barucci, & L. Young (Amsterdam: Elsevier) 205

Ortiz, J. L., Santos-Sanz, P., Sicardy, B., et al. 2017, Natur, 550, 219

Poppe, A. R., & Horányi, M. 2018, A&A, 617, L5

Robuchon, G., & Nimmo, F. 2011, Icar, 216, 426

Ruiz, J., Montoya, L., López, V., & Amils, R. 2007, OLEB, 37, 287

Rymer, A., Runyon, K. D., Clyde, B., et al. 2021, PSJ, 2, 184

Schaller, E. L., & Brown, M. E. 2007, ApJL, 659, L61

Schenk, P. M., & Zahnle, K. 2007, Icar, 192, 135

Shock, E. L., & McKinnon, W. B. 1993, Icar, 106, 464

Sittler Jr, E. C., & Hartle, R. E. 1996, JGRA, 101, 10863

Steffl, A. J., Young, L. A., Strobel, D. F., et al. 2020, AJ, 159, 274

Stern, S. A. 1991, Icar, 90, 271

Stern, S. A., Bagenal, F., Ennico, K., et al. 2015, Sci, 350, aad1815

Strobel, D. F., Summers, M. E., Herbert, F., & Sandel, B. R. 1990, GeoRL, 17, 1729

Strobel, D. F., & Zhu, X. 2017, Icar, 291, 55

Tan, S. P., & Kargel, J. S. 2018, MNRAS, 474, 4254

Tyler, R. 2009, GeoRL, 36, L15205

Tyler, G. L., Sweetnam, D. N., Anderson, J. D., et al. 1989, Sci, 246, 1466

Vokrouhlický, D., Nesvorný, D., & Levison, H. F. 2008, AJ, 136, 1463

White, O. L., Moore, J. M., Howard, A. D., et al. 2021, The Pluto System After New Horizons (Tuscon, AZ: Univ. Arizona Press) 55

Young, L. A., Kammer, J. A., Steffl, A. J., et al. 2018, Icar, 300, 174

Young, L. A., Bertrand, T., Trafton, L. M., et al. 2021, The Pluto System After New Horizons (Tuscon, AZ: Univ. Arizona Press) 321

Zhang, X., Strobel, D. F., & Imanaka, H. 2017, Natur, 551, 352

Zolotov, M. Y. 2007, GeoRL, 34, L23203